MEMOIRS

of the
American Mathematical Society

ISSN 0065-9266 (print)
ISSN 1947-6221 (online)

Number 1338

Naturality and Mapping Class Groups in Heegard Floer Homology

András Juhász
Dylan P. Thurston
Ian Zemke

September 2021 • Volume 273 • Number 1338 (first of 5 numbers)

AMS AMERICAN MATHEMATICAL SOCIETY

Library of Congress Cataloging-in-Publication Data

Cataloging-in-Publication Data has been applied for by the AMS.
See http://www.loc.gov/publish/cip/.
DOI: https://doi.org/10.1090/memo/1338

Memoirs of the American Mathematical Society

This journal is devoted entirely to research in pure and applied mathematics.

Subscription information. Beginning with the January 2010 issue, *Memoirs* is accessible from `www.ams.org/journals`. The 2021 subscription begins with volume 269 and consists of six mailings, each containing one or more numbers. Subscription prices for 2021 are as follows: for paper delivery, US$1085 list, US$868 institutional member; for electronic delivery, US$955 list, US$764 institutional member. Upon request, subscribers to paper delivery of this journal are also entitled to receive electronic delivery. If ordering the paper version, add US$20 for delivery within the United States; US$80 for outside the United States. Subscription renewals are subject to late fees. See `www.ams.org/help-faq` for more journal subscription information. Each number may be ordered separately; *please specify number* when ordering an individual number.

Back number information. For back issues see `www.ams.org/backvols`.

Subscriptions and orders should be addressed to the American Mathematical Society, P. O. Box 845904, Boston, MA 02284-5904 USA. *All orders must be accompanied by payment.* Other correspondence should be addressed to 201 Charles Street, Providence, RI 02904-2213 USA.

Memoirs of the American Mathematical Society (ISSN 0065-9266 (print); 1947-6221 (online)) is published bimonthly (each volume consisting usually of more than one number) by the American Mathematical Society at 201 Charles Street, Providence, RI 02904-2213 USA. Periodicals postage paid at Providence, RI. Postmaster: Send address changes to Memoirs, American Mathematical Society, 201 Charles Street, Providence, RI 02904-2213 USA.

This publication is indexed in *Mathematical Reviews*®, *Zentralblatt MATH*, *Science Citation Index*®, *Science Citation Index*™*-Expanded*, *ISI Alerting Services*℠, *SciSearch*®, *Research Alert*®, *CompuMath Citation Index*®, *Current Contents*®*/Physical, Chemical & Earth Sciences*.
This publication is archived in *Portico* and *CLOCKSS*.
Printed in the United States of America.

∞ The paper used in this book is acid-free and falls within the guidelines established to ensure permanence and durability.
Visit the AMS home page at `https://www.ams.org/`

10 9 8 7 6 5 4 3 2 1 26 25 24 23 22 21

Contents

Abstract

We show that all versions of Heegaard Floer homology, link Floer homology, and sutured Floer homology are natural. That is, they assign concrete groups to each based 3-manifold, based link, and balanced sutured manifold, respectively. Furthermore, we functorially assign isomorphisms to (based) diffeomorphisms, and show that this assignment is isotopy invariant.

The proof relies on finding a simple generating set for the fundamental group of the "space of Heegaard diagrams," and then showing that Heegaard Floer homology has no monodromy around these generators. In fact, this allows us to give sufficient conditions for an arbitrary invariant of multi-pointed Heegaard diagrams to descend to a natural invariant of 3-manifolds, links, or sutured manifolds.

Received by the editor June 17, 2015, and, in revised form, August 18, 2018.

Article electronically published on November 8, 2021.

DOI: https://doi.org/10.1090/memo/1338

2020 *Mathematics Subject Classification.* Primary 57M27, 57R58; Secondary 57R25, 57R65.

Key words and phrases. Heegaard Floer homology, 3-manifold, Heegaard diagram.

AJ is affiliated with the Mathematical Institute, University of Oxford, Woodstock Road, Oxford, OX2 6GG, United Kingdom. Email: juhasza@maths.ox.ac.uk.

AJ was supported by a Royal Society Research Fellowship and OTKA grant NK81203. This project has received funding from the European Research Council (ERC) under the European Union's Horizon 2020 research and innovation programme (grant agreement No 674978).

DPT is affiliated with the Department of Mathematics, Indiana University, Bloomington, 831 E. Third Street, Bloomington, Indiana 47405. Email: dpthurst@indiana.edu.

DPT was supported by NSF grants number DMS-1008049 and DMS-1507244.

IZ is affiliated with the Department of Mathematics, Princeton University, Princeton, New Jersey 08544. Email: izemke@math.princeton.edu.

IZ was supported by NSF Postdoctoral Research Fellowship DMS-1703685.

CHAPTER 1

Introduction

The Heegaard Floer homology groups, defined by Ozsváth and Szabó [**26, 27**], are powerful invariants. They come in four different versions – plus, minus, infinity, and hat – each of which associates a graded abelian group to a closed oriented 3-manifold. These were later extended to knots by Ozsváth and Szabó [**25**] and independently by Rasmussen [**32**]. Furthermore, the hat and minus versions were extended to links by Ozsváth and Szabó [**29**], and the hat version to sutured manifolds by Juhász [**16**]. These groups are initially well-defined only up to isomorphism, but, in order to get more powerful invariants that admit diffeomorphism actions, cobordism maps, and to be able to consider specific elements such as contact invariants, we need a naturally associated group, not just a group up to isomorphism. We address that issue in this paper.

1.1. Motivation

To better understand what it means to have a naturally associated group, we explain some of the naturality issues that arise in topology. Even though the examples considered here are classical, they have strong analogies with the case of Heegaard Floer homology. The reader familiar with naturality issues should skip to Section 1.2 for the statement of our results.

When defining algebraic invariants in topology, it is essential to place them in a functorial setting. For example, suppose we construct an algebraic invariant of topological spaces that depends on various choices, and hence only assigns an isomorphism class of, say, groups to a space. We cannot talk about maps between isomorphism classes of groups, or consider specific elements of an isomorphism class.

An early example of this phenomenon is provided by the fundamental group, which depends on the choice of basepoint in an essential way. Indeed, given a space X and basepoints p, $q \in X$, there is no "canonical" isomorphism between $\pi_1(X,p)$ and $\pi_1(X,q)$; one has to specify a homotopy class of paths from p to q first. (The word "canonical" is often used in an imprecise way in the literature, we will specify its precise meaning later in this section.) It follows that π_1 can only be defined functorially on the category of pointed topological spaces.

The naturality/functoriality issues that might arise are perfectly illustrated by simplicial homology. First, one has to restrict to the category of triangulable spaces. Even settling invariance up to isomorphism took several decades. The main question was the following: Given triangulations T and T' of the space X, how do we compare the groups $H_*(T)$ and $H_*(T')$? The first attempts tried to proceed via the Hauptvermutung: Do T and T' have *isomorphic* subdivisions? We now know this is false, but even if it were true, it would not provide naturality as the choice of isomorphism is not unique. The issues of invariance and naturality were settled by Alexander's method of simplicial approximation, which provides for any pair of triangulations T and T' of X an isomorphism $\beta(T,T')\colon H_*(T) \to H_*(T')$. So how do we obtain the group $H_*(X)$ from this data? First, let us recall a definition due to Eilenberg and Steenrod [**9**, Definition 6.1].

DEFINITION 1.1. A *transitive system of groups* consists of

- a set M, and for every $\alpha \in M$, a group G_α,
- for every pair $(\alpha, \beta) \in M \times M$, an isomorphism $\pi^\alpha_\beta\colon G_\alpha \to G_\beta$ such that
  (1) $\pi^\alpha_\alpha = \mathrm{Id}_{G_\alpha}$ for every $\alpha \in M$,
  (2) $\pi^\beta_\gamma \circ \pi^\alpha_\beta = \pi^\alpha_\gamma$ for every α, β, $\gamma \in M$.

A transitive system of groups gives rise to a single group G as follows: Let G be the set of elements $g \in \prod_{\alpha\in M} G_\alpha$ for which $\pi^\alpha_\beta(g(\alpha)) = g(\beta)$ for every α, $\beta \in M$.

REMARK 1.2. For every $\alpha \in M$, let $p_\alpha\colon \prod_{\alpha \in M} G_\alpha \to G_\alpha$ be the projection. Then $p_\alpha|_G\colon G \to G_\alpha$ is an isomorphism. In fact, G is a universal object, obtained as a limit along the directed graph on M where there is a unique edge from α to β for every $(\alpha, \beta) \in M \times M$. The assignment $\alpha \mapsto G_\alpha$ is a functor from M to the category of groups, which is the diagram along which we take the limit.

In this paper, instead of the limit, we work with the colimit $\coprod_{\alpha\in M} G_\alpha/{\sim}$, where $g_\alpha \sim g_\beta$ for $g_\alpha \in G_\alpha$ and $g_\beta \in G_\beta$ if and only if $\pi^\alpha_\beta(g_\alpha) = g_\beta$. The group structure on $\coprod_{\alpha\in M} G_\alpha/{\sim}$ is given by pointwise multiplication of equivalence classes. Each embedding of G_α into $\coprod_{\alpha\in M} G_\alpha/{\sim}$ is an isomorphism. It is easy to check that this satisfies the universal property for a colimit. The map $g \mapsto [g(\alpha)]$ for $g \in G$ gives a natural isomorphism between the limit and the colimit, where $[g(\alpha)]$ is the equivalence class of $g(\alpha)$ for an arbitrary $\alpha \in M$.

We call the π^α_β *canonical isomorphisms.* So, if we are constructing some algebraic invariant, and have isomorphisms for any pair of choices, we only call these isomorphisms "canonical" if they satisfy properties (1) and (2) above. An instance of a transitive system of groups is given by taking M to be the set of all triangulations of a triangulable space X, and for any pair $(T, T') \in M \times M$, setting $\pi^T_{T'} = \beta(T, T')$. Another example of a transitive system is given in the case of Morse homology by Schwarz [**35**, Section 4.1.3], where one needs to compare homology groups defined using different Morse functions.

Classical homology was put in a functorial framework by the Eilenberg–Steenrod axioms, whereas the gauge theoretic invariants of 3- and 4-manifolds are expected to satisfy properties similar to the topological quantum field theory (TQFT) axioms of Atiyah, called a "secondary TQFT." Our motivating question is whether Heegaard Floer homology fits into such a functorial picture. Heegaard Floer homology is a package of invariants of 3- and 4-manifolds defined by Ozsváth and Szabó [**27, 28**]. It follows from our work that each version of Heegaard Floer homology individually satisfies the classical TQFT axioms, but it is important to note that the closed 4-manifold invariant is obtained by mixing the $+$, $-$, and ∞ versions, hence deviating from Atiyah's original description. Building on our work, the third author [**39**] has shown that the 4-manifold invariants are also well-defined.

In its simplest form, Heegaard Floer homology assigns an Abelian group $\widehat{HF}(Y)$ to a closed oriented 3-manifold Y, well-defined up to isomorphism. The construction depends on a choice of Heegaard diagram for Y. Given two Heegaard diagrams $\mathcal{H}$ and $\mathcal{H}'$ for Y, our goal is to construct a canonical isomorphism $\widehat{HF}(\mathcal{H}) \to \widehat{HF}(\mathcal{H}')$ such that the set of diagrams, together with these isomorphisms form a transitive system of groups, yielding a single group $\widehat{HF}(Y)$. We want to do this in a way that every diffeomorphism $d\colon Y_0 \to Y_1$ induces an isomorphism

$$d_*\colon \widehat{HF}(Y_0) \to \widehat{HF}(Y_1).$$

For this (and also to get the canonical isomorphisms), one has to consider diagrams embedded in Y, not only "abstract" ones. That is, we consider triples $(\Sigma, \boldsymbol{\alpha}, \boldsymbol{\beta})$ where Σ is a subsurface of Y that splits Y into two handlebodies, and $\boldsymbol{\alpha}$, $\boldsymbol{\beta} \subset \Sigma$ are attaching sets for the two handlebodies. Then the main question is: How do we compare $\widehat{HF}$ for diagrams that are embedded in Y differently? Note that our construction of canonical isomorphisms differs from that of Ozsváth and Szabó [**28**, Theorem 2.1] in a subtle but essential way; see Remark 2.46.

The Reidemeister-Singer theorem provides an analogue of the Hauptvermutung in the case of Heegaard splittings: Any two Heegaard splittings of Y become isotopic after stabilizations. However, this isotopy is far from being unique. In fact, the fundamental group $\pi_1(\mathcal{S}(Y,\Sigma))$ of the space of Heegaard splittings equivalent to (Y,Σ) is highly non-trivial. So we could have a loop of Heegaard diagrams $\{\, \mathcal{H}_t \colon t \in [0,1] \,\}$ of Y along which $\widehat{HF}(\mathcal{H}_t)$ has monodromy. Indeed, let us recall the following definition.

DEFINITION 1.3. Let $\Sigma \subset Y$ be a Heegaard surface. Then the *Goeritz group* of the Heegaard splitting (Y,Σ) is defined as

$$G(Y,\Sigma) = \ker\left(\mathrm{MCG}(Y,\Sigma) \to \mathrm{MCG}(Y)\right).$$

In other words, $G(Y,\Sigma)$ consists of automorphisms d of (Y,Σ) (considered up to isotopy preserving the splitting) such that d is isotopic to Id_Y if we are allowed to move Σ.

According to Johnson and McCullough [**14**], there is a short exact sequence

$$1 \to \pi_1(\mathrm{Diff}(Y)) \to \pi_1(\mathcal{S}(Y,\Sigma)) \to G(Y,\Sigma) \to 1.$$

Let $\mathcal{H} = (\Sigma, \boldsymbol{\alpha}, \boldsymbol{\beta}, z)$ be a Heegaard diagram of Y. Ignoring basepoint issues, an element of $\pi_1(\mathcal{S}(Y,\Sigma))$ coming from $\pi_1(\mathrm{Diff}(Y))$ acts trivially on the Heegaard Floer homology $\widehat{HF}(\mathcal{H})$ (as this is the action of Id_Σ, the endpoint of the loop), so this descends to an action of $G(Y,\Sigma)$ on $\widehat{HF}(\mathcal{H})$. The 3-sphere has a unique genus g Heegaard splitting for every $g \geq 0$. At the time of writing of this paper, it is unknown whether $G(S^3,\Sigma)$ is finitely generated when the genus of Σ is greater than 2. Understanding the group $G(Y,\Sigma)$ for a general 3-manifold Y and splitting Σ seems even more difficult. Heegaard Floer homology is invariant under stabilization, and the "fundamental group" of the space of Heegaard diagrams modulo stabilizations is easier to understand, as we shall see in this paper.

1.2. Statement of results

We prove that Heegaard Floer homology is an invariant of based 3-manifolds in the following strong sense. (We ignore gradings and Spin^c structures for the moment.)

DEFINITION 1.4. Let Man be the category whose class of objects $|\mathsf{Man}|$ consists of closed, connected, oriented 3-manifolds, and whose morphisms are diffeomorphisms. Let Man_* be the category whose objects are pairs (Y,p), where $Y \in |\mathsf{Man}|$

and $p \in Y$ is a choice of basepoint, and whose morphisms are basepoint-preserving diffeomorphisms.

Also, let R-Mod be the category of R-modules for any ring R, and let k-Vect be the category of vector spaces over k for any field k.

Recall that Ozsváth and Szabó defined four different versions of Heegaard Floer homology, named $\widehat{HF}$, HF^-, HF^+, and HF^∞. We will write HF without decoration to mean any of these four versions. Note that these are all modules over the polynomial ring $\mathbb{F}_2[U]$, where the U-action is trivial on $\widehat{HF}$.

THEOREM 1.5. *There are functors*

$$\widehat{HF},\ HF^-,\ HF^+,\ HF^\infty \colon \mathsf{Man}_* \to \mathbb{F}_2[U]\text{-}\mathsf{Mod},$$

such that, for a based 3-manifold (Y,p), the groups $HF(Y,p)$ are isomorphic to the various versions of Heegaard Floer homology defined by Ozsváth and Szabó [**26, 27**]. *Furthermore, isotopic diffeomorphisms induce identical maps on HF.*

As $HF_{\mathrm{red}}(Y,p)$ is defined as $HF^+(Y,p)/\mathrm{Im}(U^k)$ for k sufficiently large (see [**27**, Definition 4.7]), it immediately follows from Theorem 1.5 that HF_{red} is also functorial.

Ozsváth and Szabó [**26, 27**] showed that the *isomorphism class* of $HF(Y)$ is an invariant of the 3-manifold Y. The statement in Theorem 1.5 is stronger, in that it says that $HF(Y,p)$ is actually a well-defined group, not just an isomorphism class of groups. The first step towards naturality was made by Ozsváth and Szabó [**28**, Theorem 2.1], who constructed maps Ψ between $HF(\Sigma,\boldsymbol{\alpha},\boldsymbol{\beta})$ and $HF(\Sigma,\boldsymbol{\alpha}',\boldsymbol{\beta}')$ for a fixed Heegaard surface Σ and equivalent "abstract" diagrams $(\Sigma,\boldsymbol{\alpha},\boldsymbol{\beta})$ and $(\Sigma,\boldsymbol{\alpha}',\boldsymbol{\beta}')$. The maps Ψ are isomorphisms, and they satisfy conditions (1) and (2) of Definition 1.1. Furthermore, they also defined maps for stabilizations.

In the present paper, we explain how to canonically compare invariants of diagrams with different embeddings in Y. As it turns out, the additional checks are the following: One has to prove that HF has no monodromy around the simple handleswap loop of Figure 4, and show that the map on HF induced by a diffeomorphism $d\colon (\Sigma,\boldsymbol{\alpha},\boldsymbol{\beta}) \to (\Sigma,\boldsymbol{\alpha}',\boldsymbol{\beta}')$ isotopic to Id_Σ agrees with the canonical isomorphism Ψ.

One surprise in Theorem 1.5 is the appearance of the basepoint. To make this precise, we look at the mapping class group.

DEFINITION 1.6. For a smooth manifold M, its *mapping class group* is

$$MCG(M) = \mathrm{Diff}(M)/\mathrm{Diff}_0(M) = \pi_0(\mathrm{Diff}(M)),$$

where $\mathrm{Diff}(M)$ is the group of diffeomorphisms of X, and $\mathrm{Diff}_0(M)$ is the subgroup of diffeomorphisms isotopic to the identity, which is also the connected component of the identity in $\mathrm{Diff}(M)$. Similarly, for a based smooth manifold (M,p), its *based mapping class group* is

$$MCG(M,p) = \mathrm{Diff}(M,p)/\mathrm{Diff}_0(M,p) = \pi_0(\mathrm{Diff}(M,p)),$$

where we consider maps that preserve the basepoint.

COROLLARY 1.7. *For a based 3-manifold (Y,p), the group $MCG(Y,p)$ acts naturally on $HF(Y,p)$ for any of the four versions of Heegaard Floer homology.*

PROOF. This follows immediately from Theorem 1.5 when restricted to automorphisms of (Y,p). □

It is easy to construct examples where the action of the mapping class group is non-trivial. For instance, for a 3-manifold Y, the evident diffeomorphism that exchanges the two factors of $Y \# Y$ (preserving a basepoint) will act via $x \otimes y \mapsto y \otimes x$ on $\widehat{HF}(Y \# Y) \cong \widehat{HF}(Y) \otimes \widehat{HF}(Y)$, which is non-trivial if Y is sufficiently complicated.

Recall that, from the fibration $\mathrm{Diff}(Y,p) \to \mathrm{Diff}(Y) \to Y$, there is a Birman exact sequence for based mapping class groups for any connected manifold Y:

$$\pi_1(Y) \to MCG(Y,p) \to MCG(Y) \to 0.$$

Thus, from an action of $MCG(Y,p)$ on $\widehat{HF}$, we get an action of $\pi_1(Y)$ on $\widehat{HF}$. This action of $\pi_1(Y)$ is trivial if the action descends to an action of the unbased mapping class group $MCG(Y)$. This action of $\pi_1(Y)$ is *not* always trivial. However, it factors through an action of $H_1(Y)$. For the proof of a precise formula, originally conjectured by the first author, see the work of the third author [**40**].

For the other three versions, $HF^-(Y,p)$, $HF^+(Y,p)$, and $HF^\infty(Y,p)$, the action of $\pi_1(Y)$ on HF is always trivial, in analogy with the situation for monopole Floer homology [**18**]. This has been shown by the third author [**39**].

There is also a version of Theorem 1.5 for links. Let Link be the category of oriented links in S^3, whose morphisms are orientation preserving diffeomorphisms $d\colon (S^3, L_1) \to (S^3, L_2)$. Let Link_* be the category whose objects are *based oriented links*: pairs $(L, \mathbf{p})$, where $L \subset S^3$ is an oriented link and $\mathbf{p} = \{p_1, \dots, p_n\} \subset L$ is a set of basepoints, exactly one on each component of L. The morphisms are diffeomorphisms of S^3 preserving the based oriented link.

THEOREM 1.8. *There are functors*

$$\widehat{HFL}\colon \mathsf{Link}_* \to \mathbb{F}_2\text{-}\mathsf{Vect},$$
$$HFL^-\colon \mathsf{Link}_* \to \mathbb{F}_2[U]\text{-}\mathsf{Mod},$$

agreeing up to isomorphism with the link invariants defined by Ozsváth and Szabó [**25,29**] *and Rasmussen* [**32**]. *Isotopic diffeomorphisms induce identical maps on* $\widehat{HFL}$ *and* HFL^-.

As in Corollary 1.7, Theorem 1.8 implies that $MCG(S^3, L, \mathbf{p})$ acts on $\widehat{HFL}(L, \mathbf{p})$ and $HFL^-(L, \mathbf{p})$. Again, one can ask whether this action is non-trivial, and in particular, whether the basepoint makes a difference. For simplicity, consider knots, in which case there is an exact sequence

$$\pi_1(S^1) \to MCG(S^3, K, p) \to MCG(S^3, K) \to 0.$$

In this context, Sarkar [**34**] has constructed many examples where the action of $\pi_1(S^1)$ on $\widehat{HFL}(K,p)$ is non-trivial. More concretely, let $\sigma \in MCG(S^3, K, p)$ be the positive finger move (or Dehn twist) along K, defined in [**34**, p. 4]. Then it follows from [**34**, Theorem 6.1] that the action of σ on $\widehat{HFL}(K,p)$ for prime knots up to 9 crossings is non-trivial more often than not. He also proved a formula for the $\pi_1(S^1)$ action for knots in S^3, which was then extended to links in arbitrary 3-manifolds by the third author [**41**].

There are several variants of Theorem 1.8. For instance, Ozsváth and Szabó [**30**] have defined the group $HFK^-(Y, K, p)$ for a rationally null-homologous knot K in an oriented 3-manifold Y. They also defined $HFL^-(Y, L, \mathbf{p})$ for a link L in an integer homology sphere Y; see [**29**, Theorem 4.7]. There is also more structure that can be put on the result. In particular, there is a spectral sequence from $HFK^-(Y, K, p)$ converging to $HF^-(Y)$. These invariants are again functorial. In fact, Hendricks and Manolescu [**13**, Proposition 2.3] showed that Theorem 1.5 also holds on the chain level in the homotopy category of chain complexes of $\mathbb{F}_2[U]$-modules. A knot induces a filtration on the Heegaard Floer chain complex, and the canonical isomorphisms preserve the knot filtration; see Juhász and Marengon [**15**, Section 5.11]. Finally, for the naturality of $\widehat{HFL}$ of links in arbitrary oriented 3-manifolds and several basepoints on each link component, see Juhász [**17**, Proposition 4.10].

We will unify the proofs of Theorems 1.5 and 1.8 in the more general setting of *balanced sutured manifolds*. Let Sut be the category of sutured 3-manifolds and diffeomorphisms, and let $\mathsf{Sut}_{\mathrm{bal}}$ be the full subcategory of balanced sutured manifolds. (For definitions and details, see Definitions 2.1 and 2.25 below.)

THEOREM 1.9. *There is a functor*

$$SFH\colon \mathsf{Sut}_{\mathrm{bal}} \to \mathbb{F}_2\text{-}\mathsf{Vect},$$

agreeing up to isomorphism with the sutured manifold invariant defined by the first author [**16**]. *Isotopic diffeomorphisms induce identical maps on SFH.*

All Heegaard Floer homology groups discussed above decompose along Spin^c structures. For example,

$$SFH(M, \gamma) = \bigoplus_{\mathfrak{s} \in \mathrm{Spin}^c(M,\gamma)} SFH(M, \gamma, \mathfrak{s}).$$

In addition, each summand $SFH(M, \gamma, \mathfrak{s})$ carries a relative homological $\mathbb{Z}_{\mathfrak{d}(\mathfrak{s})}$-grading, where $\mathfrak{d}(\mathfrak{s})$ is the divisibility of the Chern class $c_1(\mathfrak{s}) \in H^2(M)$. So, for any x, $y \in SFH(M, \gamma, \mathfrak{s})$, the grading difference $\mathrm{gr}(x, y)$ is an element of $\mathbb{Z}_{\mathfrak{d}(\mathfrak{s})}$.

These gradings are natural in the following sense. Suppose that $d\colon (M, \gamma) \to (N, \nu)$ is a diffeomorphism. Then the induced map

$$d_*\colon SFH(M, \gamma) \to SFH(N, \nu)$$

restricts to an isomorphism

$$d_*|_{SFH(M,\gamma,\mathfrak{s})}\colon SFH(M, \gamma, \mathfrak{s}) \to SFH(N, \nu, d(\mathfrak{s}))$$

that preserves the relative homological grading. Completely analogous results hold for the other versions of Heegaard Floer homology.

Now we outline the main technical tools behind the above results; for further details, we refer the reader to Chapter 2. To be able to treat the various versions of Heegaard Floer homology simultaneously, we consider an arbitrary algebraic invariant F of abstract (i.e., not necessarily embedded) diagrams of sutured manifolds in a given class $\mathcal{S}$ of sutured manifolds (e.g., knot complements in case F is knot Floer homology). An *isotopy diagram* is a sutured diagram with attaching sets taken up to isotopy. We work with these to avoid admissibility issues. Let $\mathcal{G}(\mathcal{S})$ be the directed graph whose vertices are isotopy diagrams of sutured manifolds in $\mathcal{S}$, and the vertices H and H' are connected by an edge if either the α-curves or the β-curves differ by a sequence of isotopies and handleslides (called an α- or β-equivalence), or, if H' is obtained from H by a stabilization or a destabilization, and there is an edge for every diffeomorphism $d\colon H \to H'$. We say that F is a *weak Heegaard invariant* if, for every edge e from H to H' in $\mathcal{G}(\mathcal{S})$, there is an induced isomorphism

$$F(e)\colon F(H) \to F(H').$$

A weak Heegaard invariant then gives rise to an invariant of sutured manifolds in the class $\mathcal{S}$, well-defined up to isomorphism.

To assign a concrete algebraic object to each sutured manifold in the class $\mathcal{S}$, we then define the notion of a *strong Heegaard invariant.* Such an F has to commute along certain distinguished loops in $\mathcal{G}(\mathcal{S})$. These loops include rectangles where opposite edges are of the same type (i.e., they are both α-equivalences, β-equivalences, stabilizations, or diffeomorphisms; see Definition 2.29), and the aforementioned simple handleswap triangles of Figure 4 (involving an α-handleslide, a β-handleslide, and a diffeomorphism). Furthermore, a strong Heegaard invariant has to satisfy the property that, if $e\colon H \to H$ is a diffeomorphism isotopic to the identity of the Heegaard surface, then $F(e) = \mathrm{Id}_{F(H)}$.

Given a strong Heegaard invariant F and a sutured manifold (M,γ) in the class $\mathcal{S}$, we obtain the invariant $F(M,\gamma)$ as follows. We take the subgraph $\mathcal{G}_{(M,\gamma)}$ of $\mathcal{G}(\mathcal{S})$ whose vertices are isotopy diagrams *embedded* in (M,γ), and where we only consider diffeomorphisms that are isotopic to the identity in M. Then our main result is Theorem 2.38, which states that, given any two paths in $\mathcal{G}_{(M,\gamma)}$ from H to H', the composition of F along these paths coincide. The proof of this occupies most of the paper, and relies on a careful analysis of the bifurcations occurring in generic 2-parameter families of gradient vector fields on 3-manifolds. It easily follows that these compositions give a canonical isomorphism $F(H) \to F(H')$, and we obtain $F(M,\gamma)$ via Definition 1.1. We show that the different versions of Heegaard Floer homology are strong Heegaard invariants in Sections 10.2, 10.3, and 10.4.

One consequence of naturality is that we can now talk about specific elements of $HF(Y)$, which is necessary for the definition of the contact class. Another consequence is that we can define maps on Heegaard Floer homology induced by diffeomorphisms and cobordisms. The paper might also be of interest to 3-manifold topologists, as it sheds more light on the space of Heegaard splittings and diagrams, potentially telling more about the structure of the Goeritz group.

1.3. Outline

In Chapter 2, we study Heegaard invariants. In Section 2.1, we recall the definition of *sutured manifolds*, and explain how to assign sutured manifolds to based 3-manifolds and links. This allows us to unify our treatment of naturality by focusing on sutured manifolds. In Section 2.2, we discuss *abstract* and *embedded Heegaard diagrams* of sutured manifolds. This distinction plays an important role in our construction of natural 3-manifold invariants.

In Section 2.3, we define *moves* on abstract sutured diagrams, namely, α-equivalence, β-equivalence, stabilization, destabilization, and diffeomorphism. We show that any two diagrams defining the same sutured manifold can be connected by a sequence of such moves. We define *weak Heegaard invariants* (Definition 2.24), and state that the different versions of Heegaard Floer homology are weak Heegaard invariants (Theorems 2.26, 2.27, and 2.28). They assign an object in a category to every sutured diagram in a given class (e.g., diagrams of based 3-manifolds or link complements), and a morphism to every move. A weak Heegaard invariant gives rise to an invariant of based 3-manifolds (or based links, or sutured manifolds) that is well-defined up to isomorphism.

A key new notion is that of a *strong Heegaard invariant* (Definition 2.32), which we introduce in Section 2.4. This is a weak Heegaard invariant F that has to satisfy four axioms: functoriality (under compositions of moves of the same type), commutativity (which implies that F remains the same if we swap the order of two moves), continuity (isotopic diffeomorphisms induce the same morphism), and *simple handleswap invariance* (this requires that F commutes along a certain triangle of moves).

In Section 2.5, we explain how a strong Heegaard invariant F gives rise to a natural invariant of a given class of sutured manifolds. In Theorem 2.38, which is one of the deepest results of this work, we show that, if we connect two *embedded* diagrams of a sutured manifold (M, γ) by a sequence of moves such that we only allow diffeomorphisms that are *isotopic to the identity in* M (Definition 2.34), and we compose the morphisms induced by F, then the result does not depend on the choice of moves. This gives rise to a transitive system (Definition 1.1) that allows us to assign an invariant to every sutured manifold in the given class that is functorial under diffeomorphisms (Definition 2.42). We then prove Theorems 1.5, 1.8, and 1.9 stated in the introduction.

In Chapter 3, we present some examples that highlight some of the issues that arise when trying to define functorial Heegaard Floer invariants.

Most of the remainder of this paper is aimed at proving Theorem 2.38. In Chapter 4, we give a brief overview of the types of *singularities* that appear in generic 2-parameter families of smooth functions. A generic smooth function is Morse. In a generic 1-parameter family, one can have A_2 (birth-death) singularities at isolated parameter values. In a generic 2-parameter family, $A_3^{\pm}$ (birth-death-birth) points appear at isolated parameter values.

As a Heegaard diagram arises from a generic gradient vector field, not just a Morse function, we review the bifurcations of generic 1-parameter and 2-parameter families of *gradient vector fields* in Chapter 5. In Section 5.1, we recall some classical results from the theory of dynamical systems, such as the center manifold theorem (Theorem 5.3) and the reduction principle (Theorem 5.4). In Section 5.2, we state that a generic gradient vector field is Morse–Smale. In a generic 1-parameter family, a non-hyperbolic singular point might appear, or a stable and an unstable manifold might intersect non-transversely. The classification of bifurcations of generic 2-parameter families of gradients is already very complicated in dimension 3, and was carried out by Vegter [**37**]. In Section 5.3, we define the space

$$\mathcal{FV}(M, \gamma)$$

of gradient-like vector fields on the sutured manifold (M, γ) that we are going to work with. This is weakly contractible by Corollary 5.20.

In Chapter 6, we convert Morse–Smale vector fields to (overcomplete) Heegaard diagrams, codimension-1 bifurcations of gradients to Heegaard moves, and codimension-2 bifurcations to loops of diagrams.

We introduce the notion of *separable gradients* in Section 6.1. These have at most codimension-2 bifurcations, and the most important criterion is that there is no gradient flow-line from an index 2 to an index 1 critical point. A surface *separates* such a gradient vector field essentially if it is transverse to it, and contains all index 0 and 1 critical points on one side, and all index 2 and 3 critical points on the other side. The space of such surfaces is contractible, and they divide the sutured manifold into two sutured compression bodies (Proposition 6.7). Furthermore, we can continuously deform the separating surface if we deform the gradient vector field without introducing a birth-death bifurcation (Proposition 6.11).

In Section 6.2, we show how to obtain an *overcomplete* diagram (Definition 6.13) from a codimension-0 gradient vector field by intersecting the separating surface Σ with the unstable manifolds of index-1 critical points and the stable manifolds of index-2 critical points. The diagram is overcomplete because the attaching curves might not be linearly independent in $H_1(\Sigma)$. We assign a certain graph to a separable gradient, and we obtain a sutured diagram once we choose a spanning forest for this graph by deleting attaching curves corresponding to edges that lie outside the spanning forest (Definition 6.15). In the opposite direction, given a diagram of a sutured manifold such that the α- and β-curves are transverse, the space of *simple* Morse–Smale gradients (Definition 6.16) that induce it is non-empty and connected (Propositions 6.17 and 6.18).

In Section 6.3, we translate codimension-1 bifurcations of gradients to moves on overcomplete diagrams. We show that a 1-parameter family of Morse–Smale gradients induces a diffeomorphism of the diagram isotopic to the identity in M (Lemma 6.21). In Proposition 6.28, we show that birth-death bifurcations correspond to *generalized (de)stabilizations* (Definition 6.26) or the appearance of a null-homotopic α- or β-curve, and a non-transverse intersection of stable and unstable manifolds to a generalized handleslide (Definition 6.27) or a tangency between an α- and a β-curve. In Proposition 6.32, we show that, given a generic 2-parameter family of gradients and a rectangle in the parameter space that intersects a codimension-1 stratum in two points on opposite sides, the diagrams in the corners form a distinguished rectangle appearing in the Commutativity Axiom of strong Heegaard invariants.

In Section 6.4, we show how to choose spanning trees appropriately in Propositions 6.28 and 6.30 to pass from overcomplete to actual Heegaard diagrams, without altering the relationship of the diagrams before and after the bifurcation in an essential way. In Section 6.5, we show that, given a move on sutured diagrams, it can be converted to a path of gradients (Proposition 6.35).

In Section 6.6, we translate codimension-2 bifurcations to loops of diagrams. Given a codimension-2 bifurcation of gradients at a parameter value p, the codimension-1 strata divide a neighborhood of p in the parameter space into chambers. Taking a point in each chamber, we obtain a loop of diagrams such that neighboring diagrams are related by generalized Heegaard moves. We describe these loops in Theorem 6.37. There are many more loops than what appear in the definition of strong Heegaard invariants.

In Chapter 7, we break down generalized stabilizations, generalized handleslides, and the loops of diagrams appearing in Theorem 6.37 into the simpler moves and loops that feature in the definition of strong Heegaard invariants. During the simplification procedure, we work with overcomplete diagrams, and will only later choose spanning trees to pass to actual sutured diagrams. We pass from 2-parameter families of gradients to certain combinatorial structures, namely, *polyhedral decompositions* of the parameter space D^2 (Definition 7.1). These are certain regular CW decompositions of D^2, and we label the vertices with (overcomplete) diagrams. Neighboring diagrams are related by generalized moves. From now on, we only modify the polyhedral decomposition, and forget about the family of gradients.

In Section 7.1, we show how to replace a generalized stabilization with a simple stabilization, a number of α-handleslides, and a number of β-handleslides. There are several choices one can make during this resolution process. If we use different choices at the two ends of a codimension-1 generalized stabilization stratum, then we can interpolate between them (Lemma 7.8). We can also replace a generalized handleslide by a simple handleslide and an isotopy. In Section 7.2, we apply this resolution process to the loops described in Theorem 6.6. We then break them down into loops of diagrams appearing in the definition of strong Heegaard invariants and (k, l)-handleswaps (Definition 7.9). We write (k, l)-handleswaps in terms of simple handleswaps in Chapter 8. The above procedure can be thought of as finding a simple generating set for the space of Heegaard diagrams of a given sutured manifold.

We prove Theorem 2.38 in Chapter 9. The idea is the following: Given a loop of diagrams, we lift them to a loop of gradients such that the bifurcations induce the given moves. We then extend this loop from S^1 to a generic 2-parameter family of gradients over D^2. We choose a polyhedral decomposition of D^2 compatible with the bifurcation stratification given by this family. Each vertex is labeled by a diagram. We then apply the resolution procedure to obtain the simple loops appearing in the definition of strong Heegaard invariants. Finally, a simple combinatorial argument gives that, since a strong Heegaard invariant commutes along the simple loops, it also commutes along S^1.

In Chapter 10, we prove Theorem 2.33, which states that Heegaard Floer homology *HF* is a strong Heegaard invariant. We define the isomorphisms associated to Heegaard moves in Section 10.1, where we reprove that *HF* is a weak Heegaard invariant. For this, we only use triangle maps that are more computable than continuation maps. We verify that *HF* is a strong Heegaard invariant in Section 10.2. We prove the Continuity Axiom in Proposition 10.28, and simple handleswap invariance in Section 10.3. We put the above pieces together in Section 10.4 to prove Theorem 2.33.

Finally, in Appendix A, we sketch a description of strong Heegaard invariants for classical (i.e., not sutured) single pointed Heegaard diagrams that is equivalent to Definition 2.32, and instead of α-equivalences and β-equivalences, uses more elementary moves: α-isotopies, β-isotopies, α-handleslides, and β-handleslides.

Acknowledgements

We are extremely grateful to Peter Ozsváth for numerous helpful conversations. We would also like to thank Valentin Afraimovich, Ryan Budney, Boris Hasselblatt, Matthew Hedden, Michael Hutchings, Martin Hyland, Jesse Johnson, Robert Lipshitz, Saul Schleimer, Zoltán Szabó, and the referee for their guidance and suggestions.

This project would not have been possible without the hospitality of the Tambara Institute of Mathematical Sciences, the Mathematical Sciences Research Institute, and the Isaac Newton Institute. Most of the work was carried out while the first author was at the University of Cambridge and the second author was at Barnard College, Columbia University.

CHAPTER 2

Heegaard Invariants

2.1. Sutured manifolds

Sutured manifolds were introduced by Gabai [**11**]. The following definition is slightly less general, in that it excludes toroidal sutures.

DEFINITION 2.1. A *sutured manifold* (M, γ) is a compact oriented 3-manifold M with boundary, together with a set $\gamma \subset \partial M$ of pairwise disjoint annuli. Furthermore, the interior of each component of γ contains a *suture*; i.e., a homologically nontrivial oriented simple closed curve. We denote the union of the sutures by $s(\gamma)$. In addition, every component of $R(\gamma) = \partial M \setminus \operatorname{Int}(\gamma)$ is oriented. Define $R_+(\gamma)$ (respectively $R_-(\gamma)$) to be those components of $\partial M \setminus \operatorname{Int}(\gamma)$ whose orientations agree (respectively disagree) with the orientation of ∂M, or equivalently, whose normal vectors point out of (respectively into) M. The orientation on $R(\gamma)$ must be coherent with respect to $s(\gamma)$; i.e., if δ is a component of $\partial R(\gamma)$ and is given the boundary orientation, then δ must represent the same homology class in $H_1(\gamma)$ as some suture.

A sutured manifold (M, γ) is called *proper* if the map $\pi_0(\gamma) \to \pi_0(\partial M)$ is surjective and M has no closed components (i.e., the map $\pi_0(\partial M) \to \pi_0(M)$ is surjective).

Note that, for a proper sutured manifold (M, γ), the orientations of $s(\gamma)$ and M completely determine the orientation on $R(\gamma)$, and hence the decomposition $R(\gamma) = R_+(\gamma) \cup R_-(\gamma)$.

CONVENTION 2.2. In this paper, we will assume that all sutured manifolds are proper, in addition to not having any toroidal sutures.

To see the connection between sutured manifolds and closed 3-manifolds, observe that, if (M, γ) is a sutured manifold such that ∂M is a sphere with a single suture (dividing ∂M into two disks), then the quotient of M, where ∂M is identified with a point, is a closed 3-manifold with a distinguished basepoint given by the equivalence class of ∂M. For the other direction, we introduce the following definitions.

DEFINITION 2.3. Suppose that M is a smooth manifold, and let $L \subset M$ be a properly embedded submanifold. For each point $p \in L$, let $N_pL = T_pM/T_pL$ be the fiber of the normal bundle of L over p, and let $UN_pL = (N_pL \setminus \{0\})/\mathbb{R}_+$ be the fiber of the unit normal bundle of L over p. Then the *(spherical) blowup* of M along L, denoted by $\operatorname{Bl}_L(M)$, is a manifold with boundary obtained from M by replacing each point $p \in L$ by $UN_p(L)$. There is a natural projection $\operatorname{Bl}_L(M) \to M$. For further details, see Arone and Kankaanrinta [**4**].

For instance, if $L \subset M$ is a submanifold of codimension 1, then $\operatorname{Bl}_L(M)$ is the usual operation of cutting M open along L.

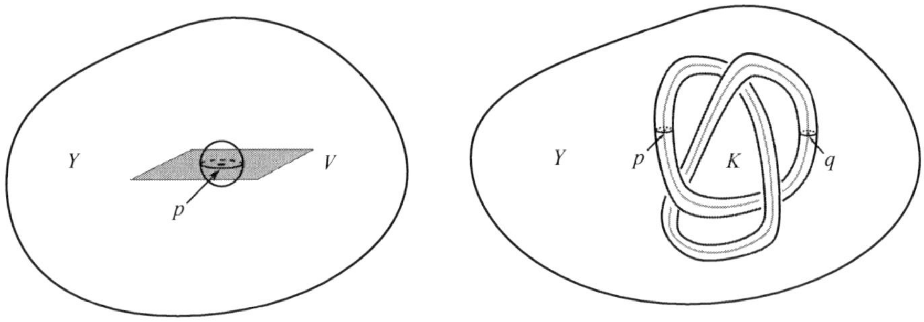

FIGURE 1. The sutured manifolds $Y(p, V)$ and $Y(K, p, q)$.

DEFINITION 2.4. Let Y be a closed, connected, oriented 3-manifold, together with a basepoint p and an oriented tangent 2-plane $V < T_pM$. Then $Y(p, V) = (M, \gamma)$ is the sutured manifold with $M = \mathrm{Bl}_p(Y)$ and suture $s(\gamma) = (V \setminus \{0\})/\mathbb{R}_+$ in the resulting S^2 boundary component of M. See the left-hand side of Figure 1. We orient $s(\gamma)$ such that, if we lift it to V, then the lift goes around the origin in the positive direction.

There is a similar construction for links, as well.

DEFINITION 2.5. Let (Y, K, p, q) be an oriented knot with two basepoints. Then $Y(K, p, q) = (M, \gamma)$ is the sutured manifold with $M = \mathrm{Bl}_K(Y)$ and $s(\gamma) = UN_pK \cup UN_qK$, sitting inside the torus ∂M, as on the right-hand side of Figure 1. The orientation of K induces an orientation of NK. We orient UN_pK coherently with N_pK, while UN_qK is oriented incoherently with N_qK.

Similarly, if $(Y, L, \mathbf{p}, \mathbf{q})$ is a based oriented link with exactly one p and one q basepoint on each component of L, then we define $Y(L, \mathbf{p}, \mathbf{q})$ to be the sutured manifold (M, γ) with $M = \mathrm{Bl}_L(Y)$, and sutures obtained for each component of L as above.

2.2. Sutured diagrams

With these examples in mind, we turn to definitions for sutured Heegaard diagrams. Since, in this paper, we need to be careful about naturality of the constructions, we are careful in our definitions, distinguishing, for instance, between attaching sets and isotopy classes of attaching sets.

DEFINITION 2.6. Let Σ be a compact oriented surface with boundary. An *attaching set in* Σ is a one-dimensional smooth submanifold $\boldsymbol{\delta} \subset \mathrm{Int}(\Sigma)$ such that each component of $\Sigma \setminus \boldsymbol{\delta}$ contains at least one component of $\partial\Sigma$. We will denote the isotopy class of $\boldsymbol{\delta}$ by $[\boldsymbol{\delta}]$.

DEFINITION 2.7. The sutured manifold (M, γ) is a *sutured compression body* if either there is an attaching set $\boldsymbol{\delta} \subset R_+(\gamma)$ such that, if we compress $R_+(\gamma)$ inside M along all the components of $\boldsymbol{\delta}$, we get a surface that is isotopic to $R_-(\gamma)$ relative to γ, or, there is an attaching set $\boldsymbol{\delta} \subset R_-(\gamma)$ such that, if we compress $R_-(\gamma)$ inside M along all the components of $\boldsymbol{\delta}$, we get a surface that is isotopic to $R_+(\gamma)$ relative to γ. We call $\boldsymbol{\delta}$ an *attaching set* for (M, γ).

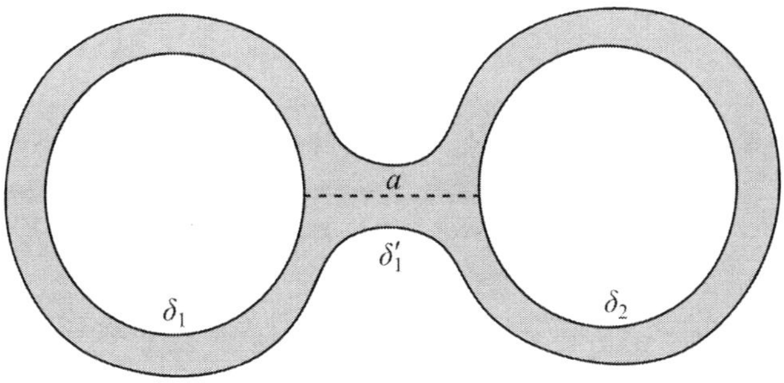

FIGURE 2. Handlesliding the curve δ_1 over δ_2 along the arc a gives δ_1'.

DEFINITION 2.8. Given an attaching set $\boldsymbol{\delta}$ in Σ, let $C(\boldsymbol{\delta}) = (M, \gamma)$ be the sutured compression body obtained by taking M to be $\Sigma \times [0,1]$ and attaching 3-dimensional 2-handles along $\boldsymbol{\delta} \times \{1\}$, while $\gamma = \partial\Sigma \times [0,1]$. In addition, let $C_-(\boldsymbol{\delta}) = R_-(M,\gamma) = \Sigma \times \{0\}$ and

$$C_+(\boldsymbol{\delta}) = R_+(M,\gamma) = \partial C(\boldsymbol{\delta}) \setminus \operatorname{Int}(C_-(\boldsymbol{\delta}) \cup \gamma).$$

If $\boldsymbol{\delta}$ and $\boldsymbol{\delta}'$ are both attaching sets in Σ, then we say they are *compression equivalent*, and we write $\boldsymbol{\delta} \sim \boldsymbol{\delta}'$, if there is a diffeomorphism $d \colon C(\boldsymbol{\delta}) \to C(\boldsymbol{\delta}')$ such that $d|_{C_-(\boldsymbol{\delta})}$ is the identity. This is an equivalence relation that descends to isotopy classes of attaching sets. So we will write $[\boldsymbol{\delta}] \sim [\boldsymbol{\delta}']$ if $\boldsymbol{\delta} \sim \boldsymbol{\delta}'$.

Observe that $\chi(C_+(\boldsymbol{\delta})) = \chi(C_-(\boldsymbol{\delta})) + 2|\boldsymbol{\delta}|$. So $\boldsymbol{\delta} \sim \boldsymbol{\delta}'$ implies that $|\boldsymbol{\delta}| = |\boldsymbol{\delta}'|$.

LEMMA 2.9. *Let $\boldsymbol{\delta} \subset \Sigma$ be an attaching set in a compact oriented surface with boundary, and let $C(\boldsymbol{\delta}) = (M,\gamma)$ be the corresponding sutured compression body. Then $\pi_2(M) = 0$.*

PROOF. Consider the Mayer–Vietoris sequence for the pair $(\Sigma \times I, H)$, where H is the union of the handles attached to $\Sigma \times \{1\}$ along $\boldsymbol{\delta} \times \{1\}$:

$$0 = H_2(\Sigma \times I) \oplus H_2(H) \to H_2(M) \to H_1((\Sigma \times I) \cap H) \xrightarrow{i} H_1(\Sigma \times I) \oplus H_1(H).$$

Of course, $H_i(H) = 0$ for $i \in \{1,2\}$, and $H_2(\Sigma \times I) = 0$ as Σ has no closed components. Since $\boldsymbol{\delta}$ is an attaching set, the map $\pi_0(\partial\Sigma) \to \pi_0(\Sigma \setminus \boldsymbol{\delta})$ is surjective. Hence, the components of $\boldsymbol{\delta}$ are linearly independent in $H_1(\Sigma)$, and so the map i is injective. It follows that $H_2(M) = 0$. In particular, every smoothly embedded 2-sphere S in M is null-homologous; i.e., there is a compact submanifold-with-boundary N of M such that $\partial N = S$. If we attach 2-handles to M along the components of γ, we obtain a compression body, which embeds into a handlebody, and hence also into $\mathbb{R}^3$. In $\mathbb{R}^3$, the sphere S bounds a ball, hence N is diffeomorphic to D^3, and S is null-homotopic. □

DEFINITION 2.10. Let δ_1 and δ_2 be two disjoint simple closed curves in Σ, and fix an embedded arc a from δ_1 to δ_2 whose interior is disjoint from $\delta_1 \cup \delta_2$ and from $\partial\Sigma$. Then a regular neighborhood of the graph $\delta_1 \cup a \cup \delta_2$ is a planar surface with three boundary components: one is isotopic to δ_1, the other is isotopic to δ_2, and the third is a new curve δ_1' that we call the curve obtained by *handlesliding δ_1 over δ_2 along the arc a*; see Figure 2.

Suppose $\boldsymbol{\delta}$ and $\boldsymbol{\delta}'$ are two systems of attaching circles. We say that $\boldsymbol{\delta}$ and $\boldsymbol{\delta}'$ are *related by a handleslide* if there are components δ_1 and δ_2 of $\boldsymbol{\delta}$ and a component δ_1' of $\boldsymbol{\delta}'$ such that δ_1' can be obtained by handle-sliding δ_1 over δ_2 along some arc whose interior is disjoint from $\boldsymbol{\delta}$, and $\boldsymbol{\delta}' = (\boldsymbol{\delta} \setminus \delta_1) \cup \delta_1'$. If D and D' are isotopy classes of attaching sets, then they are related by a handleslide if they have representatives $\boldsymbol{\delta}$ and $\boldsymbol{\delta}'$, respectively, such that $\boldsymbol{\delta}$ and $\boldsymbol{\delta}'$ are related by a handleslide.

LEMMA 2.11. *If $\boldsymbol{\delta}$ and $\boldsymbol{\delta}'$ are related by a handleslide, then $\boldsymbol{\delta} \sim \boldsymbol{\delta}'$. Conversely, if $\boldsymbol{\delta} \sim \boldsymbol{\delta}'$, then $[\boldsymbol{\delta}]$ and $[\boldsymbol{\delta}']$ are related by a sequence of handleslides.*

PROOF. The first part is immediate. For the second part, the proof of Bonahon [**5**, Proposition B.1] for ordinary compression bodies can be adapted to this context. □

DEFINITION 2.12. A *sutured diagram* is a triple $(\Sigma, \boldsymbol{\alpha}, \boldsymbol{\beta})$, where Σ is a compact oriented surface with boundary, and $\boldsymbol{\alpha}$ and $\boldsymbol{\beta}$ are two attaching sets in Σ. An *isotopy diagram* is a triple $(\Sigma, [\boldsymbol{\alpha}], [\boldsymbol{\beta}])$, where $(\Sigma, \boldsymbol{\alpha}, \boldsymbol{\beta})$ is a sutured diagram.

DEFINITION 2.13. Let (M, γ) be a sutured manifold. Then we say that $(\Sigma, \boldsymbol{\alpha}, \boldsymbol{\beta})$ is an *(embedded) diagram of* (M, γ) if the following hold:

(1) $\Sigma \subset M$ is an oriented surface with $\partial\Sigma = s(\gamma)$ as oriented 1-manifolds.
(2) The components of $\boldsymbol{\alpha}$ bound disjoint disks to the negative side of Σ, and the components of $\boldsymbol{\beta}$ bound disjoint disks to the positive side of Σ.
(3) If we compress Σ along $\boldsymbol{\alpha}$, we get a surface isotopic to $R_-(\gamma)$ relative to γ.
(4) If we compress Σ along $\boldsymbol{\beta}$, we get a surface isotopic to $R_+(\gamma)$ relative to γ.

In other words, Σ cuts (M, γ) into two sutured compression bodies, with attaching sets $\boldsymbol{\alpha}$ and $\boldsymbol{\beta}$, respectively (see Definition 2.7).

Note that, if $[\boldsymbol{\alpha}'] = [\boldsymbol{\alpha}]$ and $[\boldsymbol{\beta}'] = [\boldsymbol{\beta}]$, then $(\Sigma, \boldsymbol{\alpha}', \boldsymbol{\beta}')$ is also a sutured diagram of (M, γ). So we say that (Σ, A, B) is an *isotopy diagram of* (M, γ) if there is a sutured diagram $(\Sigma, \boldsymbol{\alpha}, \boldsymbol{\beta})$ of (M, γ) such that $A = [\boldsymbol{\alpha}]$ and $B = [\boldsymbol{\beta}]$.

LEMMA 2.14. *Every sutured manifold (M, γ) has a diagram.*

PROOF. The proof of Juhász [**16**, Proposition 2.13] provides a sutured Heegaard diagram $(\Sigma, \boldsymbol{\alpha}, \boldsymbol{\beta})$ such that $\Sigma \subset M$. □

DEFINITION 2.15. Let (M, γ) be a sutured manifold. We say that the oriented surface $\Sigma \subset M$ is a *Heegaard surface of* (M, γ) if $\partial\Sigma = s(\gamma)$ and Σ divides (M, γ) into two sutured compression bodies.

DEFINITION 2.16. A sutured diagram $(\Sigma, \boldsymbol{\alpha}, \boldsymbol{\beta})$ *defines* a sutured manifold (M, γ) as follows. To obtain M, take $\Sigma \times [-1, 1]$, and attach 3-dimensional 2-handles to $\Sigma \times \{-1\}$ along $\boldsymbol{\alpha} \times \{-1\}$ and to $\Sigma \times \{1\}$ along $\boldsymbol{\beta} \times \{1\}$. The annuli are taken to be $\gamma = \partial\Sigma \times [-1, 1]$, with the sutures $s(\gamma) = \Sigma \times \{0\}$. Then (M, γ) is well-defined up to diffeomorphism relative to Σ. (Note that, if we think of Σ as the middle level $\Sigma \times \{0\} \subset M$, then $(\Sigma, \boldsymbol{\alpha}, \boldsymbol{\beta})$ is a sutured diagram of M.)

If $\boldsymbol{\alpha}'$ is isotopic to $\boldsymbol{\alpha}$ and $\boldsymbol{\beta}'$ is isotopic to $\boldsymbol{\beta}$, then the sutured manifold (M', γ') defined by $(\Sigma, \boldsymbol{\alpha}', \boldsymbol{\beta}')$ is diffeomorphic (relative to Σ) to the sutured manifold (M, γ) defined by $(\Sigma, \boldsymbol{\alpha}, \boldsymbol{\beta})$. So an isotopy diagram H defines a diffeomorphism type of sutured manifolds that we will denote by $S(H)$.

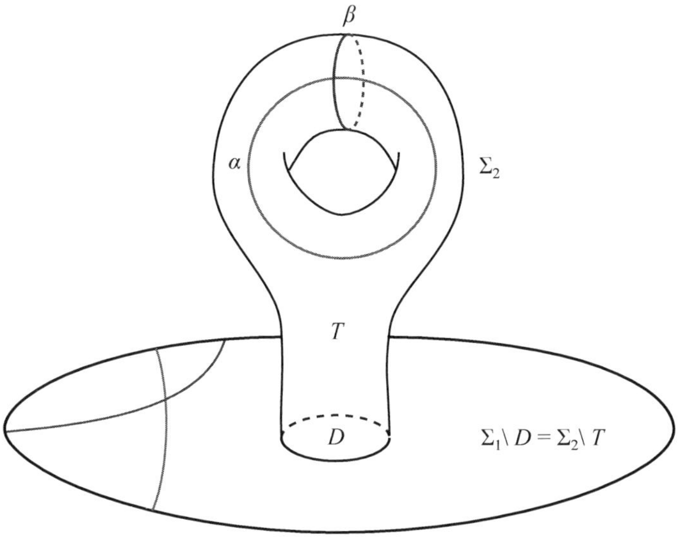

FIGURE 3. The diagram $(\Sigma_2, \boldsymbol{\alpha}_2, \boldsymbol{\beta}_2)$ is obtained from $(\Sigma_1, \boldsymbol{\alpha}_1, \boldsymbol{\beta}_1)$ by a stabilization.

2.3. Moves on diagrams and weak Heegaard invariants

DEFINITION 2.17. The isotopy diagrams (Σ_1, A_1, B_1) and (Σ_2, A_2, B_2) are α-*equivalent* if $\Sigma_1 = \Sigma_2$ and $B_1 = B_2$, while $A_1 \sim A_2$. Similarly, they are β-*equivalent* if $\Sigma_1 = \Sigma_2$ and $A_1 = A_2$, while $B_1 \sim B_2$.

DEFINITION 2.18. The sutured diagram $(\Sigma_2, \boldsymbol{\alpha}_2, \boldsymbol{\beta}_2)$ is obtained from the diagram $(\Sigma_1, \boldsymbol{\alpha}_1, \boldsymbol{\beta}_1)$ by a *stabilization* if

- there is a disk $D \subset \Sigma_1$ and a punctured torus $T \subset \Sigma_2$ such that
$$\Sigma_1 \setminus D = \Sigma_2 \setminus T,$$
- $\boldsymbol{\alpha}_1 = \boldsymbol{\alpha}_2 \cap (\Sigma_2 \setminus T)$,
- $\boldsymbol{\beta}_1 = \boldsymbol{\beta}_2 \cap (\Sigma_2 \setminus T)$,
- $\boldsymbol{\alpha}_2 \cap T$ and $\boldsymbol{\beta}_2 \cap T$ are simple closed curves that intersect each other transversely in a single point.

In this case, we also say that $(\Sigma_1, \boldsymbol{\alpha}_1, \boldsymbol{\beta}_1)$ is obtained from $(\Sigma_2, \boldsymbol{\alpha}_2, \boldsymbol{\beta}_2)$ by a *destabilization*. For an illustration, see Figure 3.

Let H_1 and H_2 be isotopy diagrams. Then H_2 is obtained from H_1 by a *(de)stabilization* if they have representatives $(\Sigma_2, \boldsymbol{\alpha}_2, \boldsymbol{\beta}_2)$ and $(\Sigma_1, \boldsymbol{\alpha}_1, \boldsymbol{\beta}_1)$, respectively, such that $(\Sigma_2, \boldsymbol{\alpha}_2, \boldsymbol{\beta}_2)$ is obtained from $(\Sigma_1, \boldsymbol{\alpha}_1, \boldsymbol{\beta}_1)$ by a (de)stabilization.

If $d\colon \Sigma \to \Sigma'$ is a diffeomorphism of surfaces and C is an isotopy class of attaching sets in Σ, then $d(C)$ is defined as $[d(\boldsymbol{\gamma})]$, where $\boldsymbol{\gamma}$ is an arbitrary attaching set representing C.

DEFINITION 2.19. Let $H_1 = (\Sigma_1, A_1, B_1)$ and $H_2 = (\Sigma_2, A_2, B_2)$ be isotopy diagrams. A *diffeomorphism* $d\colon H_1 \to H_2$ is an orientation preserving diffeomorphism $d\colon \Sigma_1 \to \Sigma_2$ such that $d(A_1) = A_2$ and $d(B_1) = B_2$.

We now recall the notion of a graph, from a rather categorical viewpoint.

DEFINITION 2.20. A *graph* G consists of

(1) a class $|G|$ whose elements are called the objects (or vertices) of the graph,
(2) for each pair $(A, B) \in |G| \times |G|$, a set $G(A, B)$ whose elements are called the morphisms (or arrows) from A to B.

DEFINITION 2.21. A *morphism of graphs* $F\colon G \to K$ between two graphs G and K consists of

(1) a map $F\colon |G| \to |K|$,
(2) for each pair $(A, B) \in |G| \times |G|$ of objects, a map

$$F\colon G(A, B) \to K(F(A), F(B)).$$

Notice that every category is a graph, and every functor between categories is a morphism of graphs.

DEFINITION 2.22. Let $\mathcal{G}$ be the graph whose class of vertices $|\mathcal{G}|$ consists of isotopy diagrams and, for H_1, $H_2 \in |\mathcal{G}|$, the edges $\mathcal{G}(H_1, H_2)$ can be written as a union of four sets

$$\mathcal{G}(H_1, H_2) = \mathcal{G}_\alpha(H_1, H_2) \cup \mathcal{G}_\beta(H_1, H_2) \cup \mathcal{G}_{\text{stab}}(H_1, H_2) \cup \mathcal{G}_{\text{diff}}(H_1, H_2).$$

The set $\mathcal{G}_\alpha(H_1, H_2)$ consists of a single arrow if H_1 and H_2 are α-equivalent, and is empty otherwise. The set $\mathcal{G}_\beta(H_1, H_2)$ is defined similarly using β-equivalence. The set $\mathcal{G}_{\text{stab}}(H_1, H_2)$ consists of a single arrow if H_2 is obtained from H_1 by a stabilization or a destabilization, and is empty otherwise. Finally, $\mathcal{G}_{\text{diff}}(H_1, H_2)$ consists of all diffeomorphisms from H_1 to H_2. The graph $\mathcal{G}$ is thus the union of four subgraphs, namely $\mathcal{G}_\alpha$, $\mathcal{G}_\beta$, $\mathcal{G}_{\text{stab}}$, and $\mathcal{G}_{\text{diff}}$.

Note that the graphs $\mathcal{G}_\alpha$, $\mathcal{G}_\beta$, and $\mathcal{G}_{\text{diff}}$ are in fact categories when endowed with the obvious compositions. We have a version of the Reidemeister–Singer theorem.

PROPOSITION 2.23. *The isotopy diagrams H_1, $H_2 \in |\mathcal{G}|$ can be connected by an oriented path if and only if they define diffeomorphic sutured manifolds. Furthermore, the existence of an unoriented path from H_1 to H_2 implies the existence of an oriented one.*

PROOF. By Juhász [**16**, Proposition 2.15], if two diagrams define diffeomorphic sutured manifolds, then they become diffeomorphic after a sequence of isotopies, handleslides, stabilizations, and destabilizations. (Actually, the above mentioned result is stated for balanced diagrams, but the same proof works for arbitrary ones.) Lemma 2.11 implies that every handleslide is an α- or β-equivalence, which concludes the proof of the first claim.

For the second claim, observe that, if $* \in \{\, \alpha, \beta, \text{stab}, \text{diff} \,\}$, then $\mathcal{G}_*(H_1, H_2) \neq \emptyset$ if and only if $\mathcal{G}_*(H_2, H_1) \neq \emptyset$. □

DEFINITION 2.24. Let $\mathcal{S}$ be a set of diffeomorphism types of sutured manifolds, and let $\mathcal{C}$ be any category. We denote by $\mathcal{G}(\mathcal{S})$ the full subgraph of $\mathcal{G}$ spanned by those isotopy diagrams H for which $S(H) \in \mathcal{S}$. A *weak Heegaard invariant of $\mathcal{S}$* is a morphism of graphs $F\colon \mathcal{G}(\mathcal{S}) \to \mathcal{C}$ such that, for every arrow e of $\mathcal{G}(\mathcal{S})$, the image $F(e)$ is an isomorphism.

Observe that, if $F\colon \mathcal{G}(\mathcal{S}) \to \mathcal{C}$ is a weak Heegaard invariant and H_1 and H_2 lie in the same path-component of $\mathcal{G}(\mathcal{S})$ (i.e., if $S(H_1) = S(H_2)$), then $F(H_1)$ and $F(H_2)$ are isomorphic objects of the category $\mathcal{C}$. In this language, we can state the main invariance results previously known. We first introduce some important sets of diffeomorphism types of sutured manifolds. We denote by $[(M, \gamma)]$ the diffeomorphism type of (M, γ).

DEFINITION 2.25. A *balanced* sutured manifold is a proper sutured manifold (M, γ) such that $\chi(R_+(\gamma)) = \chi(R_-(\gamma))$. Equivalently, by Juhász [**16**, Proposition 2.9], it is a proper sutured manifold that has a diagram $(\Sigma, \boldsymbol{\alpha}, \boldsymbol{\beta})$ with $|\boldsymbol{\alpha}| = |\boldsymbol{\beta}|$. Then we define the following types of sutured manifolds:

(1) Let $\mathcal{S}_{\text{man}}$ be the set of all $[Y(p, V)]$, where Y is a closed, oriented, based 3-manifold, $p \in Y$, and $V < T_pM$ is an oriented tangent 2-plane.
(2) Let $\mathcal{S}_{\text{link}}$ be the set of all $[S^3(L, \mathbf{p}, \mathbf{q})]$, where $(L, \mathbf{p}, \mathbf{q})$ is a based oriented link in S^3 with exactly one $\mathbf{p}$ and one $\mathbf{q}$ marking on each component of L.
(3) Let $\mathcal{S}_{\text{bal}}$ be the set of all $[(M, \gamma)]$, where (M, γ) is a balanced sutured manifold.

THEOREM 2.26 ([**27**]). *The morphisms*

$$\widehat{HF},\ HF^-,\ HF^+,\ HF^\infty \colon \mathcal{G}(\mathcal{S}_{\text{man}}) \to \mathbb{F}_2[U]\text{-}\mathsf{Mod}$$

are weak Heegaard invariants of $\mathcal{S}_{\text{man}}$ (where the U-action is trivial on $\widehat{HF}$).

THEOREM 2.27 ([**25, 29, 32**]). *The morphisms*

$$\widehat{HFL},\ HFL^- \colon \mathcal{G}(\mathcal{S}_{\text{link}}) \to \mathbb{F}_2[U]\text{-}\mathsf{Mod}$$

are weak Heegaard invariants of $\mathcal{S}_{\text{link}}$.

THEOREM 2.28 ([**16**]). *The morphism*

$$SFH \colon \mathcal{G}(\mathcal{S}_{\text{bal}}) \to \mathbb{F}_2\text{-}\mathsf{Vect}$$

is a weak Heegaard invariant of $\mathcal{S}_{\text{bal}}$.

However, these theorems are not enough to give invariants in the stronger sense of Theorems 1.5, 1.8, and 1.9, which assign to a based 3-manifold, a based link, or a balanced sutured manifold an object of $\mathbb{F}_2[U]$-Mod, rather than an isomorphism class of objects of $\mathbb{F}_2[U]$-Mod. For that, we look for further structure in the graph $\mathcal{G}$.

2.4. Strong Heegaard invariants

DEFINITION 2.29. Let $H_i = (\Sigma_i, [\boldsymbol{\alpha}_i], [\boldsymbol{\beta}_i])$ be isotopy diagrams for $1 \le i \le 4$. A *distinguished rectangle* in $\mathcal{G}$ is a subgraph

$$\begin{array}{ccc} H_1 & \xrightarrow{e} & H_2 \\ \big\downarrow{\scriptstyle f} & & \big\downarrow{\scriptstyle g} \\ H_3 & \xrightarrow{h} & H_4 \end{array}$$

of $\mathcal{G}$ that satisfies one of the following properties:

(1) Both e and h are α-equivalences, while f and g are β-equivalences.
(2) Either both e and h are α-equivalences, or both e and h are β-equivalences, while f and g are both stabilizations.
(3) Either both e and h are α-equivalences, or both e and h are β-equivalences, while f and g are both diffeomorphisms. In this case, we necessarily have $\Sigma_1 = \Sigma_2$ and $\Sigma_3 = \Sigma_4$. We require, in addition, that the diffeomorphisms $f \colon \Sigma_1 \to \Sigma_3$ and $g \colon \Sigma_2 \to \Sigma_4$ are the same.

(4) The maps e, f, g, and h are all stabilizations, such that there are disjoint discs D_1, $D_2 \subset \Sigma_1$ and disjoint punctured tori T_1, $T_2 \subset \Sigma_4$ satisfying $\Sigma_1 \setminus (D_1 \cup D_2) = \Sigma_4 \setminus (T_1 \cup T_2)$, and such that $\Sigma_2 = (\Sigma_1 \setminus D_1) \cup T_1$ and $\Sigma_3 = (\Sigma_1 \setminus D_2) \cup T_2$.

(5) The maps e and h are stabilizations, while f and g are diffeomorphisms. Furthermore, there are disks $D \subset \Sigma_1$ and $D' \subset \Sigma_3$, and punctured tori $T \subset \Sigma_2$ and $T' \subset \Sigma_4$, such that $\Sigma_1 \setminus D = \Sigma_2 \setminus T$ and $\Sigma_3 \setminus D' = \Sigma_4 \setminus T'$, and the diffeomorphisms f and g satisfy $f(D) = D'$, $g(T) = T'$, and $f|_{\Sigma_1 \setminus D} = g|_{\Sigma_2 \setminus T}$.

REMARK 2.30. In case (1), we have $\Sigma_i = \Sigma_j$ for $i, j \in \{1, \dots, 4\}$, so a distinguished rectangle in this case is of the form

$$\begin{array}{ccc} (\Sigma, A, B) & \longrightarrow & (\Sigma, A', B) \\ \downarrow & & \downarrow \\ (\Sigma, A, B') & \longrightarrow & (\Sigma, A', B'). \end{array}$$

In case (2), we necessarily have $\Sigma_1 = \Sigma_2$ and $\Sigma_3 = \Sigma_4$. Without loss of generality, consider the situation when both e and h are α-equivalences. Then we have a rectangle

$$\begin{array}{ccc} (\Sigma, [\boldsymbol{\alpha}], [\boldsymbol{\beta}]) & \longrightarrow & (\Sigma, [\overline{\boldsymbol{\alpha}}], [\boldsymbol{\beta}]) \\ \downarrow & & \downarrow \\ (\Sigma', [\boldsymbol{\alpha}'], [\boldsymbol{\beta}']) & \longrightarrow & (\Sigma', [\overline{\boldsymbol{\alpha}}'], [\boldsymbol{\beta}']) \end{array}$$

such that there is a disk $D \subset \Sigma$ and a punctured torus $T \subset \Sigma'$ with $\Sigma \setminus D = \Sigma' \setminus T$. Furthermore, we can assume that $\boldsymbol{\alpha} = \boldsymbol{\alpha}' \cap (\Sigma' \setminus T)$ and $\boldsymbol{\beta} = \boldsymbol{\beta}' \cap (\Sigma' \setminus T)$, while $\overline{\boldsymbol{\alpha}} = \overline{\boldsymbol{\alpha}}' \cap (\Sigma' \setminus T)$. Since $\boldsymbol{\alpha}' \sim \overline{\boldsymbol{\alpha}}'$, the curves $\boldsymbol{\alpha}' \cap T$ and $\overline{\boldsymbol{\alpha}}' \cap T$ are isotopic.

In case (4), the fact that all four diagrams contain

$$S = \Sigma_1 \setminus (D_1 \cup D_2)$$

implies that $\boldsymbol{\alpha}_i \cap S = \boldsymbol{\alpha}_j \cap S$ and $\boldsymbol{\beta}_i \cap S = \boldsymbol{\beta}_j \cap S$ for every $i, j \in \{1, \dots, 4\}$.

DEFINITION 2.31. A *simple handleswap* is a subgraph of $\mathcal{G}$ of the form

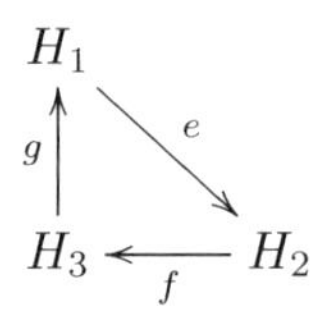

such that

(1) $H_i = (\Sigma \# \Sigma_0, [\boldsymbol{\alpha}_i], [\boldsymbol{\beta}_i])$ are isotopy diagrams for $i \in \{1, 2, 3\}$, where Σ_0 is a genus two surface,

(2) e is an α-equivalence, f is a β-equivalence, and g is a diffeomorphism,

(3) in the punctured genus two surface $P = (\Sigma \# \Sigma_0) \setminus \Sigma$, the above triangle is conjugate to the triangle in Figure 4; i.e., there is a diffeomorphism that throws $P \cap H_i$ onto the pictures in the green circles, sending the α-circles in P to the two red circles, and the β-circles in P to the two blue circles,

(4) in Σ, the diagrams H_1, H_2, and H_3 are identical.

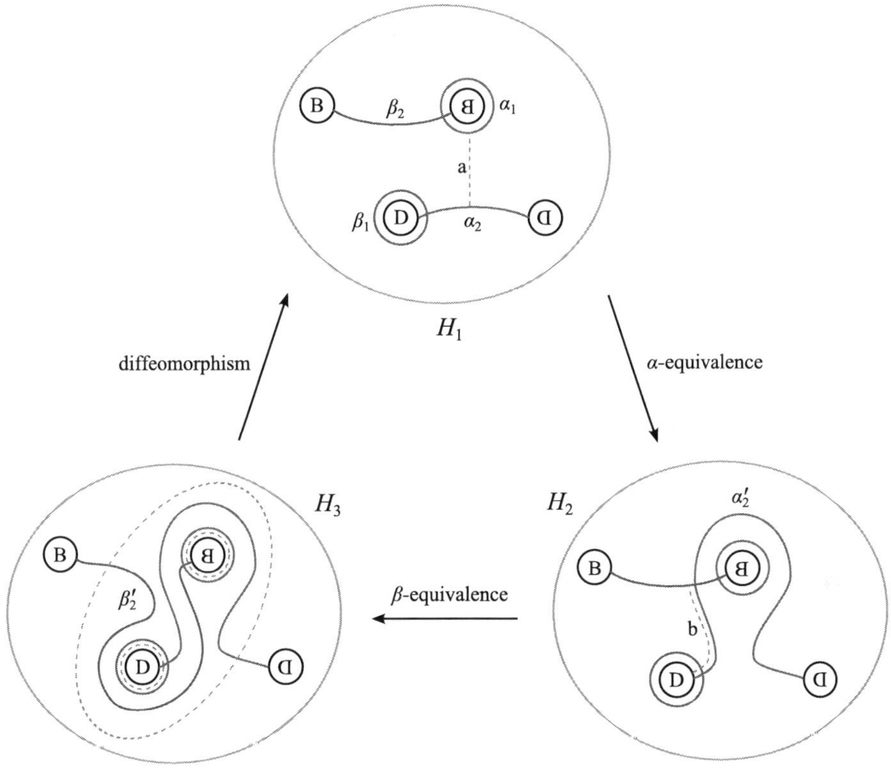

FIGURE 4. A simple handleswap, which is a loop of diagrams consisting of an α-handleslide, a β-handleslide, and a diffeomorphism. The green curve is the boundary of the punctured genus two surface P that is obtained by identifying the circles marked with corresponding letters (namely, B and D). We draw the α-curves in red and the β-curves in blue.

So $P \cap \boldsymbol{\alpha}_1$ consists of closed curves α_1 and α_2 and $P \cap \boldsymbol{\beta}_1$ consists of closed curves β_1 and β_2 such that $\alpha_i \cap \beta_i$ consists of a single point for $i \in \{1, 2\}$, while both $\alpha_1 \cap \beta_2$ and $\alpha_2 \cap \beta_1$ are empty. The arrow e from H_1 to H_2 corresponds to handle-sliding α_1 over α_2 along the dashed arc a. The arrow f from H_2 to H_3 corresponds to handle-sliding β_2 over β_1 along the dashed arc b. Finally, the diffeomorphism g maps H_3 to H_1 by performing Dehn twists around the dashed curves depicted in the lower left corner of Figure 4; namely a left-handed Dehn twist along the large dashed curve and right-handed Dehn twists around the smaller ones.

DEFINITION 2.32. Let $\mathcal{S}$ be a set of diffeomorphism types of sutured manifolds. A *strong Heegaard invariant of* $\mathcal{S}$ is a weak Heegaard invariant $F\colon \mathcal{G}(\mathcal{S}) \to \mathcal{C}$ that satisfies the following axioms:

(1) **Functoriality**: The restriction of F to $\mathcal{G}_\alpha(\mathcal{S})$, $\mathcal{G}_\beta(\mathcal{S})$, and $\mathcal{G}_{\text{diff}}(\mathcal{S})$ are functors to $\mathcal{C}$. Furthermore, if $e\colon H_1 \to H_2$ is a stabilization and $e'\colon H_2 \to H_1$ is the corresponding destabilization, then $F(e') = F(e)^{-1}$.

(2) **Commutativity**: For every distinguished rectangle

$$\begin{array}{ccc} H_1 & \xrightarrow{e} & H_2 \\ \downarrow{\scriptstyle f} & & \downarrow{\scriptstyle g} \\ H_3 & \xrightarrow{h} & H_4 \end{array}$$

in $\mathcal{G}(\mathcal{S})$, we have $F(g) \circ F(e) = F(h) \circ F(f)$.

(3) **Continuity**: If $H \in |\mathcal{G}(\mathcal{S})|$ and $e \in \mathcal{G}_{\text{diff}}(H, H)$ is a diffeomorphism isotopic to Id_Σ, then $F(e) = \text{Id}_{F(H)}$.

(4) **Handleswap invariance**: For every simple handleswap

$$\begin{array}{ccc} H_1 & & \\ {\scriptstyle g}\uparrow & \searrow{\scriptstyle e} & \\ H_3 & \xleftarrow[f]{} & H_2 \end{array}$$

in $\mathcal{G}(\mathcal{S})$, we have $F(g) \circ F(f) \circ F(e) = \text{Id}_{F(H_1)}$.

Note that, in axiom (3), if $H = (\Sigma, \boldsymbol{\alpha}, \boldsymbol{\beta})$ and $e_t \colon \Sigma \to \Sigma$ for $t \in [0,1]$ is an isotopy from e to Id_Σ, then $(\Sigma, e_t(\boldsymbol{\alpha}), e_t(\boldsymbol{\beta}))$ represents the same isotopy diagram as H. Hence $e_t \in \mathcal{G}_{\text{diff}}(H, H)$ for every $t \in [0,1]$.

Axiom (2) of Definition 2.32 and the second part of axiom (1) imply commutativity for any distinguished rectangle involving destabilizations. That is why we only considered stabilizations in Definition 2.29.

THEOREM 2.33. *The following are strong Heegaard invariants:*

(1) *Sutured Floer homology, SFH, is a strong Heegaard invariant of $\mathcal{S}_{\text{bal}}$.*
(2) *The Heegaard Floer homology invariants $\widehat{HF}$, HF^+, HF^-, and HF^∞ are strong Heegaard invariants of $\mathcal{S}_{\text{man}}$.*
(3) *The link Floer homology groups $\widehat{HFL}$ and HFL^- are strong Heegaard invariants of $\mathcal{S}_{\text{link}}$.*

We will prove Theorem 2.33 in Chapter 10.

2.5. Construction of the Heegaard Floer functors

We next explain how Theorem 2.33 lets us associate, for instance, a group $SFH(M, \gamma)$ to a balanced sutured manifold (M, γ).

DEFINITION 2.34. Suppose that H_1 and H_2 are two isotopy diagrams of (M, γ) with $H_i = (\Sigma_i, A_i, B_i)$, and let $\iota_i \colon \Sigma_i \to M$ be the inclusion for $i \in \{1, 2\}$. Then a diffeomorphism $d \colon H_1 \to H_2$ is *isotopic to the identity in M* if $\iota_2 \circ d \colon \Sigma_1 \to M$ is isotopic to $\iota_1 \colon \Sigma_1 \to M$ relative to $s(\gamma)$.

DEFINITION 2.35. Let (M, γ) be a sutured manifold. Then $\mathcal{G}_{(M,\gamma)}$ is the subgraph of $\mathcal{G}$ whose vertex set [1] $|\mathcal{G}_{(M,\gamma)}|$ consists of all isotopy diagrams of (M, γ). The set of edges between H_1, $H_2 \in |\mathcal{G}_{(M,\gamma)}|$ is defined to be

$$\mathcal{G}_{(M,\gamma)}(H_1, H_2) = \mathcal{G}_\alpha(H_1, H_2) \cup \mathcal{G}_\beta(H_1, H_2) \cup \mathcal{G}_{\text{stab}}(H_1, H_2) \cup \mathcal{G}^0_{\text{diff}}(H_1, H_2),$$

[1] Observe that $|\mathcal{G}_{(M,\gamma)}|$ is a set, not a proper class, as we defined a sutured diagram for (M, γ) to be a submanifold of M.

where $\mathcal{G}_\alpha$, $\mathcal{G}_\beta$, and $\mathcal{G}_{\text{stab}}$ are as before, and $\mathcal{G}^0_{\text{diff}}(H_1, H_2)$ is the set of diffeomorphisms from H_1 to H_2 isotopic to the identity in M.

We will prove the following stronger version of Proposition 2.23 in Section 7.1.

PROPOSITION 2.36. *Let (M,γ) be sutured manifold. In the graph $\mathcal{G}_{(M,\gamma)}$, any two vertices can be connected by an oriented path.*

DEFINITION 2.37. Given a weak Heegaard invariant $F\colon \mathcal{G}(\mathcal{S}) \to \mathcal{C}$ and an oriented path η in $\mathcal{G}(\mathcal{S})$ of the form

$$H_0 \xrightarrow{e_1} H_1 \xrightarrow{e_2} \cdots \xrightarrow{e_n} H_n,$$

define $F(\eta)$ to be the isomorphism

$$F(e_n) \circ \cdots \circ F(e_1) : F(H_0) \to F(H_n).$$

For a weak Heegaard invariant, the isomorphism $F(\eta)$ from $F(H_0)$ to $F(H_n)$ might depend on the choice of the path η. However, according to the following theorem, this ambiguity disappears if we require F to be a strong Heegaard invariant and we restrict our attention to the subgraph $\mathcal{G}_{(M,\gamma)}$.

THEOREM 2.38. *Let $\mathcal{S}$ be a set of diffeomorphism types of sutured manifolds containing $[(M,\gamma)]$. Furthermore, let $F\colon \mathcal{G}(\mathcal{S}) \to \mathcal{C}$ be a strong Heegaard invariant. Given isotopy diagrams H, $H' \in |\mathcal{G}_{(M,\gamma)}|$ and any two oriented paths η and ν in $\mathcal{G}_{(M,\gamma)}$ from H to H', we have*

$$F(\eta) = F(\nu).$$

REMARK 2.39. Another interpretation of Theorem 2.38 is that, if we extend $\mathcal{G}_{(M,\gamma)}$ to a 2-complex with 2-cells corresponding to the various polygons in Definition 2.32, the result is simply-connected.

Theorem 2.38 is one of the most important and deepest results of this paper. We will prove it in Chapter 9, and develop the necessary technical tools in Chapters 4–7.

DEFINITION 2.40. Let $\mathcal{S}$ be a set of diffeomorphism types of balanced sutured manifolds containing $[(M,\gamma)]$, and let $F\colon \mathcal{G}(\mathcal{S}) \to \mathcal{C}$ be a strong Heegaard invariant. If H and H' are isotopy diagrams of (M,γ), then let

$$F_{H,H'} = F(\eta),$$

where η is an arbitrary oriented path connecting H to H'. By Theorem 2.38, the map $F_{H,H'}$ does not depend on the choice of η.

COROLLARY 2.41. *Suppose that H, H', $H'' \in |\mathcal{G}_{(M,\gamma)}|$. Then*

$$F_{H,H''} = F_{H',H''} \circ F_{H,H'}.$$

Motivated by Definition 1.1 and Remark 1.2, we obtain a natural invariant of sutured manifolds from a strong Heegaard invariant as follows. As usual, we denote the category of abelian groups by Ab.

DEFINITION 2.42. Let $\mathcal{S}$ be a set of diffeomorphism types of balanced sutured manifolds, and let $F\colon \mathcal{G}(\mathcal{S}) \to \mathsf{Ab}$ be a strong Heegaard invariant. Fix a balanced sutured manifold (M,γ) with $[(M,\gamma)] \in \mathcal{S}$, and suppose that H and H' are isotopy diagrams of (M,γ). We say that the elements $x \in F(H)$ and $y \in F(H')$ are *equivalent*, in short $x \sim y$, if $y = F_{H,H'}(x)$. By Theorem 2.38, this is an equivalence relation on the disjoint union $\coprod_{H\in|\mathcal{G}_{(M,\gamma)}|} F(H)$. The equivalence class of an element

$x \in F(H)$ is denoted by $[x]$. Under the natural addition operation, these equivalence classes form an abelian group that we call $F(M,\gamma)$. Furthermore, let $I_H\colon F(H) \to F(M,\gamma)$ be the isomorphism that maps x to $[x]$.

If $\phi\colon (M,\gamma) \to (M',\gamma')$ is a diffeomorphism, then we define

$$F(\phi)\colon F(M,\gamma) \to F(M',\gamma')$$

as follows. Pick an isotopy diagram $H = (\Sigma, A, B)$ of (M,γ), and let $d = \phi|_\Sigma$. Then $H' = d(H)$ is an isotopy diagram of (M',γ'), and d is a diffeomorphism from H to H', so it induces a map $F(d)\colon F(H) \to F(H')$. We define the isomorphism $F(\phi)$ as $I_{H'} \circ F(d) \circ (I_H)^{-1}$. So we have a commutative diagram

$$\begin{array}{ccc} F(H) & \xrightarrow{F(d)} & F(H') \\ \big\downarrow{\scriptstyle I_H} & & \big\downarrow{\scriptstyle I_{H'}} \\ F(M,\gamma) & \xrightarrow{F(\phi)} & F(M',\gamma'). \end{array}$$

PROPOSITION 2.43. *In the above definition, the isomorphism $F(\phi)$ does not depend on the choice of isotopy diagram H of (M,γ).*

PROOF. Suppose that $H_1 = (\Sigma_1, A_1, B_1)$ and $H_2 = (\Sigma_2, A_2, B_2)$ are isotopy diagrams of (M,γ). Let $d_1 = d|_{\Sigma_1}$ and $d_2 = d|_{\Sigma_2}$, and write $H_1' = d_1(H_1)$ and $H_2' = d_2(H_2)$. Then we have to show that

$$I_{H_1'} \circ F(d_1) \circ (I_{H_1})^{-1} = I_{H_2'} \circ F(d_2) \circ (I_{H_2})^{-1}.$$

Since $(I_{H_2})^{-1} \circ I_{H_1} = F_{H_1,H_2}$ and $(I_{H_2'})^{-1} \circ I_{H_1'} = F_{H_1',H_2'}$, this amounts to proving that

$$F_{H_1',H_2'} \circ F(d_1) = F(d_2) \circ F_{H_1,H_2}. \tag{2.44}$$

Pick a path η in $\mathcal{G}_{(M,\gamma)}$ of the form

$$D_0 \xrightarrow{e_1} D_1 \xrightarrow{e_2} \cdots \xrightarrow{e_n} D_n,$$

such that $D_0 = H_1$ and $D_n = H_2$. There is a corresponding path η' in $\mathcal{G}_{(M,\gamma)}$ from H_1' to H_2' of the form

$$D_0' \xrightarrow{e_1'} D_1' \xrightarrow{e_2'} \cdots \xrightarrow{e_n'} D_n',$$

obtained as follows. For every $i \in \{1,\dots,n\}$, let $D_i' = \phi(D_i)$, and let $h_i\colon D_i \to D_i'$ be the restriction of ϕ to D_i. If e_i is an α-equivalence, β-equivalence, or stabilization, then we denote by e_i' the corresponding α-equivalence, β-equivalence, or stabilization from D_{i-1}' to D_i'. Furthermore, if e_i is a diffeomorphism isotopic to the identity, then we take

$$e_i' = h_i \circ e_i \circ h_{i-1}^{-1};$$

this is also isotopic to the identity. Consider the following subgraph of $\mathcal{G}(\mathcal{S})$:

$$\begin{array}{ccccccc} D_0 & \xrightarrow{e_1} & D_1 & \xrightarrow{e_2} & \cdots & \xrightarrow{e_n} & D_n \\ \big\downarrow{\scriptstyle h_0=d_1} & & \big\downarrow{\scriptstyle h_2} & & & & \big\downarrow{\scriptstyle h_n=d_2} \\ D_0' & \xrightarrow{e_1'} & D_1' & \xrightarrow{e_2'} & \cdots & \xrightarrow{e_n'} & D_n'. \end{array}$$

By construction, each small square is either a distinguished rectangle, or a commuting rectangle of diffeomorphisms. The functor F commutes along the former by the Commutativity Axiom of strong Heegaard invariants, and along the latter by the Functoriality Axiom for $\mathcal{G}_{\text{diff}}(\mathcal{S})$. Hence F also commutes along the large rectangle, giving equation (2.44). □

Let $\mathsf{Sut}_{\text{bal}}$, $\mathsf{Sut}_{\text{man}}$, and $\mathsf{Sut}_{\text{link}}$ denote the full subcategories of Sut whose objects have diffeomorphism types lying in $\mathcal{S}_{\text{bal}}$, $\mathcal{S}_{\text{man}}$, and $\mathcal{S}_{\text{link}}$, respectively.

Proof of Theorem 1.9. By Theorem 2.33, the morphism $F = SFH$ is a strong Heegaard invariant of $\mathcal{S}_{\text{bal}}$. Given isotopy diagrams H and H' of the balanced sutured manifold (M, γ), Theorem 2.38 gives an isomorphism $F_{H,H'}\colon F(H) \to F(H')$. These isomorphisms are canonical according to Corollary 2.41. Hence, the groups $F(H)$ and the isomorphisms $F_{H,H'}$ form a transitive system, and we obtain the limit

$$SFH(M, \gamma) = F(M, \gamma)$$

as in Definition 2.42. A diffeomorphism ϕ between balanced sutured manifolds induces an isomorphism $F(\phi)$ as in Definition 2.42, and these are well-defined according to Proposition 2.43. So we have all the ingredients for a functor $SFH\colon \mathsf{Sut}_{\text{bal}} \to \mathbb{F}_2\text{-}\mathsf{Vect}$.

What remains to show is that isotopic diffeomorphisms induce identical maps on SFH, or equivalently, that, for any diffeomorphism $\phi\colon (M, \gamma) \to (M, \gamma)$ isotopic to $\text{Id}_{(M,\gamma)}$, we have $F(\phi) = \text{Id}_{SFH(M,\gamma)}$. Let H be an isotopy diagram of (M, γ), and write $d = \phi|_H$ and $H' = \phi(H)$. By definition, $F(\phi) = I_{H'} \circ F(d) \circ (I_H)^{-1}$. So this is the identity if and only if

$$F(d) = (I_{H'})^{-1} \circ I_H = F_{H,H'}.$$

This is true since d is isotopic to the identity, hence it corresponds to an edge of $\mathcal{G}_{(M,\gamma)}$ between H and H', and so, if we take the path η from H to H' to be the single edge d, then $F(d) = F(\eta) = F_{H,H'}$. □

Lemma 2.45. *Let (Y, p) be a based 3-manifold, and let V_0 and V_1 be oriented 2-planes in T_pY. Suppose that ϕ, $\psi\colon (Y, p) \to (Y, p)$ are diffeomorphisms such that $d\phi(V_0) = V_1$ and $d\psi(V_0) = V_1$ in an oriented sense. Suppose furthermore that both ϕ and ψ are isotopic to Id_Y through diffeomorphisms fixing p. Then ϕ and ψ are isotopic to each other through diffeomorphisms fixing p and mapping V_0 to V_1.*

Proof. This follows from the fact that the Grassmannian M of oriented 2-planes in T_pY is homeomorphic to S^2, and is hence simply-connected, together with an isotopy extension argument as follows.

Let $\{\, \phi_t : t \in I \,\}$ and $\{\, \psi_t : t \in I \,\}$ be isotopies from Id_Y to ϕ and ψ, respectively, through diffeomorphisms fixing p. Since the Grassmannian M is simply-connected, there is a 2-parameter family of 2-planes $V_{t,u} < T_pY$ for $(t, u) \in I \times I$ such that $V_{t,0} = d\phi_t(V_0)$ and $V_{t,1} = d\psi_t(V_0)$ for every $t \in I$, while $V_{0,u} = V_0$ and $V_{1,u} = V_1$ for every $u \in I$. The 2-planes $V_{t,u}$ form a vector bundle ν over $I \times I$. Since ν is trivial, there is a family of isomorphisms $i_{t,u}\colon V_0 \to V_{t,u}$ for $(t, u) \in I \times I$ such that $i_{0,u} = \text{Id}_{V_0}$ for every $u \in I$, and $i_{t,0} = (d\phi_t)|_{V_0}$ and $i_{t,1} = (d\psi_t)|_{V_0}$ for every $t \in I$. We can extend this to a 2-parameter family of isomorphisms $j_{t,u}\colon T_pY \to T_pY$ such that $j_{t,u}|_{V_0} = i_{t,u}$ for every $(t, u) \in I \times I$, while $j_{0,u} = \text{Id}_{T_pY}$ for every $u \in I$, and $j_{t,0} = d\phi_t$ and $j_{t,1} = d\psi_t$ for every $t \in I$. By Gromov's parametric h-principle for open manifolds [**10**, Theorem 7.2.3], there is a neighborhood U of p and a family of diffeomorphisms $h_{t,u}\colon (U, p) \to (Y, p)$ such that $dh_{t,u} = j_{t,u}$ for every $(t, u) \in I \times I$, and $h_{t,0} = \phi_t|_U$ and $h_{t,1} = \psi_t|_U$ for every $t \in I$. Using the relative isotopy extension theorem, we obtain a 2-parameter family of diffeomorphism $g_{t,u}\colon (Y,p) \to (Y,p)$ such that $g_{0,u} = \mathrm{Id}_Y$ for every $u \in I$, while $g_{t,0} = \phi_t$ and $g_{t,1} = \psi_t$ for every $t \in I$; furthermore, $g_{t,u}|_U = h_{t,u}$ for every $(t,u) \in I \times I$. Then the family $\{\, g_{1,u}\colon u \in I \,\}$ provides the desired isotopy from $g_{1,0} = \phi$ to $g_{1,1} = \psi$. □

Proof of Theorem 1.5. Let HF be one of the four versions of Heegaard Floer homology, and let $\mathsf{Man}_{*,V}$ be the category of based 3-manifolds with a choice of oriented tangent 2-plane at the basepoint. A morphism from the object (Y,p,V) to (Y',p',V') is a diffeomorphism $\phi\colon (Y,p) \to (Y',p')$ such that $d\phi(V) = V'$ in an oriented sense. As in the proof of Theorem 1.9, by Theorem 2.33, HF induces a functor $HF_1\colon \mathsf{Sut}_{\mathrm{man}} \to \mathsf{Ab}$. Composing with the functor $(Y,p,V) \mapsto Y(p,V)$ from Definition 2.4 gives a functor $HF_2\colon \mathsf{Man}_{*,V} \to \mathsf{Ab}$. As in the proof of Theorem 1.9, we obtain that isotopic morphisms induce identical maps, where we say that two morphisms from (Y,p,V) to (Y,p,V') are isotopic if they can be connected by a path of morphisms from (Y,p,V) to (Y,p,V').

Each fiber of the forgetful functor $\mathsf{Man}_{*,V} \to \mathsf{Man}_*$ is a sphere, which is simply-connected, so $HF_2(Y,p,V)$ has no monodromy along any loop of oriented 2-planes in T_pY. More precisely, fix a based manifold $(Y,p) \in \mathsf{Man}_*$, and let M be the Grassmannian of oriented 2-planes in T_pY. Our goal is to construct a canonical isomorphism from $HF_2(Y,p,V_0)$ to $HF_2(Y,p,V_1)$ for any pair $(V_0,V_1) \in M \times M$. Take an arbitrary morphism ϕ from (Y,p,V_0) to (Y,p,V_1), and such that ϕ is isotopic to Id_Y through diffeomorphisms fixing p. Then we claim that the isomorphism

$$HF_2(\phi)\colon HF_2(Y,p,V_0) \to HF_2(Y,p,V_1)$$

is independent of the choice of ϕ. Indeed, by Lemma 2.45, if ψ is another choice, then ϕ and ψ are isotopic through diffeomorphisms fixing p and mapping V_0 to V_1, and hence $HF_2(\phi) = HF_2(\psi)$. We denote this isomorphism by i_{V_0,V_1}. So the groups $HF_2(Y,p,V)$ for $V \in M$ and the isomorphisms I_{V_0,V_1} for $(V_0,V_1) \in M \times M$ form a transitive system, and hence we can take the limit $HF(Y,p)$. We have shown that HF_2 factors through a functor $HF\colon \mathsf{Man}_* \to \mathsf{Ab}$.

Since the forgetful morphism $\mathbb{F}_2[U]$-$\mathsf{Mod} \to \mathsf{Ab}$ is faithful, for each of $\widehat{HF}$, HF^-, HF^+, and HF^∞, Theorem 2.33 gives a functor with target category $\mathbb{F}_2[U]$-Mod, as in the statement of the theorem. □

Proof of Theorem 1.8. By Theorem 2.33, the link Floer homology groups $\widehat{HFL}$ and HFL^- are strong Heegaard invariants of $\mathcal{S}_{\mathrm{link}}$. Let HFL be one of $\widehat{HFL}$ and HFL^-. As in the proof of Theorem 1.9, HFL induces a functor

$$HFL_1\colon \mathsf{Sut}_{\mathrm{link}} \to \mathsf{Ab}$$

for both versions of link Floer homology. Composing with the map

$$(S^3, L, \mathbf{p}, \mathbf{q}) \mapsto S^3(L, \mathbf{p}, \mathbf{q})$$

introduced in Definition 2.5 gives a functor $HFL_2\colon \mathsf{Link}_{**} \to \mathsf{Ab}$, where Link_{**} is the category of oriented links with two (distinguished) basepoints on each component. The fiber of the forgetful map $\mathsf{Link}_{**} \to \mathsf{Link}_*$ over a based link $(L,\mathbf{p})$ is homeomorphic to $\mathbb{R}^{|L|}$, and hence contractible, so – as in the proof of Theorem 1.5 – the morphism HFL_2 factors through a functor $HFL\colon \mathsf{Link}_* \to \mathsf{Ab}$. Again, the invariant takes values in a somewhat richer category than Ab. □

Finally, we indicate how to obtain Spin^c-refined versions of the above results. Let F be a strong Heegaard invariant defined on a set $\mathcal{S}$ of diffeomorphism types of balanced sutured manifolds. Fix a sutured manifold (M, γ) such that $[(M, \gamma)] \in \mathcal{S}$. Suppose that, for every isotopy diagram H of (M, γ) and every $\mathfrak{s} \in \text{Spin}^c(M, \gamma)$, we are given an abelian group $F(H, \mathfrak{s})$ such that

$$F(H) = \bigoplus_{\mathfrak{s} \in \text{Spin}^c(M,\gamma)} F(H, \mathfrak{s}).$$

In addition, we assume that if e is an edge of $\mathcal{G}_{(M,\gamma)}$ from H to H', then $F(e)|_{F(H,\mathfrak{s})}$ is an isomorphism between $F(H, \mathfrak{s})$ and $F(H', \mathfrak{s})$. Then the limit $F(M, \gamma)$ will split as a direct sum $\bigoplus_{\mathfrak{s} \in \text{Spin}^c(M,\gamma)} F(M, \gamma, \mathfrak{s})$. Relative homological gradings on the summands $F(M, \gamma, \mathfrak{s})$ can be treated in a similar manner.

REMARK 2.46. Suppose that $\mathcal{H}$ and $\mathcal{H}'$ are admissible diagrams of the same Spin^c 3-manifold $(Y, \mathfrak{s})$. Ozsváth and Szabó [**28**, Theorem 2.1] constructed an isomorphism $\Psi^\circ_{\mathfrak{s}} \colon HF^\circ(\mathcal{H}) \to HF^\circ(\mathcal{H}')$ by composing maps associated to α-equivalences, β-equivalences, and stabilizations. As the Reidemeister–Singer theorem only implies two diagrams become isotopic after a sequence of such moves, implicit in their construction is a non-unique diffeomorphism map (see [**28**, Lemma 2.10] and [**27**, Proposition 2.2]), and hence $\Psi^\circ_{\mathfrak{s}}$ is not well-defined. The isomorphism we construct in Definition 2.37 is different in that it involves a new move, namely a diffeomorphism isotopic to the identity. As we shall see in Section 10.1, the isomorphisms we associate to α-equivalences, β-equivalences, and stabilizations agree with the isomorphisms defined by Ozsváth and Szabó [**28**, Section 2.5], but they are defined in a more computable manner; see Remark 10.23.

CHAPTER 3

Examples

In this chapter, we give several examples that illustrate some of the issues that arise when one tries to define Heegaard Floer homology in a functorial manner.

EXAMPLE 3.1. This example shows why it does not suffice to work with abstract (i.e., non-embedded) Heegaard diagrams to obtain canonical isomorphisms, and hence a functorial invariant of 3-manifolds. See the diagrams

$$\mathcal{H} = (\Sigma, \{\alpha_1, \alpha_2\}, \{\beta_1, \beta_2\}, z) \quad \text{and} \quad \mathcal{H}' = (\Sigma', \{\alpha_1', \alpha_2'\}, \{\beta_1', \beta_2'\}, z')$$

in Figure 5. Both define sutured manifolds diffeomorphic to

$$(S^1 \times S^2) \# (S^1 \times S^2).$$

The diagrams $\mathcal{H}$ and $\mathcal{H}'$ are clearly diffeomorphic. Choose a diffeomorphism $d\colon \mathcal{H} \to \mathcal{H}'$. Observe that there is an involution $f\colon \mathcal{H} \to \mathcal{H}$ such that $f(\alpha_1) = \alpha_2$, $f(\beta_1) = \beta_2$, and $f(z) = z$, obtained by π rotation about the axis perpendicular to the surface and passing through z. Then $d \circ f$ is also an identification between $\mathcal{H}$ and $\mathcal{H}'$. However, the diffeomorphisms d and $d \circ f$ induce different isomorphisms between $\widehat{HF}(\mathcal{H})$ and $\widehat{HF}(\mathcal{H}')$. Indeed, $\widehat{HF}(\mathcal{H}) \cong (\mathbb{Z}_2)^4$, and f_* swaps the two $\mathbb{Z}_2$ terms lying in the "middle" homological grading. This is why, in the graph $\mathcal{G}_{(M,\gamma)}$, we only consider diagrams embedded in (M, γ), and edges corresponding only to diffeomorphisms isotopic to the identity in M. Otherwise, Theorem 2.38 would not hold.

EXAMPLE 3.2. Consider the diagram $\mathcal{H} = (\Sigma, \boldsymbol{\alpha}, \boldsymbol{\beta}, z)$ of $S^1 \times S^2$ shown in Figure 6. Here, $S^1 \times S^2$ is represented by the region bounded by the two concentric spheres with common center O, and we identify the points of the outer and inner spheres that lie on a ray through O. The Heegaard surface Σ is represented by the horizontal annulus; after gluing the outer and inner boundary circles we get a torus. There is a single α-circle and a single β-circle; they intersect in two points a and b. In

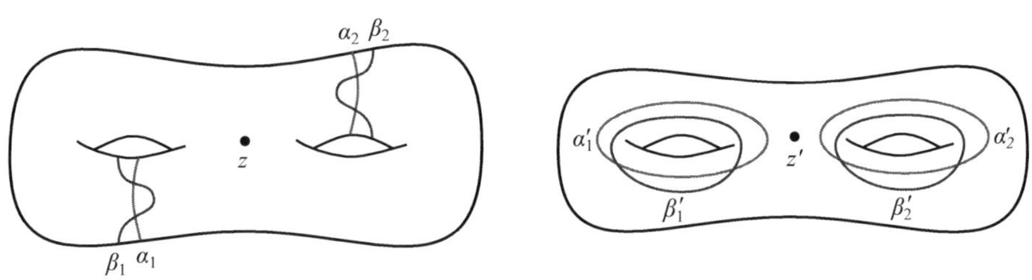

FIGURE 5. Two diffeomorphic diagrams, both defining manifolds diffeomorphic to $(S^1 \times S^2) \# (S^1 \times S^2)$, for which different identifications induce different maps on $\widehat{HF}$.

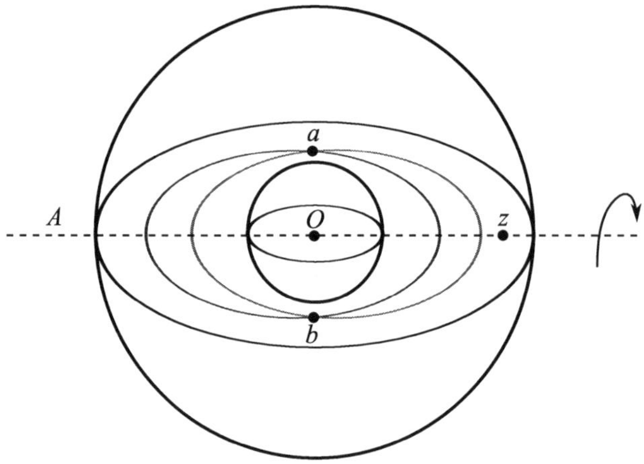

FIGURE 6. A diagram of $S^1 \times S^2$ for which an orientation reversing isotopy swaps the two generators of $\widehat{HF}$.

the diagram, the dashed line represents an axis A passing through the basepoint z. If we rotate Σ about A by an angle πt for some $t \in [0,1]$, we get an automorphism d_t of $S^1 \times S^2$. Notice that $d_1(\Sigma, \boldsymbol{\alpha}, \boldsymbol{\beta}, z) = (\Sigma, \boldsymbol{\alpha}, \boldsymbol{\beta}, z)$ and $d_1(a) = b$ and $d_1(b) = a$; furthermore, d_t fixes the basepoint z for every $t \in [0,1]$. Since $\widehat{HF}(\Sigma, \boldsymbol{\alpha}, \boldsymbol{\beta}, z)$ is generated by a and b, it appears that $\widehat{HF}$ has non-trivial monodromy around the loop of diagrams $d_t(\mathcal{H})$. However, $d_1|_\Sigma$ is orientation reversing. This shows that we need to consider oriented Heegaard surfaces in $\mathcal{G}_{(M,\gamma)}$ to obtain naturality.

EXAMPLE 3.3. Consider the diagram $\mathcal{H} = (\Sigma, \boldsymbol{\alpha}, \boldsymbol{\beta})$ of the lens-space $L(p,1)$ shown in Figure 7 for $p = 3$. In particular, Σ is the torus obtained by identifying the opposite edges of the rectangle $[0,1] \times [0,1]$, the curve α is a line of slope 0 and β is a line of slope p. Then $\alpha \cap \beta$ consists of p points $a_0, \dots, a_{p-1}$ that generate $\widehat{HF}(\mathcal{H})$. For $t \in [0,1]$, let $\mathcal{H}_t$ be the diagram of $L(p,1)$ obtained by translating β horizontally by t/p. Then $\mathcal{H}_0 = \mathcal{H}_1$, so we obtain a loop of diagrams for $L(p,1)$. Notice that $\widehat{HF}(\mathcal{H}_t)$ has non-trivial monodromy, as it maps a_i to a_{i+1} for $0 \le i \le p-1$, where $a_p = a_0$. Non-trivial monodromy makes it impossible to assign a Heegaard Floer group to $L(p,1)$ independent of the choice of diagram. This example is ruled out by requiring that there is at least one basepoint, and isotopies of the $\boldsymbol{\alpha}$ and $\boldsymbol{\beta}$ curves cannot pass through the basepoints. Note that a choice of basepoint is necessary to assign a Spin^c structure to each generator.

EXAMPLE 3.4. Here, we also consider a genus one Heegaard diagram $\mathcal{H} = (\Sigma, \boldsymbol{\alpha}, \boldsymbol{\beta}, \boldsymbol{z})$ of $S^1 \times S^2$, but with two basepoints $\boldsymbol{z} = \{z_1, z_2\}$; see Figure 8. Heegaard Floer homology for multi-pointed Heegaard diagrams was introduced in [**29**, Section 4]. Again, we draw the diagram on $[0,1] \times [0,1]$. We have two α-curves: $\alpha_1 = \{1/4\} \times [0,1]$ and $\alpha_2 = \{3/4\} \times [0,1]$. The curve β_i is a small Hamiltonian translate of α_i such that $\alpha_1 \cap \beta_1$ consists of two points that we label a, b, and $\alpha_2 \cap \beta_2$ consists of two points x, y. We also arrange that β_2 is a translate of β_1 by the vector $(1/2, 0)$. We choose two basepoints, namely $z_1 = (0, 1/2)$ and $z_2 = (1/2, 1/2)$. For $t \in [0,1]$, let d_t be the diffeomorphism of Σ given by $d_t(u,v) = (u + t/2, v)$. (This extends to $S^1 \times S^2$ as rotation by πt in the S^1-direction.) Let $\mathcal{H}_t = d_t(\mathcal{H})$ for

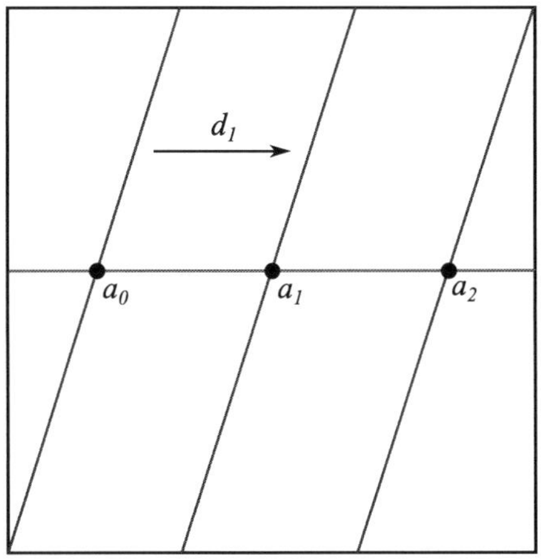

FIGURE 7. A diagram of $L(3,1)$ with no basepoint, together with an isotopy that permutes the 3 generators of $\widehat{HF}$.

$t \in [0,1]$, then $\mathcal{H}_1 = \mathcal{H}_0$, so we have a loop of doubly pointed diagrams of $S^1 \times S^2$. Notice that $\widehat{HF}(\Sigma, \boldsymbol{\alpha}, \boldsymbol{\beta}, z_1, z_2)$ is generated by the pairs $\{a, x\}$, $\{a, y\}$, $\{b, x\}$, and $\{b, y\}$. The diffeomorphism d_1 swaps the generators $\{a, y\}$ and $\{b, x\}$, and swaps the basepoints z_1 and z_2. Hence, to have naturality for $\widehat{HF}$, we need to work with based 3-manifolds and based diffeomorphisms. However, if $\mathfrak{s}_0$ denotes the torsion Spin^c-structure on $S^1 \times S^2$, a straightforward computation shows that

$$HF^-(\mathcal{H}, \mathfrak{s}_0) \cong \mathbb{Z}[U_1, U_2]/(U_1 - U_2)\langle\, \{a, x\}, \{a, y\} + \{b, x\}\,\rangle$$

as a $\mathbb{Z}[U_1, U_2]$-module, and d_1 acts trivially on it. Compare this with our discussion in the introduction that the basepoint moving map can be non-trivial on $\widehat{HF}$ but is trivial on HF^-.

EXAMPLE 3.5. Even if we isotope the α- and β-curves in the complement of the basepoint, one might obtain a loop of diagrams such that $\widehat{CF}$ has non-trivial holonomy around it. However, as we shall prove, there is no holonomy if we pass to homology.

We describe a diagram $\mathcal{H}$ of $S^1 \times S^2$ as follows; see Figure 9. Let Σ be the torus represented by $[0,1] \times [0,1]$, take α to be $\{1/2\} \times [0,1]$, and let β be a Hamiltonian translate of α such that $\alpha \cap \beta$ consists of four points $a_0, \dots, a_3$, and β is invariant under translation by $(0, 1/2)$. Let d_t be translation by $(0, t/2)$ for $t \in [0,1]$, and let $\mathcal{H}_t = (\Sigma, \alpha, d_t(\beta), z)$. By construction, $\mathcal{H}_0 = \mathcal{H}_1$. Since $d_1(a_i) = a_{i+2}$ (where $i+2$ is to be considered modulo 4), we see that d_1 acts non-trivially on $\widehat{CF}(\mathcal{H})$. However, as $\widehat{HF}(\mathcal{H})$ is generated by $a_0 + a_2$ and $a_1 + a_3$, the induced action on homology is trivial.

More generally, suppose that $(\Sigma, \boldsymbol{\alpha}, \boldsymbol{\beta}, z)$ is a Heegaard diagram, $\alpha \in \boldsymbol{\alpha}$ and $\beta \in \boldsymbol{\beta}$. Furthermore, suppose that there is a regular neighborhood $N \approx \alpha \times [-1,1]$ of α such that $\beta \subset N$ and β is transverse to the fibers $\{p\} \times [-1,1]$ for every $p \in \alpha$. Then we can apply a "finger move" inside N that is the identity outside N and preserves $\alpha \cup \beta$ setwise, and hence permutes the points of $\alpha \cap \beta$. Even though this isotopy acts non-trivially on the chain level, it is trivial on the homology level.

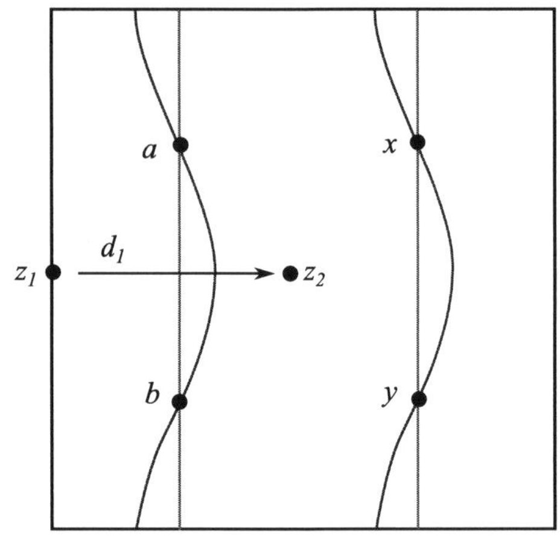

FIGURE 8. A doubly-pointed diagram of $S^1 \times S^2$. A π-rotation in the S^1-direction swaps the basepoints and induces a non-trivial automorphism of $\widehat{HF}$, while being trivial on HF^+ and HF^- in the torsion Spin^c-structure.

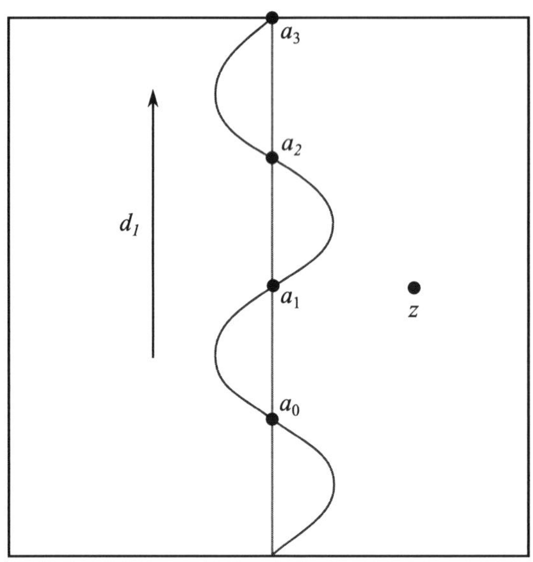

FIGURE 9. A Heegaard diagram of $S^1 \times S^2$. If we translate the β-curve in the vertical direction by 1/2, we get a non-trivial automorphism of the chain complex that is trivial on the homology.

CHAPTER 4

Singularities of Smooth Functions

In this chapter, we recall some classical results about singularities of smooth real valued functions following Arnold, Goryunov, Lyashko, and Vasil'ev [**3**]. The reader familiar with singularity theory can safely skip to Chapter 5. This part is the beginning of the proof of Theorem 2.38 on strong Heegaard invariants that culminates in Chapter 9. The reader interested in the proof of Theorem 1.5, the application of Theorem 2.38 to Heegaard Floer homology, should skip to Chapter 10.

DEFINITION 4.1. Let f be a smooth function on the manifold M. A point $p \in M$ is said to be a *critical point* of f if $df_p = 0$.

DEFINITION 4.2. Let $\mathcal{E}_n$ be the set of germs at $0 \in \mathbb{R}^n$ of smooth functions $f\colon \mathbb{R}^n \to \mathbb{R}$. Let $\mathcal{D}_n$ be the group of germs of diffeomorphisms $g\colon (\mathbb{R}^n, 0) \to (\mathbb{R}^n, 0)$. The group $\mathcal{D}_n$ acts on $\mathcal{E}_n$ by the rule $g(f) = f \circ g^{-1}$. Two function-germs f, $f' \in \mathcal{E}_n$ are said to be *equivalent* if they lie in the same $\mathcal{D}_n$-orbit.

The equivalence class of a function germ at a critical point is called a *singularity*. A *class of singularities* is any subset of the space $\mathcal{E}_n$ that is invariant under the action of the group $\mathcal{D}_n$.

DEFINITION 4.3. A critical point is said to be *nondegenerate* (or a *Morse critical point*) if the second differential (or Hessian) of the function at that point is a nondegenerate quadratic form. The class of non-degenerate critical points is called A_1.

THEOREM 4.4 (Morse Lemma). *In a neighborhood of a nondegenerate critical point $a \in M^n$ of the function $f\colon M^n \to \mathbb{R}$, there exists a coordinate system in which f has the form*

$$f(x) = -x_1^2 - \cdots - x_k^2 + x_{k+1}^2 + \cdots + x_n^2 + f(a).$$

In the above theorem, k is called the *index* of the nondegenerate critical point a, and will be denoted by $\mathcal{I}(a)$. If two elements of $\mathcal{E}_n$ have nondegenerate critical points at zero, then they are equivalent if and only if they have the same value and the same index. More generally, for an arbitrary critical point, $\mathcal{I}(a)$ is the index of the second differential of the function at a.

The most important characteristic of a class of singularities is its codimension c in the space $\mathcal{E}_n$ of function-germs. A generic function has only nondegenerate critical points. So, the codimension c of a nondegenerate critical point is zero, while any class of degenerate critical points has positive codimension. Degenerate critical points occur in an irremovable manner only in families of functions depending on parameters. In a family of functions depending on l parameters, there may occur (in such a manner that it cannot be removed through a small perturbation of the family) only a class of singularities for which $c \leq l$.

DEFINITION 4.5. A *deformation* with base $\Lambda = \mathbb{R}^l$ of the germ $f \in \mathcal{E}_n$ is the germ at zero of a smooth map $F\colon \mathbb{R}^n \times \mathbb{R}^l \to \mathbb{R}$ such that $F(x,0) \equiv f(x)$.

A deformation $F'\colon \mathbb{R}^n \times \mathbb{R}^l \to \mathbb{R}$ is *equivalent* to F if

$$F'(x,\lambda) = F(g(x,\lambda),\lambda),$$

where g is the germ at zero of a smooth map $(\mathbb{R}^n \times \mathbb{R}^l, 0) \to (\mathbb{R}^n, 0)$ with $g(x,0) \equiv x$, representing a family of diffeomorphisms depending on $\lambda \in \mathbb{R}^l$.

The deformation $F'\colon \mathbb{R}^n \times \mathbb{R}^{l'} \to \mathbb{R}$ is *induced* from F if

$$F'(x,\lambda') = F(x,\theta(\lambda')),$$

where $\theta\colon (\mathbb{R}^{l'},0) \to (\mathbb{R}^l,0)$ is a smooth germ of a mapping of the bases.

DEFINITION 4.6. A deformation $F(x,\lambda)$ of the germ $f(x)$ is said to be *versal* if every deformation $F'\colon \mathbb{R}^n \times \mathbb{R}^{l'} \to \mathbb{R}$ of $f(x)$ can be represented in the form

$$F'(x,\lambda') = F(g(x,\lambda'),\theta(\lambda')),$$

where g is a germ at zero of an l'-parameter family of diffeomorphisms $\mathbb{R}^n \times \mathbb{R}^{l'} \to \mathbb{R}^n$ such that $g(x,0) \equiv x$, and $\theta\colon (\mathbb{R}^{l'},0) \to (\mathbb{R}^l,0)$ is such that $\theta(0) = 0$; i.e., every deformation of $f(x)$ is equivalent to a deformation induced from F.

A versal deformation for which the base Λ has the smallest possible dimension is called *miniversal.*

PROPOSITION 4.7. [**3**, p. 18] *Let $f(x) \in \mathcal{E}_n$ be a germ of a critical point. We denote by $I_{\nabla f}$ the ideal of $\mathcal{E}_n$ generated by all partial derivatives $f_i = \partial f/\partial x_i$ of f, and let $Q_f = \mathcal{E}_n/I_{\nabla f}$. If $\varphi_1, \ldots, \varphi_l \in \mathcal{E}_n$ project to a basis of Q_f, then*

$$F(x,\lambda) = f(x) + \sum_{j=1}^{l} \lambda_j \varphi_j$$

is a miniversal deformation of the germ $f(x)$.

A versal deformation is unique in the following sense.

THEOREM 4.8. [**3**, p. 18] *Every ℓ-parameter versal deformation of a germ f is equivalent to a deformation induced from any other ℓ-parameter versal deformation by a suitable diffeomorphism of their bases.*

Let K be a subset of $\mathcal{E}_n$ invariant under the action of $\mathcal{D}_n$; i.e., a class of singularities. Furthermore, let $P \subset \mathcal{E}_n$ be the set of germs at 0 of all elements of the polynomial ring $\mathbb{R}[x_1, \ldots, x_n]$. A *normal form* for the class K is a map $\Phi\colon B \to P$ from a finite dimensional linear "parameter space" B to the space of polynomial germs satisfying three conditions:

- $\Phi(B)$ intersects all $\mathcal{D}_n$-orbits in K,
- the preimage in B of every $\mathcal{D}_n$-orbit is finite, and
- $\Phi^{-1}(\mathcal{E}_n \setminus K)$ lies in some proper algebraic hypersurface in B.

For more detail, see [**3**, p. 22].

THEOREM 4.9. *(*[**3**, p. 33] *and* [**8**, p. 78]*) In a generic 1-parameter family of smooth functions, the only degenerate critical points that appear have normal form*

$$f(x) = -x_1^2 - \cdots - x_k^2 + x_{k+1}^2 + \cdots + x_{n-1}^2 + x_n^3 + f(a).$$

The class of such singularities is called A_2.

In addition, in 2-parameter families, singularities of the form

$$f(x) = -x_1^2 - \cdots - x_k^2 + x_{k+1}^2 + \cdots + x_{n-1}^2 \pm x_n^4 + f(a)$$

might also appear. The class of such singularities is called $A_3^{\pm}$*.*

As a corollary of Proposition 4.7, a miniversal deformation of a singularity of type A_1 is given by

$$F(x,\lambda) = -x_1^2 - \cdots - x_k^2 + x_{k+1}^2 + \ldots x_n^2 + \lambda,$$

where $\lambda \in \mathbb{R}$.

Miniversal deformations of type A_2 singularities are given by the formula

$$F(x,\lambda) = -x_1^2 - \cdots - x_k^2 + x_{k+1}^2 + \cdots + x_{n-1}^2 + x_n^3 + \lambda_1 x_n + \lambda_2,$$

where the parameter $\lambda = (\lambda_1, \lambda_2) \in \mathbb{R}^2$. By Definition 4.6, every generic 1-parameter deformation of a type A_2 singularity (with parameter t) is equivalent to one induced from this via a suitable map $t \mapsto (\lambda_1(t), \lambda_2(t))$, and so has normal form

$$-x_1^2 - \cdots - x_k^2 + x_{k+1}^2 + \cdots + x_{n-1}^2 + x_n^3 + \lambda_1(t)x_n + \lambda_2(t).$$

The concrete value of the constant term $\lambda_2(t)$ does not affect the types of singularities appearing in the family, so from a qualitative point of view, we can assume that $\lambda_2(t) \equiv 0$. Then, in this family, for $\lambda_1 < 0$, we have two nondegenerate critical points of indices k and $k+1$ that cancel each other at $\lambda_1 = 0$, and the germs have no critical points for $\lambda_1 > 0$. Hence, we will sometimes refer to a type A_2 singularity of index k as an index k–$(k+1)$ birth-death singularity (death if $\lambda_1(t)$ is decreasing, and birth if $\lambda_1(t)$ is increasing). In 2-parameter families, type A_2 singularities appear along curves in the parameter space Λ (in the above normal form given by the formula $\lambda_1 = 0$).

Finally, type $A_3^{\pm}$ singularities have miniversal deformation

$$F(x,\lambda) = -x_1^2 - \cdots - x_k^2 + x_{k+1}^2 + \cdots + x_{n-1}^2 \pm x_n^4 + \lambda_1 x_n^2 + \lambda_2 x_n + \lambda_3$$

with parameter $\lambda \in \mathbb{R}^3$. These first appear in a non-removable manner in 2-parameter families. By versality, every generic 2-parameter deformation can be induced from this by a generic map $\lambda\colon \mathbb{R}^2 \to \mathbb{R}^3$ of their bases. To study the bifurcation set in $\Lambda = \mathbb{R}^2$ for a generic 2-parameter deformation $\Lambda \to \mathcal{E}_n$, we again disregard the constant term, and consider

$$-x_1^2 - \cdots - x_k^2 + x_{k+1}^2 + \cdots + x_{n-1}^2 \pm x_n^4 + \lambda_1(t_1,t_2)x_n^2 + \lambda_2(t_1,t_2)x_n,$$

where $(t_1,t_2) \mapsto (\lambda_1(t_1,t_2), \lambda_2(t_1,t_2))$ has non-vanishing differential at the origin. For generic values of (λ_1, λ_2), this may have

- three nondegenerate critical points, of indices k, $k+1$, and k for A_3^+ and $k+1$, k, and $k+1$ for A_3^-; or
- one nondegenerate critical point, of index k for A_3^+ and $k+1$ for A_3^-.

When the polynomial $\pm 4x_n^3 + 2\lambda_1 x_n + \lambda_2$ has multiple roots, the behavior is different. The discriminant is the cuspidal curve

$$\Delta = \{\, \lambda \in \Lambda : \ 8\lambda_1^3 \pm 27\lambda_2^2 = 0 \,\}.$$

For $\lambda \in \Delta \setminus \{0\}$, the germ $F(x,\lambda)$ has an A_1 and an A_2 singularity, while, for $\lambda = 0$, it has an $A_3^{\pm}$ singularity. Sometimes, we will refer to an A_3^+ singularity of index k as an index k–$(k+1)$–k birth-death-birth, while an A_3^- singularity of index k as an index $(k+1)$–k–$(k+1)$ birth-death-birth. For the bifurcation diagrams of the

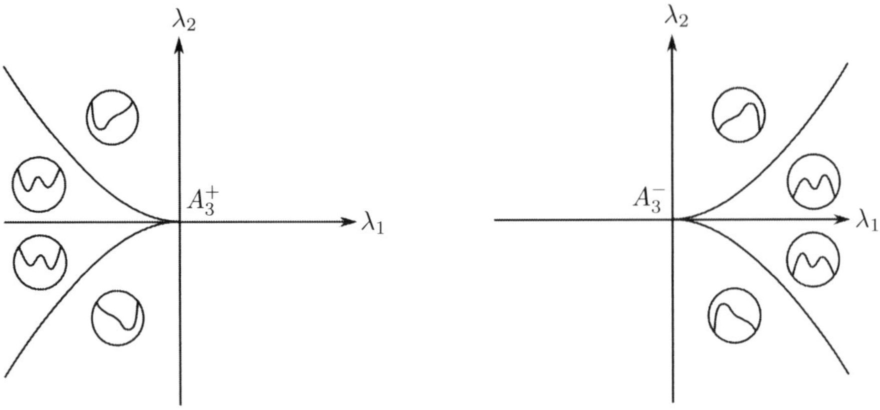

FIGURE 10. Bifurcation diagrams of the singularities A_3^+ and A_3^- for $n = 1$.

singularities A_3^+ and A_3^-, see Figure 10. Note that, in case of A_3^+, for $\lambda_1 < 0$ and $\lambda_2 = 0$, the values of two critical points coincide, which is a type of bifurcation that we disregard in this paper as we will only work with the gradient flows of functions, and not with their values. There is an analogous bifurcation for A_3^- singularities in case $\lambda_1 > 0$ and $\lambda_2 = 0$.

We now apply the above discussion to global 1- and 2-parameter families of smooth real valued functions on manifolds; see Cerf [**8**, p. 78]. For a generic 1-parameter family of smooth functions $\{ f_\lambda \colon \lambda \in \Lambda \}$, there is a discrete subset $D \subset \Lambda$ such that for every $\lambda \in \Lambda \setminus D$ the function f_λ has only nondegenerate critical points, while, for $\lambda \in D$, it has a single degenerate critical point of type A_2, where two nondegenerate critical points of neighboring indices collide.

For a generic 2-parameter family $\{ f_\lambda \colon \lambda \in \Lambda \}$, there is a subset $D \subset \Lambda$ such that f_λ has only nondegenerate critical points for $\lambda \in \Lambda \setminus D$. In addition, D is a union of embedded curves that have only cusp singularities and intersect each other in transverse double points. At a regular point $\lambda \in D$, the function f_t has a single degenerate critical point of type A_2. If λ is a double point of D, then f_t has two degenerate critical points of type A_2. Finally, at each cusp of D, the function f_t has a single degenerate critical point of type $A_3^\pm$.

CHAPTER 5

Generic One- and Two-Parameter Families of Gradients

Next, we summarize the results of Palis and Takens [**31**] and Vegter [**37**] on the classification of global bifurcations that appear in generic 1- and 2-parameter families of gradient vector fields on 3-manifolds. In Chapter 6, we will translate the codimension-1 bifurcations to moves on Heegaard diagrams, while codimension-2 bifurcations translate to loops of Heegaard diagrams. Note that the bifurcation theory of gradients is much richer than the corresponding theory for smooth functions, due to the tangencies that can appear between invariant manifolds of singular points.

5.1. Invariant manifolds

First, we review some classical definitions and results from Anosov, Aranson, Arnold, Bronshtein, Grines, and Il'yashenko [**2**, Section 4].

DEFINITION 5.1. An *invariant manifold* of a vector field is a submanifold that is tangent to the vector field at each of its points.

If v is a smooth vector field on a manifold M with a singularity at p (i.e., v is zero at p), then the linear part L_pv of v at p is an endomorphism of T_pM. In local coordinates $x = (x_1, \dots, x_n)$ around p, the linear part of v is Ax, where $A = \frac{\partial v}{\partial x}\big|_{x=0}$ and $\frac{\partial v}{\partial x}$ is the Jacobian matrix whose (i,k) entry is $\frac{\partial v_i}{\partial x_k}$.

The space T_pM can be written as a direct sum of three L_pv-invariant subspaces, namely T^s, T^u, and T^c, such that every eigenvalue of $L_pv|_{T^s}$ has negative real part, every eigenvalue of $L_pv|_{T^u}$ has positive real part, and every eigenvalue of $L_pv|_{T^c}$ has real part zero. Indeed, T^s, T^u, and T^c are spanned by the generalized eigenvectors of L_pv corresponding to eigenvalues with negative, positive, and zero real parts, respectively. Here, the superscripts s, u, and c correspond to "stable," "unstable," and "center."

DEFINITION 5.2. We say that v has a *hyperbolic singularity* at p if none of the eigenvalues of L_pv are purely imaginary; i.e., if $T^c = 0$.

THEOREM 5.3 (Center manifold theorem [**2**, Subsection 4.2]). *Let v be a C^{r+1} vector field on M with a singular point at p. Let T^s, T^u, and T^c be the L_pv-invariant subspaces of T_pM defined above.*

Then the vector field v has invariant manifolds $\mathcal{W}^s$, $\mathcal{W}^u$, and $\mathcal{W}^c$ of class C^{r+1}, C^{r+1}, and C^r, respectively, that go through p and are tangent to T^s, T^u, and T^c, respectively, at p. Integral curves of v with initial point on $\mathcal{W}^s$ (respectively, $\mathcal{W}^u$) tend exponentially to p as $t \to +\infty$ (respectively, $t \to -\infty$).

Here, $\mathcal{W}^s$ is called the (strong) stable manifold and $\mathcal{W}^u$ the (strong) unstable manifold of the singular point p. The behavior of the integral curves on the center manifold $\mathcal{W}^c$ is determined by the nonlinear terms. If v is C^∞, then $\mathcal{W}^s$ and $\mathcal{W}^u$ can be chosen to be C^∞, whereas the center manifold is only finitely smooth. In addition, while $\mathcal{W}^s$ and $\mathcal{W}^u$ are well-defined, the choice of $\mathcal{W}^c$ might not be unique.

We say that the flows ϕ on M and ψ on N are *topologically equivalent* if there is a homeomorphism $h\colon M \to N$ mapping orbits of ψ to orbits of ϕ homeomorphically, and preserving the orientation of the orbits.

THEOREM 5.4 (Reduction Principle [**2**, Subsection 4.3]). *Suppose that a C^2 vector field v has a singular point at p. Let T^s, T^u, and T^c be the invariant subspaces corresponding to the map $L_p v$. Then, in a neighborhood of the singular point p, the flow of the vector field v is topologically equivalent to the flow of the direct product of two vector fields: the restriction of v to the center manifold, and the "standard saddle" $(-a, b)$ for $a \in T^s$ and $b \in T^u$.*

This theorem can be used to study both individual vector fields and families of vector fields; by replacing a family $v(x, \varepsilon)$ on M with $(v(x, \varepsilon), 0)$ on $M \times \mathbb{R}^l$, where $x \in M$ and $\varepsilon \in \mathbb{R}^l$. The following discussion has been taken from Palis and Takens [**31**].

DEFINITION 5.5. We say that a smooth vector field v on M has a *saddle-node* at p (or a *quasi-hyperbolic singularity of type 1*) if $\dim T^c = 1$ and $v|_{\mathcal{W}^c}$ has the form

$$v = ax^2 \frac{\partial}{\partial x} + O(|x|^3)$$

with $a \neq 0$ for some center manifold $\mathcal{W}^c$ through p, where x is a local coordinate on $\mathcal{W}^c$ around p.

If $v^{\overline{\mu}}$, belonging to a one-parameter family $\{v^\mu\}$ of vector fields, has a saddle-node at p, we say that it *unfolds generically* if there is a center manifold for the family $\{v^\mu\}$ passing through p (at $\mu = \overline{\mu}$) such that v_μ, restricted to this center manifold, has the form

$$v_\mu = \left(ax^2 + b(\mu - \overline{\mu})\right) \frac{\partial}{\partial x} + O\left(|x^3| + |x \cdot (\mu - \overline{\mu})| + |\mu - \overline{\mu})|^2\right),$$

with a, $b \neq 0$.

For example, if $f : M \to \mathbb{R}$ has an A_2 singularity at p, then ∇f has a saddle-node at p.

DEFINITION 5.6. A point $p \in M$ is called a *quasi-hyperbolic singularity of type 2* of a vector field v if $\dim T^c = 1$ and there is a center manifold $\mathcal{W}^c$ of class C^m such that, on $\mathcal{W}^c$, there is a local C^m-coordinate x with

$$v|_{\mathcal{W}^c} = x^3 \cdot v_1(x) \frac{\partial}{\partial x}$$

with $v_1(0) \neq 0$.

For example, if $f : M \to \mathbb{R}$ has an $A_3^\pm$ singularity at p, then ∇f has a quasi-hyperbolic singularity of type 2 at p.

DEFINITION 5.7. Let p be a singular point of the vector field v on M. Furthermore, let the maximal flow of v be $\varphi\colon D \to M$, where $D \subset M \times \mathbb{R}$ is the flow

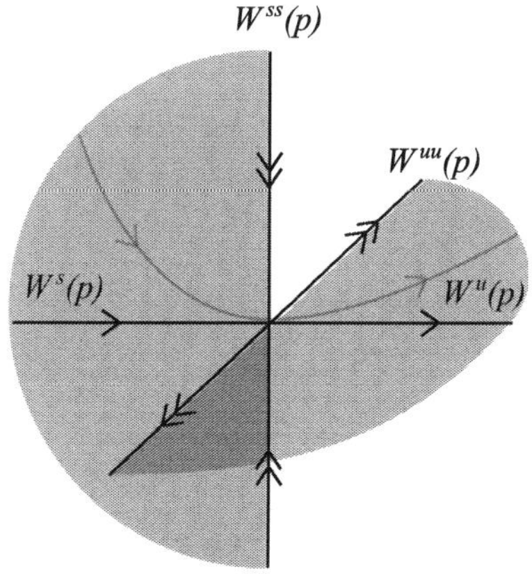

FIGURE 11. The stable and unstable sets of a saddle-node singularity are manifolds with boundary. This figure also illustrates the non-uniqueness of the center manifold; the black and the light grey curves with single arrows show two possible center manifolds.

domain. Then the *stable set* of p is

$$W^s(p) = \{\, x \in M : \lim_{t\to\infty} \varphi(x,t) = p \,\},$$

and the *unstable set* of p is

$$W^u(p) = \{\, x \in M : \lim_{t\to-\infty} \varphi(x,t) = p \,\}.$$

If p is a hyperbolic singular point of v, then both $W^s(p)$ and $W^u(p)$ are injectively immersed submanifolds of M with tangent spaces $T_pW^s(p) = T^s$ and $T_pW^u(p) = T^u$, respectively. So, in this case, the stable and unstable sets $W^s(p)$ and $W^u(p)$ coincide with the stable and unstable manifolds $\mathcal{W}^s$ and $\mathcal{W}^u$ of the singular point p, respectively.

If p is a saddle-node, $W^s(p)$ is an injectively immersed submanifold with boundary. This boundary is the *strong stable manifold* of p, and is denoted by $W^{ss}(p)$. Note that $T_pW^{ss}(p) = T^s$. Similarly, $W^u(p)$ is an injectively immersed submanifold with boundary the *strong unstable manifold* $W^{uu}(p)$; see Figure 11. In the terminology of Theorem 5.3, we have $W^{ss}(p) = \mathcal{W}^s$ and $W^{uu}(p) = \mathcal{W}^u$.

5.2. Bifurcations of gradient vector fields on 3-manifolds

Let M be compact 3-manifold. A gradient vector field $X = \mathrm{grad}_g(f)$ on M is associated with a Riemannian metric g and a smooth function $f\colon M \to \mathbb{R}$ by the relation $g(X,Y) = df(Y)$ for all smooth vector fields Y on M. Since f is strictly increasing along regular orbits of X, the vector field $\mathrm{grad}_g(f)$ has no periodic orbits or other kinds of recurrence. The singular points of X coincide with the critical points of f. Since the linear part L_pX of X at a singularity p is symmetric with respect to g, all eigenvalues of L_pX are real. (In suitable coordinates, L_pX is the Hessian matrix of f at the singular point.)

DEFINITION 5.8. A gradient vector field X on M is *Morse–Smale* if

(H) all singular points of X are hyperbolic, and
(T) all stable and unstable manifolds are transversal.

The Morse–Smale vector fields constitute an open and dense subset of the set $X^g(M)$ of all gradient vector fields on a closed manifold M. In addition, a gradient vector field is structurally stable if and only if it is Morse–Smale. Recall that a C^1 vector field v on M is *structurally stable* if the following holds: For any C^0 neighborhood U of Id_M in $\mathrm{Homeo}(M)$, there is a C^1 neighborhood V of v such that for every $v' \in V$ there is a homeomorphism $h \in U$ that maps the oriented trajectories of v to the oriented trajectories of v' (in particular, the flows of v and v' are topologically conjugate).

DEFINITION 5.9. A *k-parameter family of gradients* on a compact manifold M is a family of pairs (g^μ, f^μ) for $\mu \in \mathbb{R}^k$, where $\{g^\mu\}$ is a k-parameter family of Riemannian metrics and $\{f^\mu\}$ is a k-parameter family of smooth real-valued functions on M. Let $X^\mu = \mathrm{grad}_{g^\mu}(f^\mu)$ for $\mu \in \mathbb{R}^k$. By a slight abuse of notation, we will only write $\{X^\mu\}$ to denote a k-parameter family of gradients.

We assume that both g^μ and f^μ, and hence X^μ, depend smoothly on $(x, \mu) \in M \times \mathbb{R}^k$. The set of such pairs is endowed with the strong Whitney topology; i.e., the topology of uniform convergence of g^μ, f^μ and all their derivatives on compact sets. The resulting topological space of k-parameter families is denoted by $X^g_k(M)$.

DEFINITION 5.10. A parameter value $\overline{\mu} \in \mathbb{R}^k$ is called a *bifurcation value* for the family $\{X^\mu\}$ in $X^g_k(M)$ if $X^{\overline{\mu}}$ fails to be a Morse–Smale system. Hence $X^{\overline{\mu}}$ has at least one *orbit of tangency* between stable and unstable manifolds (i.e., an integral curve along which a stable and an unstable manifold are tangent), or at least one non-hyperbolic singular point.

The *bifurcation set* (or *bifurcation diagram*) of a family $\{X^\mu\}$ in $X^g_k(M)$ is the subset of $\mathbb{R}^k$ consisting of all bifurcation values of $\{X^\mu\}$.

Note that, in dimension three, if a stable manifold and an unstable manifold do not intersect transversely at a point x, then they are actually tangent at x. Indeed, stable and unstable manifolds always contain the flow direction, and any two linear subspaces of $\mathbb{R}^3$ containing a fixed vector are either transverse or tangent (in the sense that one is contained inside the other).

5.2.1. Generic 1-parameter families of gradients in dimension 3. We now turn to the results of Palis and Takens [**31**], following the notation of Vegter [**37**]. We will translate the bifurcations occurring in generic 1-parameter families of gradients to moves on Heegaard diagrams in Proposition 6.28. For the following definition, see Vegter [**37**, p. 122] and Palis–Takens [**31**, Section 2.a].

DEFINITION 5.11. Let v be a vector field on a 3-manifold M. Two invariant submanifolds A and B of v have a *quasi-transversal tangency* if their intersection has a connected integral curve γ that is not a single point, and, at some (and hence every) point $r \in \gamma$, the following two conditions hold:

(QT-1) $\dim(T_rA + T_rB) = 2$; so we have three cases:
  (a) $\dim A = 2$ and $\dim B = 2$,
  (b) $\dim A = 1$ and $\dim B = 2$,
  (c) $\dim A = 2$ and $\dim B = 1$.

(QT-2) In case (QT-1a), we impose the condition that the tangency between A and B is as generic as possible in the following sense. Let S be a smooth 2-dimensional cross-section for v, containing r. Take coordinates (x_1, x_2) on a neighborhood of r in S, in which $A \cap S = \{x_2 = 0\}$, while $B \cap S$ is of the form $\{x_2 = F(x_1)\}$ for some smooth function F. Condition (QT-1) amounts to $F(0) = 0$ and $\frac{dF}{dx_1}(0) = 0$. In addition, we require that

$$\frac{d^2F}{dx_1^2}(0) \neq 0.$$

By Palis and Takens [**31**, Main Theorem] and Vegter [**37**, p. 124], for an open and dense set of 1-parameter families of gradients, it is easy to describe the bifurcation diagram: It consists of isolated points in the parameter space $\mathbb{R}$ at which one of the conditions appearing in the characterization of Morse–Smale gradients is violated "in the mildest possible manner." For a generic family $\{X^\mu\} \in \mathcal{X}_1^g(M)$, at each bifurcation value $\overline{\mu} \in \mathbb{R}$ exactly one of the following two possibilities holds:

(NH) Failure of condition (H). The vector field $X^{\overline{\mu}}$ has exactly one generically unfolding saddle-node p, while all other singular points are hyperbolic, and all stable and unstable manifolds are transversal. By convention (here and later), at saddle-nodes, the set of stable and unstable manifolds required to be transverse includes $W^{ss}(p)$ and $W^{uu}(p)$, as well as $W^s(p)$ and $W^u(p)$.

(NT) Failure of condition (T). All singular points of $X^{\overline{\mu}}$ are hyperbolic, and there is a single non-transversal orbit of intersection γ of the unstable manifold $W^u(p_1^{\overline{\mu}})$ of $p_1^{\overline{\mu}}$ and the stable manifold $W^s(p_2^{\overline{\mu}})$ of $p_2^{\overline{\mu}}$ that is quasi-transversal and satisfies the additional non-degeneracy conditions below.

(ND-1) This condition expresses the "crossing at non-zero speed" of $W^u(p_1^\mu)$ and $W^s(p_2^\mu)$ as the parameter passes the value $\overline{\mu}$, where p_1^μ and p_2^μ are the saddle points of X^μ near $p_1^{\overline{\mu}}$ and $p_2^{\overline{\mu}}$, respectively. For this, we choose paths $\sigma^u, \sigma^s \colon \mathbb{R} \to M$ with $\sigma^u(\mu) \in W^u(p_1^\mu)$, $\sigma^s(\mu) \in W^s(p_2^\mu)$, and $\sigma^u(\overline{\mu}) = \sigma^s(\overline{\mu}) = r \in \gamma$. We require that

$$\dot{\sigma}^u(\overline{\mu}) - \dot{\sigma}^s(\overline{\mu}) \notin T_r W^u(p_1^{\overline{\mu}}) + T_r W^s(p_2^{\overline{\mu}}).$$

When $\dim W^u(p_1^{\overline{\mu}}) = 1$ and $\dim W^s(p_2^{\overline{\mu}}) = 2$, we impose the following additional conditions. When $\dim W^u(p_1^{\overline{\mu}}) = 2$ and $\dim W^s(p_2^{\overline{\mu}}) = 1$, we have the same conditions, but with the sign of X^μ reversed.

(ND-2) The contracting (i.e., negative) eigenvalues of the linear part of $X^{\overline{\mu}}$ at $p_1^{\overline{\mu}}$ are distinct (for gradients only real eigenvalues occur). This implies that there is a unique 1-dimensional invariant manifold

$$W^{ss}(p_1^{\overline{\mu}}) \subset W^s(p_1^{\overline{\mu}})$$

such that $T_{p_1^{\overline{\mu}}} W^{ss}(p_1^{\overline{\mu}})$ is the eigenspace of $L_{p_1^{\overline{\mu}}} X^{\overline{\mu}}$ corresponding to the strongest contracting (i.e., smallest negative) eigenvalue. We call $W^{ss}(p_1^{\overline{\mu}})$ the strong stable manifold of $p_1^{\overline{\mu}}$.

(ND-3) For some $r \in \gamma$, let $E_r \subset T_r W^s(p_2^{\overline{\mu}})$ be a 1-dimensional subspace complementary to $X^{\overline{\mu}}(r)$. Let $\phi_t^{\overline{\mu}}$ for $t \in \mathbb{R}$ be the flow of $X^{\overline{\mu}}$. Then we require that

$$\lim_{t \to -\infty} (d\phi_t^{\overline{\mu}})_r(E_r) = T_{p_1^{\overline{\mu}}} W^{ss}(p_1^{\overline{\mu}}).$$

(ND-4) The stable and unstable manifolds of any singularity $p^* \notin \left\{p_1^{\overline{\mu}}, p_2^{\overline{\mu}}\right\}$ are transversal to $W^{ss}(p_1^{\overline{\mu}})$.

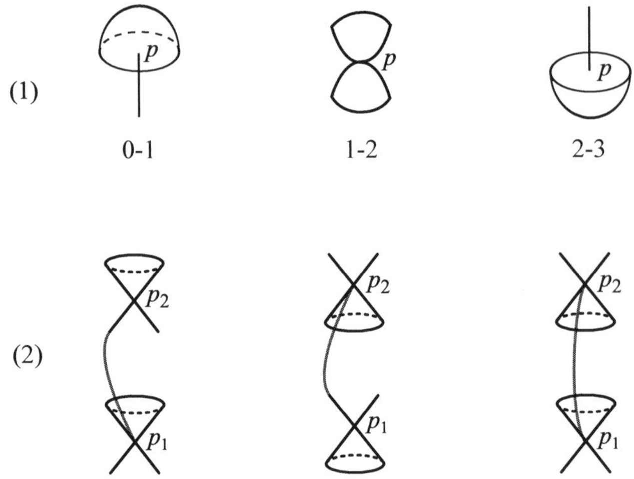

FIGURE 12. Generic codimension-1 singularities of gradient vector fields. The vertical direction in these pictures is the direction of the Morse function, so the gradient flow always goes upwards. The critical points are shown schematically, indicating only the stable and unstable manifolds. The top row shows the possibilities in case (NH): an index 0-1, an index 1-2, and an index 2-3 saddle-node. Here, the stable and unstable manifolds are manifolds with boundary. In the bottom row, we see a quasi-transversal orbit of tangency, shown in red, between $W^u(p_1)$ and $W^s(p_2)$. There are three cases, from left to right: $\mathcal{I}(p_1) = \mathcal{I}(p_2) = 1$, or $\mathcal{I}(p_1) = \mathcal{I}(p_2) = 2$, or $\mathcal{I}(p_1) = 1$ and $\mathcal{I}(p_2) = 2$.

The possibilities occurring in cases (NH) and (NT) are shown schematically in Figure 12. Note that, in case (NT), we have $\mathcal{I}(p_1^{\overline{\mu}})$, $\mathcal{I}(p_2^{\overline{\mu}}) \in \{1,2\}$ and $\mathcal{I}(p_1^{\overline{\mu}}) \leq \mathcal{I}(p_2^{\overline{\mu}})$ by condition (QT-1) of quasi-transversality.

The above conditions are labeled (i)–(vii) in Vegter [**37**, pp. 121–123]; see also conditions (i)–(v) in Palis and Takens [**31**, pp. 384–385].

5.2.2. Generic 2-parameter families of gradients in dimension 3. This section summarizes results of Vegter [**37**]; also see Carneiro and Palis [**7**] for the classification of generic 2-parameter families of gradients in arbitrary dimensions.

The instabilities (NH) and (NT) of Section 5.2.1 may also occur in an open and dense class of 2-parameter gradient families on M. The corresponding parameter values form smooth curves in the parameter space $\mathbb{R}^2$. Moreover, by the work of Vegter [**37**, p. 108], for a generic family $\{X^\mu\} \in X_2^g(M)$, at isolated values $\overline{\mu}$ of the parameter, exactly one of the following situations may occur. (These cases are described in more detail shortly.)

(A) The vector field $X^{\overline{\mu}}$ has exactly two quasi-transversal orbits of tangency between stable and unstable manifolds, satisfying analogues of conditions (ND-1)–(ND-4), while all singularities are hyperbolic.

(B) The vector field $X^{\overline{\mu}}$ has exactly one non-hyperbolic singularity, which is a saddle-node, and exactly one quasi-transversal orbit of tangency between stable and unstable manifolds that satisfies analogues of conditions (ND-1)–(ND-4).

(C) The vector field $X^{\overline{\mu}}$ has exactly two non-hyperbolic singularities, which are saddle-nodes, while all stable and unstable manifolds intersect transversely.

(D) The vector field $X^{\overline{\mu}}$ has exactly one non-hyperbolic singularity, which is quasi-hyperbolic of type 2, while all stable and unstable manifolds are transversal.

(E) All singular points of $X^{\overline{\mu}}$ are hyperbolic, and a single degenerate orbit of tangency occurs between $W^u(p_1^{\overline{\mu}})$ and $W^s(p_2^{\overline{\mu}})$ that violates exactly one of the conditions (QT-1), (QT-2), (ND-3), or (ND-4) in the "mildest possible manner" (for more detail, see (E1)–(E4)). Observe that it does not make sense to consider violation of condition (ND-1); it can be replaced by a similar condition for 2-parameter families. Condition (ND-2) also holds for generic 2-parameter families, since the set of linear *gradients* on $\mathbb{R}^2$ having two equal eigenvalues has codimension 2. Hence, generically, a pair of equal contracting eigenvalues at $p_1^{\overline{\mu}}$ does not occur together with an orbit of tangency.

If $X^{\overline{\mu}}$ has a non-hyperbolic singular point $p \in M$, as in cases (B)–(D), the set of stable and unstable manifolds also includes $W^{ss}(p)$ and $W^{uu}(p)$, respectively. Conditions (A)–(D) correspond to conditions I–IV of Vegter [**37**, pp. 111–112]. For condition (E), see [**37**, p. 125].

Next, we consider the bifurcation sets in $\mathbb{R}^2$ near parameter values $\overline{\mu}$ for which we have one of the situations described above. Such a parameter value is in the closure of smooth curves in $\mathbb{R}^2$ that correspond to the occurrence of bifurcations that may also occur in 1-parameter families. The curves limiting on $\overline{\mu}$ corresponding to tangencies with invariant manifolds of far-away singularities (i.e., singular points of $X^{\overline{\mu}}$ not directly contributing to the failure of conditions (NH) or (NT) in cases (A)–(E)) are called *secondary bifurcations.* We do not list the cases that arise from the ones below by reversing the sign of X^{μ}, which simply amounts to swapping superscripts "u" and "s."

(A1) There are four hyperbolic singular points $p_1^{\overline{\mu}}, \dots, p_4^{\overline{\mu}}$ of $X^{\overline{\mu}}$ such that the orbits of tangency are contained in $W^u(p_1^{\overline{\mu}}) \cap W^s(p_2^{\overline{\mu}})$ and $W^u(p_3^{\overline{\mu}}) \cap W^s(p_4^{\overline{\mu}})$, respectively. We allow $p_1^{\overline{\mu}} = p_3^{\overline{\mu}}$ or $p_2^{\overline{\mu}} = p_4^{\overline{\mu}}$, or both. Note that $\mathcal{I}(p_i^{\overline{\mu}}) \in \{1, 2\}$ for every $i \in \{1, \dots, 4\}$, and necessarily $\mathcal{I}(p_1^{\overline{\mu}}) \leq \mathcal{I}(p_2^{\overline{\mu}})$ and $\mathcal{I}(p_3^{\overline{\mu}}) \leq \mathcal{I}(p_4^{\overline{\mu}})$. See the top row of Figure 13 for schematic drawings of the possibilities when each $p_i^{\overline{\mu}}$ has index 1. Generically, the bifurcation set consists of two smooth curves that intersect transversely at $\overline{\mu}$; see Figure 21.

(A2) There are three hyperbolic singular points $p_1^{\overline{\mu}}$, $p_2^{\overline{\mu}}$, and $p_3^{\overline{\mu}}$ of $X^{\overline{\mu}}$ such that the orbits of tangency are contained in $W^u(p_1^{\overline{\mu}}) \cap W^s(p_2^{\overline{\mu}})$ and $W^u(p_2^{\overline{\mu}}) \cap W^s(p_3^{\overline{\mu}})$, respectively. Again, each $p_i^{\overline{\mu}}$ has index 1 or 2, and

$$\mathcal{I}(p_1^{\overline{\mu}}) \leq \mathcal{I}(p_2^{\overline{\mu}}) \leq \mathcal{I}(p_3^{\overline{\mu}}).$$

See the bottom row of Figure 13 for an illustration. The bifurcation set consists of five codimension-1 strata meeting at $\overline{\mu}$; see Figure 22.

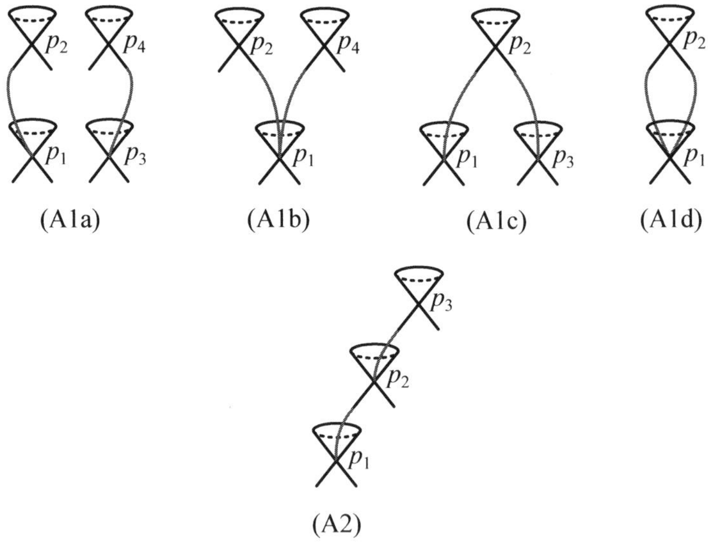

FIGURE 13. **The possibilities for two nondegenerate orbits of tangency between hyperbolic singular points from case (A).** The top row shows the possibilities in case (A1), while the bottom row illustrates case (A2). In this figure, each critical points is index 1, with stable manifold consisting of a curve and unstable manifold consisting of a disk. The orbits of tangencies are shown in red.

(B1) The vector field $X^{\overline{\mu}}$ has one saddle-node $p^{\overline{\mu}}$ and one quasi-transverse orbit of tangency between $W^u(p_1^{\overline{\mu}})$ and $W^s(p_2^{\overline{\mu}})$, where $p_1^{\overline{\mu}}$ and $p_2^{\overline{\mu}}$ are hyperbolic saddle-points of $X^{\overline{\mu}}$; see Figure 14 for one case. The bifurcation set consists of two curves that intersect transversely at $\overline{\mu}$; see Figure 23.

(B2) The vector field $X^{\overline{\mu}}$ has a saddle-node $p^{\overline{\mu}}$ and a hyperbolic saddle-point $\overline{p}^{\overline{\mu}}$ whose stable manifold has one quasi-transverse orbit of tangency with the unstable manifold of $p^{\overline{\mu}}$. Secondary bifurcations are due to the occurrence of tangencies between $W^u(p_*^{\mu})$ and $W^s(\overline{p}^{\mu})$ for each saddle point $p_*^{\mu} \neq \overline{p}^{\mu}$ such that $W^u(p_*^{\overline{\mu}}) \cap W^s(p^{\overline{\mu}}) \neq \emptyset$; suppose there are s of these. For an illustration of one case, see Figure 14. Then the bifurcation diagram consists of $s+3$ codimension-1 strata meeting at $\overline{\mu}$; see Figure 24.

(B3) The vector field $X^{\overline{\mu}}$ has a saddle-node $p^{\overline{\mu}}$ and a quasi-transverse orbit of tangency between $W^{ss}(p^{\overline{\mu}})$ and $W^u(\overline{p}^{\overline{\mu}})$, where $\overline{p}^{\overline{\mu}}$ is a saddle-point of $X^{\overline{\mu}}$; see Figure 14. The bifurcation set consists of 3 codimension-1 strata meeting at $\overline{\mu}$; see Figure 25.

(C) For an open and dense class of 2-parameter families $\{X^\mu\}$ of $X_2^g(M)$, we have a pair p_1, p_2 of saddle-nodes occurring at isolated values $\overline{\mu}$ of the parameter. For an illustration, see Figure 14. There are two curves Γ_1 and Γ_2 in the parameter plane corresponding to the occurrence of exactly one saddle-node of X^μ near p_1 and p_2, respectively. Generically, these curves are transversal; see Figure 26.

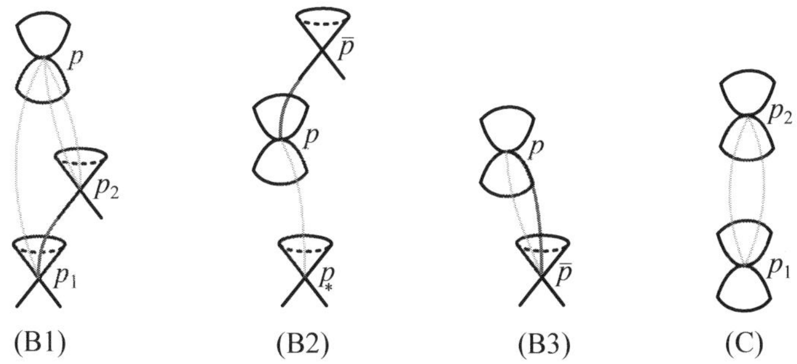

FIGURE 14. Codimension-2 bifurcations that involve a saddle-node. As in Figure 13, we have drawn schematic possibilities showing only the stable and unstable manifolds. In the case of an index 1-2 saddle-node, the stable and unstable manifolds are each half-planes. The red lines show quasi-transverse orbits of tangency between different singularities. The green lines are flows that are transverse intersections between stable and unstable manifolds.

(D) In a neighborhood of $\overline{\mu}$, the bifurcation diagram consists of parameter values μ for which X^μ, and hence f^μ, has a degenerate singular point near p. For an open and dense class of 2-parameter families $\{f^\mu\}$ for which $\operatorname{grad}(f^\mu)$ has a quasi-hyperbolic singularity of type 2, there are μ-dependent local coordinates (x, y, z) in which f^μ can be written as

$$\pm x^4 + \mu_1 x^2 + \mu_2 x \pm y^2 \pm z^2,$$

having a singularity of type $A_3^{\pm}$. So the bifurcation diagram near $\overline{\mu}$ is the well-known cusp; see Figure 28. The pair of curves having $\overline{\mu}$ in their closure corresponds to the occurrence of exactly one saddle-node near p. For an illustration, see Figure 15.

(E1) We have $\dim W^u(p_1^{\overline{\mu}}) = \dim W^s(p_2^{\overline{\mu}}) = 1$, violation of (QT-1). Secondary bifurcations may be present due to occurrence of an orbit of tangency between $W^s(p_2^\mu)$ and an unstable manifold (of dimension 2) intersecting $W^s(p_1^\mu)$, or between $W^u(p_1^\mu)$ and a stable manifold (of dimension 2) intersecting $W^u(p_2^\mu)$; see Figure 16. For μ close to $\overline{\mu}$, let D_r^μ be a continuous family of smooth discs contained in a level set of f^μ such that $W^u(p_1^\mu) \cap D_r^\mu = \{r\}$. Let $U_1^\mu, \ldots, U_n^\mu$ be the intersections of D_r^μ with unstable manifolds having non-empty intersections with $W^s(p_1^\mu)$. Similarly, $S_1^\mu, \ldots, S_m^\mu$ denote intersections of D_r^μ and stable manifolds meeting $W^u(p_2^\mu)$. The corresponding bifurcation diagram consists of $n + m$ curves in the parameter plane, having $\overline{\mu}$ in their closure; see Figure 29. For parameter values μ on these curves, we have either $W^s(p_2^\mu) \cap D_r^\mu \in U_i^\mu$ for some $i \in \{1, \ldots, n\}$, or $W^u(p_1^\mu) \cap D_r^\mu \in S_j^\mu$ for some $j \in \{1, \ldots, m\}$.

(E2) We have $d^2F^{\overline{\mu}}/dx^2 = 0$, violation of (QT-2), where $F^{\overline{\mu}}$ is the function F defined in (QT-2) for the vector field $v^{\overline{\mu}}$. For an open and dense class of 2-parameter families, we have $(d^3F^{\overline{\mu}}/dx^3)(0) \neq 0$, while the family $\{F^\mu\}$ is a versal unfolding of $F^{\overline{\mu}}$. The latter condition implies the existence of local coordinates (μ, x) near $(\overline{\mu}, r)$ in which $(\overline{\mu}, r)$ corresponds to $(0, 0) \in \mathbb{R}^2 \times \mathbb{R}$,

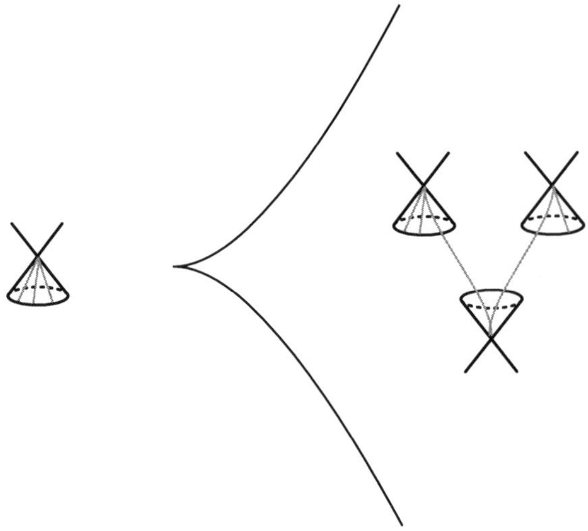

FIGURE 15. Local codimension-2 bifurcation of type (D). We have drawn the bifurcation diagram for an A_3^- singularity, and indicated the dynamics on the two sides of the bifurcation set.

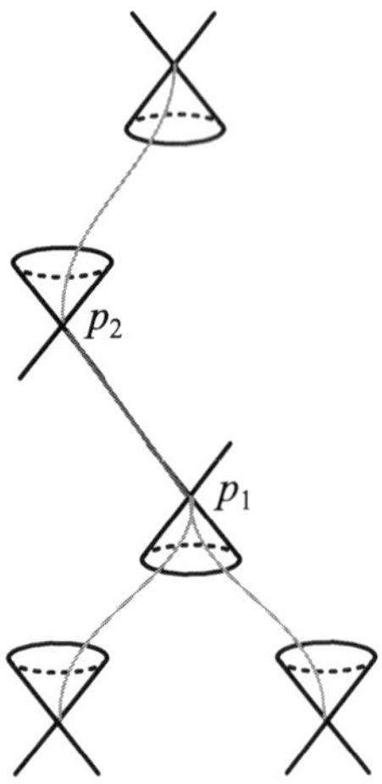

FIGURE 16. The dynamics at a bifurcation of type (E1).

such that

$$F^\mu(x) = x^3 + \mu_1 x + \mu_2.$$

The bifurcation values form a cusp in the parameter space $\mathbb{R}^2$.

(E3) Situation of case (QT-1b), where $\lim_{t\to-\infty} d\phi_t^{\overline{\mu}}(E_r)$ is the eigenspace of the linear part at $p_1^{\overline{\mu}}$ corresponding to the *weakest* contracting eigenvalues, which is violation of (ND-3). Let D_r^μ be as in case (E1). Then D_r^μ contains $U_1^\mu, \ldots, U_n^\mu$ that are intersections of D_r^μ and unstable manifolds meeting $W^s(p_1^\mu)$. Secondary bifurcations occur for parameter values μ for which $W^s(p_2^\mu) \cap D_2^\mu$ is tangent to U_i^μ for some $i \in \{1, \ldots, n\}$.

(E4) The vector field X^μ has an orbit of tangency as in case (QT-1b), and exactly one hyperbolic saddle $p_*^{\overline{\mu}}$ different from $p_1^{\overline{\mu}}$ and $p_2^{\overline{\mu}}$ such that $W^u(p_*^{\overline{\mu}})$ and $W^{ss}(p_1^{\overline{\mu}})$ are not transversal, which is violation of (ND-4). In this case, we have secondary bifurcations for parameter values μ for which one of the following occur:
 (a) $W^u(p_1^\mu) \cap D_r^\mu \in W^s(p_2^\mu) \cap D_r^\mu$,
 (b) $W^u(p_*^\mu) \cap D_r^\mu$ is tangent to $W^s(p_2^\mu) \cap D_r^\mu$.

We remark that conditions (A1) and (A2) correspond to conditions I.a and I.b of Vegter [**37**, p. 113]. Conditions (B1)–(B3) correspond to conditions II.a–II.c of Vegter [**37**, p. 114], while conditions (E1)–(E4) appear as conditions 1–4 in Vegter [**37**, pp. 125–128].

5.3. Sutured functions and gradient-like vector fields

In this section, we introduce sutured functions, which are smooth functions on a sutured manifold with prescribed boundary behavior. Then we define and study gradient-like vector fields for sutured functions.

DEFINITION 5.12. Let (M, γ) be a sutured manifold. A *sutured function* on (M, γ) is a smooth function $f\colon M \to [-1, 1]$ such that
 (1) $f^{-1}(\pm 1) = R_\pm(\gamma)$ and $f^{-1}(0) \supset s(\gamma)$,
 (2) f has no critical points along $R(\gamma)$,
 (3) $f|_\gamma$ has no critical points.

The space of sutured functions on (M, γ) is contractible. Indeed, the set of sutured functions for a given $f|_\gamma$ is convex, while the space of possible $f|_\gamma$ is contractible. For a sutured function f, we denote by $C(f)$ the set of critical points of f; i.e.,

$$C(f) = \{\, p \in M \colon df_p = 0 \,\}.$$

By conditions (2) and (3), the set $C(f)$ lies in the interior of M. The following definition was motivated by Milnor [**24**, Definition 3.1].

DEFINITION 5.13. Let f be a sutured function on (M, γ). A vector field v on M is a *gradient-like vector field for* f if
 (1) $v(f) > 0$ on $M \setminus C(f)$,
 (2) $C(f)$ has a neighborhood U such that $v|_U = \operatorname{grad}_g(f|_U)$ for some Riemannian metric g on U,
 (3) $v_p \in T_p\gamma$ for every $p \in \gamma$.

REMARK 5.14. Note that Milnor [**24**, Definition 3.1] defined gradient-like vector fields for a Morse function f on an n-manifold M. Instead of condition (2), he required that, for any critical point p of f, there are coordinates $(x_1, \dots, x_n)$ in a neighborhood U of p such that

$$f = f(p) - x_1^2 - \dots - x_\lambda^2 + x_{\lambda+1}^2 + \dots + x_n^2$$

and v has coordinates $(-x_1, \dots, -x_\lambda, x_{\lambda+1}, \dots, x_n)$ throughout U. When studying families of smooth functions, as we have seen, more complicated singularities can arise. We could require that, around such a singularity, v is the Euclidean gradient in a local coordinate system in which the singularity is in normal form. But then it is unclear whether the space of gradient-like vector fields is contractible, as the space of such local coordinate systems is rather complicated. Hence, we have chosen to work with condition (2), as the space of metrics is clearly contractible. As a tradeoff, one has to resort to such results as Theorems 5.3 and 5.4 to understand the invariant manifolds of v near a singular point.

Let $\mathcal{FV}(M,\gamma)$ be the space of pairs (f,v), where f is a sutured function on (M,γ) and v is a gradient-like vector field for f. We endow $\mathcal{FV}(M,\gamma)$ with the C^∞-topology.

DEFINITION 5.15. A *Morse function* on (M,γ) is a sutured function $f\colon M \to [-1,1]$ such that all critical points of f are non-degenerate. For a Morse function f and $i \in \{0,1,2,3\}$, let $C_i(f)$ be the set of critical points of f of index i.

By condition (2) of Definition 5.13, every gradient-like vector field v of a Morse function has only hyperbolic singular points. In particular, we can talk about the stable and unstable manifolds $W^s(p)$ and $W^u(p)$ of a singular point p of v. If we also want to refer to the vector field v, then we write $W^u(p,v)$ and $W^s(p,v)$. Note that the Morse index $\mathcal{I}(p)$ of the critical point $p \in C(f)$ agrees with $\dim W^s(p)$. Indeed, in a suitable coordinate system around p, the linearization $L_p v$ coincides with the Hessian of f at p. Furthermore, notice that every point

$$x \in M \setminus \bigcup_{p \in C(f)} (W^u(p) \cup W^s(p))$$

lies on a compact flow-line connecting $R_-(\gamma)$ and $R_+(\gamma)$.

DEFINITION 5.16. We say that $(f,v) \in \mathcal{FV}(M,\gamma)$ satisfies the *Morse–Smale condition* if v is Morse–Smale in the sense of Definition 5.8. We denote the subspace of Morse–Smale pairs in $\mathcal{FV}(M,\gamma)$ by $\mathcal{FV}_0(M,\gamma)$.

If $(f,v) \in \mathcal{FV}_0(M,\gamma)$, then f is a Morse function on (M,γ). Furthermore, for every p, $q \in C(f)$, the intersection $W^u(p) \cap W^s(q)$ is a manifold of dimension $\mathcal{I}(q) - \mathcal{I}(p)$ that we denote by $W(p,q)$. In particular, $W(p,q) = \emptyset$ if $\mathcal{I}(q) - \mathcal{I}(p) < 0$.

REMARK 5.17. Notice that, for $(f,v) \in \mathcal{FV}(M,\gamma)$, the condition that $W^u(p)$ and $W^s(q)$ intersect transversely is automatically satisfied if at least one of p or q has index 0 or 3. For a pair $(f,v) \in \mathcal{FV}(M,\gamma)$ where f is a Morse function, the Morse–Smale condition can be violated by having flows between critical points of index 1, flows between critical points of index 2, flows from index 2 to index 1 critical points, or an orbit of tangency (see Definition 5.10) in $W^u(p) \cap W^s(q)$ for $p \in C_1(f)$ and $q \in C_2(f)$. Flows from index 2 to index 1 critical points only appear generically in 2-parameter families, while the other possibilities already occur in generic 1-parameter families.

DEFINITION 5.18. We say the pair $(f,v) \in \mathcal{FV}(M,\gamma)$ is *codimension-1* if $(f,v) \notin \mathcal{FV}_0(M,\gamma)$, but v appears as $X^{\overline{\mu}}$ for some 1-parameter family $\{X^\mu\} \in X_1^g(M)$ that is generic in the sense of Section 5.2.1. We denote the space of codimension-1 pairs by $\mathcal{FV}_1(M,\gamma)$, and the union $\mathcal{FV}_0(M,\gamma) \cup \mathcal{FV}_1(M,\gamma)$ by $\mathcal{FV}_{\le 1}(M,\gamma)$.

In an analogous manner, we say that $(f,v) \in \mathcal{FV}(M,\gamma)$ is *codimension-2* if $(f,v) \notin \mathcal{FV}_{\le 1}(M,\gamma)$, but v appears as $X^{\overline{\mu}}$ for some 2-parameter family $\{X^\mu\} \in X_2^g(M)$ that is generic in the sense of Section 5.2.2. We denote the space of codimension-2 pairs by $\mathcal{FV}_2(M,\gamma)$. Finally, we set

$$\mathcal{FV}_{\le 2}(M,\gamma) = \bigcup_{i \in \{0,1,2\}} \mathcal{FV}_i(M,\gamma).$$

The following proposition implies that every gradient-like vector field is actually a gradient for some Riemannian metric. The advantage of gradient-like vector fields is that they are easier to manipulate than metrics, which is useful in actual constructions.

PROPOSITION 5.19. *Let $(f, v) \in \mathcal{FV}(M, \gamma)$. Then the space $G(f, v)$ of Riemannian metrics g on M for which $v = grad_g(f)$ is non-empty and contractible.*

PROOF. By definition, there is a metric g on a neighborhood U of $C(f)$ such that $v|_U = \text{grad}_g(f|_U)$. Pick a smaller neighborhood V of $C(f)$ such that $\overline{V} \subset U$. We are going to extend $g|_V$ to the whole manifold M such that $v = \text{grad}_g(f)$ everywhere. Such a metric g on M satisfies $g(v_x, v_x) = v_x(f) > 0$ and $g(v_x, w_x) = 0$ for every $x \notin C(f)$ and $w_x \in \ker(df_x)$. So the extension g on M is uniquely determined by a choice of metric on the 2-plane bundle $\ker(df)|_{M\setminus V}$ that smoothly extends the metric given on $\ker(df)|_{V\setminus C(f)}$. For this, pick an arbitrary metric on $\ker(df)|_{M\setminus V}$ and piece it together with $g|_{U\setminus C(f)}$ using a partition of unity subordinate to the covering $\{U, M \setminus V\}$ of M. Hence $G(f, v) \neq \emptyset$.

The space $G(f, v)$ is contractible because it is convex. Indeed, if g_0, $g_1 \in G(f, v)$, then $g_i(v, w) = w(f)$ for every vector field w on M and $i \in \{0, 1\}$. Let $g_t = (1-t)g_0 + tg_1$ for $t \in I$ be an arbitrary convex combination of g_0 and g_1. Then $g_t(v, w) = w(f)$ for every w on M; i.e., $v = \text{grad}_{g_t}(f)$. □

COROLLARY 5.20. *The space $\mathcal{FV}(M, \gamma)$ is weakly contractible.*

PROOF. Let $\mathcal{F}(M, \gamma)$ be the space of sutured functions and $\mathcal{G}(M, \gamma)$ the space of Riemannian metrics on (M, γ), respectively. Both $\mathcal{F}(M, \gamma)$ and $\mathcal{G}(M, \gamma)$ are contractible. Consider the projection

$$\pi\colon \mathcal{F}(M, \gamma) \times \mathcal{G}(M, \gamma) \to \mathcal{FV}(M, \gamma)$$

given by $\pi(f, g) = (f, \text{grad}_g(f))$; this is a Serre fibration. For $(f, v) \in \mathcal{FV}(M, \gamma)$, the fiber is $\pi^{-1}(f, v) = G(f, v)$, which is contractible by Proposition 5.19. Hence the base space $\mathcal{FV}(M, \gamma)$ is weakly contractible. □

CHAPTER 6

Translating Bifurcations of Gradients to Heegaard Diagrams

We will now translate the singularities of Sections 5.2.1 and 5.2.2 in terms of Heegaard diagrams. Loosely speaking, each generic gradient gives a Heegaard diagram, each codimension-1 singularity gives a move between Heegaard diagrams, and each codimension-2 singularity gives a contractible loop of Heegaard diagrams. The codimension-1 and codimension-2 singularities give moves and loops of moves, respectively, that are more complicated than the ones appearing in the definition of weak and strong Heegaard invariants. In Chapter 7, we will see how to simplify these families.

The overall idea is that, to construct a Heegaard splitting from the gradient of a generic Morse function (not necessarily self-indexing), take one compression body to be a small neighborhood of the union of all flows starting at $R_-(\gamma)$ or index 0 critical points and ending at index 1 critical points. The other compression body is then isotopic to a small neighborhood of the flows starting at index 2 critical points and ending at $R_+(\gamma)$ or at index 3 critical points. To further construct the α- and β-curves of a Heegaard diagram, we take the intersection of the Heegaard surface with the unstable manifolds of some of the index 1 critical points and with the stable manifolds of some of the index 2 critical points.

This Heegaard splitting extends naturally across codimension-1 and codimension-2 singularities, as long as there is no flow from an index 2 to an index 1 critical point. In each case, we will analyze how the corresponding Heegaard diagrams change.

6.1. Separability of gradients

We now introduce *separability*, our main technical tool for obtaining Heegaard splittings compatible with gradient-like vector fields that have at most codimension-2 degeneracies. In the sections that follow, we explain how to enhance these Heegaard splittings to Heegaard diagrams for generic gradients, to moves between diagrams for codimension-1 gradients, and to loops of diagrams for codimension-2 gradients.

DEFINITION 6.1. We say that the pair $(f, v) \in \mathcal{FV}_{\leq 2}(M, \gamma)$ is *separable* if

- it is not codimension-2 of type (E1); i.e., if for every pair of non-degenerate critical points $p \in C_2(f)$ and $q \in C_1(f)$, we have $W^u(p) \cap W^s(q) = \emptyset$; and
- if it is codimension-2 of type (C) (i.e., it has two birth-death singularities at p and at q), then $f(p) \neq f(q)$.

(This second condition is codimension-3, and hence generic for 2-parameter families.)

DEFINITION 6.2. Suppose that $(f,v) \in \mathcal{FV}_{\leq 2}(M,\gamma)$. Then we partition $C(f)$ into two subsets, namely $C_{01}(f,v)$ and $C_{23}(f,v)$, as follows. We define $C_{01}(f,v)$ to be the set of those critical points $p \in C(f)$ for which one of the following holds:

(1) $p \in C_0(f) \cup C_1(f)$, or
(2) p is an index 0-1 birth-death, or
(3) p is an index 1-2 birth-death, (f,v) is codimension-2 of type (B1), and $\mathcal{I}(p_1) = \mathcal{I}(p_2) = 1$, or
(4) p is an index 1-2 birth-death, (f,v) is codimension-2 of type (B2) or (B3), and $\mathcal{I}(\overline{p}) = 1$,
(5) $(f,v) \in \mathcal{FV}_2(M,\gamma)$ is type (C), and if q is the other birth-death critical point, then $f(p) < f(q)$, or
(6) (f,v) is codimension-2 of type (D), and p is an index 1-0-1, 0-1-0, or 1-2-1 birth-death-birth.

Finally, if p is an index 1-2 birth-death critical point and (f,v) is codimension-1 of type (NH), then we can put p in either $C_{01}(f,v)$ or $C_{23}(f,v)$. Having defined $C_{01}(f,v)$, we let $C_{23}(f,v) = C(f) \setminus C_{01}(f,v)$.

It follows from condition (5) of Definition 6.2 that $C_{01}(f,v)$ contains at most one index 1-2 birth-death critical point.

DEFINITION 6.3. Suppose that $(f,v) \in \mathcal{FV}_{\leq 2}(M,\gamma)$. We say that a properly embedded surface $\Sigma \subset M$ *separates* (f,v) if

(1) $\Sigma \pitchfork v$,
(2) $M = M_- \cup M_+$, such that $M_- \cap M_+ = \Sigma$ and $R_\pm(\gamma) \subset M_\pm$,
(3) $C_{01}(f,v) \subset M_-$ and $C_{23}(f,v) \subset M_+$, and
(4) $\partial\Sigma = s(\gamma)$.

We denote the set of surfaces that separate (f,v) by $\Sigma(f,v)$. When $(f,v) \in \mathcal{FV}_1(M,\gamma)$ is type (NH) with an index 1-2 birth-death singularity p, then there are two different choices for the partition $(C_{01}(f,v), C_{23}(f,v))$ of $C(f)$, depending on where we put p, hence $\Sigma(f,v)$ is not completely unique. If we put p into $C_{01}(f,v)$, then we denote the resulting set $\Sigma_-(f,v)$, and we write $\Sigma_+(f,v)$ when $p \in C_{23}(f,v)$. Often, we suppress this choice in our notation, and simply write $\Sigma(f,v)$ (which is then either $\Sigma_-(f,v)$ or $\Sigma_+(f,v)$).

Notice that, if Σ separates (f,v), then Σ is necessarily orientable as it is transverse to v and M is orientable. We orient Σ such that the normal orientation given by v, followed by the orientation of Σ, agrees with the orientation on M; i.e., such that Σ is oriented as the boundary of M_-.

If $(f,v) \in \mathcal{FV}_{\leq 2}(M,\gamma)$, then, for every $p \in C_{01}(f,v)$, the manifold $W^s(p) \setminus R_-(\gamma)$ is diffeomorphic to

- a single point if $p \in C_0(f)$, or p is a birth-death-birth of index 0-1-0,
- $\mathbb{R}$ if $p \in C_1(f)$, or p is an index 1-0-1 or 1-2-1 birth-death-birth,
- $[0,\infty)$ if p is an index 0-1 birth-death, or
- $\mathbb{R} \times [0,\infty)$ if p is an index 1-2 birth-death.

In addition, if (f,v) is separable and $p \in C_{01}(f,v)$ is not an index 1-2 birth-death, then $\partial W^s(p) \subset C_{01}(f,v) \cup R_-(\gamma)$ (where $\partial W^s(p)$ is the topological boundary). If $p \in C_{01}(f,v)$ is an index 1-2 birth-death, then $\partial W^{ss}(p) \subset C_{01}(f,v) \cup R_-(\gamma)$, while

$$\partial W^s(p) \subset C_{01}(f,v) \cup R_-(\gamma) \cup W^{ss}(p) \cup \bigcup \{\, W^s(p') : p' \in C_{01}(f,v) \setminus \{p\} \,\}.$$

Analogous statements hold for $C_{23}(f,v)$. The above discussion justifies the following definition.

DEFINITION 6.4. Suppose that the pair $(f,v) \in \mathcal{FV}_{\le 2}(M,\gamma)$ is separable. Then we define the relative CW complexes $(R_-(\gamma) \cup \Gamma_{01}(f,v), R_-(\gamma))$ and $(R_+(\gamma) \cup \Gamma_{23}(f,v), R_+(\gamma))$ by taking

$$\Gamma_{01}(f,v) = \bigcup_{p \in C_{01}(f,v)} W^s(p),$$

$$\Gamma_{23}(f,v) = \bigcup_{p \in C_{23}(f,v)} W^u(p).$$

The set of vertices of $\Gamma_{01}(f,v)$ is $C_{01}(f,v)$. The closed 1-cells of $\Gamma_{01}(f,v)$ are the closures of the components of the $W^{ss}(p) \setminus \{p\}$ for $p \in C_{01}(f,v)$ an index 1-2 birth-death, and the closures of the components of $W^s(p) \setminus \{p\}$ for every other $p \in C_{01}(f,v)$. Finally, $\Gamma_{01}(f,v)$ has at most one (closed) 2-cell, namely $\overline{W^s(p)}$ if $p \in C_{01}(f,v)$ is an index 1-2 birth-death. We define the cell decomposition of $\Gamma_{23}(f,v)$ in an analogous manner. Finally, we set $\Gamma(f,v) = \Gamma_{01}(f,v) \cup \Gamma_{23}(f,v)$.

Note that Γ_{01} is either a graph (i.e., a 1-complex), or obtained from a graph by an elementary expansion involving the 1-cell $W^{ss}(p)$ and the 2-cell $W^s(p)$ if $C_{01}(f,v)$ contains an index 1-2 birth-death p.

REMARK 6.5. In light of Definition 6.4, we now motivate Definition 6.2. We partitioned the set of critical points $C(f)$ into $C_{01}(f,v)$ and $C_{23}(f,v)$ precisely so that we can form the relative CW complexes $(R_-(\gamma) \cup \Gamma_{01}(f,v), R_-(\gamma))$ and $(R_+(\gamma) \cup \Gamma_{23}(f,v), R_+(\gamma))$. We would like to have $C_1(f) \subset C_{01}(f,v)$ and $C_2(f) \subset C_{23}(f,v)$ because if Σ is a separating surface, then – as we shall see in Section 6.2 – we can obtain a Heegaard diagram from it by taking α-curves to be $W^u(p) \cap \Sigma$ for some $p \in C_1(f)$ and β-curves to be $W^s(p) \cap \Sigma$ for some $p \in C_2(f)$. This also explains our rule in case (B2). For example, suppose that (f,v) has an index 1-2 birth-death critical point at p, and a non-degenerate critical point at $\overline{p}$ of index 1, such that there is a flow φ from p to $\overline{p}$. Since we have to place $\overline{p}$ in $C_{01}(f,v)$, we must also put p into $C_{01}(f,v)$, otherwise the 1-cell $\overline{\varphi} \subset W^s(\overline{p})$ would have one endpoint in $\Gamma_{23}(f,v)$.

In case (E1), we do not obtain a CW complex (whichever side we assign p to) for a similar reason, explaining why those gradients are not separable. Our choices for placing the index 1-2 birth-death critical points of a pair $(f,v) \in \mathcal{FV}_2(M,\gamma)$ in every case other than (B2) are purely conventional to make the construction more canonical, and most proofs would also work for the other choices. However, we do adhere to these conventions in Theorem 6.37.

When $(f,v) \in \mathcal{FV}_1(M,\gamma)$ has an index 1-2 birth-death critical point p, there is no canonical way to decide where to put p, and, in fact, the rule in case (B2) forces us to allow both possibilities: If $\{\, (f_\lambda, v_\lambda) : \lambda \in \mathbb{R}^2 \,\}$ is a generic 2-parameter family such that (f_0, v_0) has a type (B2) bifurcation, where we have to put the A_2 point in $C_{01}(f,v)$, then we have to do the same for (f_λ, v_λ) when λ lies in the stratum of the bifurcation set corresponding to the A_2 singularity.

LEMMA 6.6. *Suppose that $(f,v) \in \mathcal{FV}_{\le 2}(M,\gamma)$ is separable and $\Sigma \in \Sigma(f,v)$. Then the surface Σ intersects every flow line of v in $M \setminus \Gamma(f,v)$ in exactly one point.*

PROOF. Note that $M \setminus \Gamma(f, v)$ is a saturated subset of M (i.e., it is a union of complete flow lines). The closure of a non-constant flow line τ of v is diffeomorphic to I, and has both endpoints in $R(\gamma) \cup C(f)$. If the maximal open interval on which τ is defined is (a, b) (where a might be $-\infty$ and b might be $+\infty$), then let these endpoints be $\tau(a) = \lim_{t \to a+} \tau(t)$ and $\tau(b) = \lim_{t \to b-} \tau(t)$. If $\tau(a) \in C_{23}(f, v)$, then $\tau \subset \Gamma_{23}(f, v)$. Similarly, if $\tau(b) \in C_{01}(f, v)$, then $\tau \subset \Gamma_{01}(f, v)$. Consequently, every flow line τ of $v|_{M \setminus \Gamma(f,v)}$ has

$$\begin{aligned} &\tau(a) \in R_-(\gamma) \cup C_{01}(f, v) \subset M_-, \text{ and} \\ &\tau(b) \in R_+(\gamma) \cup C_{23}(f, v) \subset M_+, \end{aligned}$$

so $\tau \cap \Sigma \neq \emptyset$. Since Σ is positively transverse to v, once an integral curve of v enters M_+, it can never leave it, so $|\tau \cap \Sigma| = 1$. $\square$

Using Lemma 6.6, we can endow $\Sigma(f, v)$ with a topology as follows. Choose a smooth function $h \colon M \to I$ such that $h^{-1}(0) = R(\gamma)$, and let $w = hv$. Unlike v, the vector field w is complete, and v and w have the same phase portrait inside $M \setminus R(\gamma)$. Let $\varphi \colon \mathbb{R} \times M \to M$ be the flow of w. For surfaces Σ and Σ' in $\Sigma(f, v)$, we define the function $d_{\Sigma', \Sigma} \in C^\infty(\Sigma)$ by requiring that $\varphi(x, d_{\Sigma',\Sigma}(x)) \in \Sigma'$ for every $x \in \Sigma$. This uniquely determines $d_{\Sigma',\Sigma}(x)$ by Lemma 6.6. If we fix $\Sigma_0 \in \Sigma(f, v)$, then the map $b_{\Sigma_0} \colon \Sigma(f, v) \to C^\infty(\Sigma_0)$ given by $b_{\Sigma_0}(\Sigma) = d_{\Sigma,\Sigma_0}$ is bijective. The topology on $\Sigma(f, v)$ is the pullback of the Whitney C^∞-topology on $C^\infty(\Sigma_0)$ along b_{Σ_0}. This is independent of the choice of Σ_0, since $d_{\Sigma,\Sigma_1} = d_{\Sigma,\Sigma_0} \circ i_{\Sigma_0,\Sigma_1} + d_{\Sigma_0,\Sigma_1}$, where $i_{\Sigma_0,\Sigma_1} \colon \Sigma_1 \to \Sigma_0$ is the diffeomorphism given by $i_{\Sigma_0,\Sigma_1}(x) = \varphi(x, d_{\Sigma_0,\Sigma_1}(x))$. In addition, the map $f \mapsto f \circ i_{\Sigma_0,\Sigma_1}$ from $C^\infty(\Sigma_0)$ to $C^\infty(\Sigma_1)$ and the map $g \mapsto g + d_{\Sigma_0,\Sigma_1}$ from $C^\infty(\Sigma_1)$ to $C^\infty(\Sigma_1)$ are both homeomorphims. The function $d_{\Sigma,\Sigma'}$ depends continuously on h, hence the topology that we defined is independent of the choice of h.

PROPOSITION 6.7. *Suppose that $(f, v) \in \mathcal{FV}_{\leq 2}(M, \gamma)$ is separable. Then the space $\Sigma(f, v)$ is non-empty and contractible. Furthermore, every $\Sigma \in \Sigma(f, v)$ divides (M, γ) into two sutured compression bodies; i.e., it is a Heegaard surface of (M, γ).*

More precisely, in the indeterminate case when $(f, v) \in \mathcal{FV}_1(M, \gamma)$ and f has an index 1-2 birth-death critical point, we mean that both $\Sigma_-(f, v)$ and $\Sigma_+(f, v)$ are non-empty and contractible.

PROOF. By the above discussion, it is clear that if $\Sigma(f, v) \neq \emptyset$, then it is homeomorphic to $C^\infty(\Sigma)$, hence it is contractible.

Next, we show that $\Sigma(f, v) \neq \emptyset$. Let N_{01} be a thin regular neighborhood of $\Gamma_{01}(f, v) \cup R_-(\gamma)$, and consider the surface $\Sigma_{01} = \overline{\partial N_{01} \setminus \partial M}$. Similarly, pick a regular neighborhood N_{23} of $\Gamma_{23}(f, v) \cup R_+(\gamma)$, and define $\Sigma_{23} = \overline{\partial N_{23} \setminus \partial M}$. Choosing sufficiently small and nice regular neighborhoods, we can suppose that $\Sigma_{01} \cap \Sigma_{23} = \emptyset$, and that Σ_{01} and Σ_{23} are transverse to v. Their union $\Sigma_{01} \cup \Sigma_{23}$ separates M into three pieces. Two of them are N_{01} and N_{23}, and we call the third piece P. Now $v|_P$ is a nowhere vanishing vector field that points into P along Σ_{01}, points out of P along Σ_{23}, and is tangent to $\gamma \cap P$. In addition, $v(f) > 0$ on P, so an isotopy from Σ_{01} to Σ_{23} relative to γ is given by flowing along $v/v(f)$. In particular, $(P, \gamma \cap P)$ is a product sutured manifold, and the flow lines of $v|_P$ give an I-fibration. By isotoping Σ_{01} near γ flowing along v, we can obtain a surface Σ'_{01}

such that $\partial\Sigma'_{01} = s(\gamma)$. Hence $\Sigma'_{01} \in \Sigma(f, v)$ (with M_- isotopic to N_{01} and M_+ isotopic to $N_{23} \cup P$).

Observe that Σ_{01} divides (M, γ) into the sutured manifolds $(N_{01}, \gamma \cap N_{01})$ and $(N_{23} \cup P, \gamma \cap (N_{23} \cup P))$. Since Γ_{01} is either a graph (i.e., a 1-complex), or obtained from a graph by an elementary expansion if $C_{01}(f, v)$ contains an index 1-2 birth-death, $(N_{01}, \gamma \cap N_{01})$ is a sutured compression body (where $R_+(\gamma \cap N_{01}) = \Sigma_{01}$ can be compressed to be isotopic to $R_-(\gamma \cap N_{01})$). Similarly, $(N_{23}, \gamma \cap N_{23})$ is also a sutured compression body (where $R_-(\gamma \cap N_{23}) = \Sigma_{23}$ can be compressed to be isotopic to $R_+(\gamma \cap N_{23})$). As $(P, \gamma \cap P)$ is a product, $(N_{23} \cup P, \gamma \cap (N_{23} \cup P))$ is a sutured compression body. Every element of $\Sigma(f, v)$ is isotopic to Σ_{01} relative to γ, hence also divides (M, γ) into two sutured compression bodies. □

DEFINITION 6.8. Let $B(M, \gamma)$ be the space of pairs $(f, v) \in \mathcal{FV}_{\leq 2}(M, \gamma)$ that are separable, and let $E(M, \gamma)$ be the space of triples (f, v, Σ), where $(f, v) \in B(M, \gamma)$ and $\Sigma \in \Sigma(f, v)$. There is a projection $\pi\colon E(M, \gamma) \to B(M, \gamma)$ defined by forgetting Σ. For $(f, v) \in B(M, \gamma)$, let $\chi(f, v)$ be the Euler characteristic of Σ for any $\Sigma \in \Sigma(f, v)$ (which is independent of the choice of Σ by Proposition 6.7). For $k \in \mathbb{Z}$, we define

$$B_k(M, \gamma) = \{\, (f, v) \in B(M, \gamma)\colon \chi(f, v) = k \,\}.$$

Finally, we set $E_k(M, \gamma) = \pi^{-1}(B_k(M, \gamma))$ and $\pi_k = \pi|_{E_k(M, \gamma)}$.

The total space $E(M, \gamma)$ depends on whether $\Sigma(M, \gamma)$ stands for $\Sigma_+(M, \gamma)$ or $\Sigma_-(M, \gamma)$, but the base $B(M, \gamma)$ is independent of this choice according to the following result.

LEMMA 6.9. *For $(f, v) \in B(M, \gamma)$, we have*

$$\chi(f, v) = \chi(R_-(\gamma)) + \sum_{p \in C_{01}(f, v)} i(p),$$

where

- *$i(p) = 2$ for $p \in C_0(f)$ or p an index 0-1-0 birth-death-birth,*
- *$i(p) = 0$ for p a birth-death, and*
- *$i(p) = -2$ for $p \in C_1(f)$, or p an index 1-0-1 or 1-2-1 birth-death-birth.*

PROOF. Recall that $(R_-(\gamma) \cup \Gamma_{01}(f, v), R_-(\gamma))$ is a relative CW complex of dimension at most two. We saw in the proof of Proposition 6.7 that every $\Sigma \in \Sigma(f, v)$ is isotopic to the surface $\Sigma_{01} = \overline{\partial N_{01} \setminus \partial M}$ relative to γ, where N_{01} is a regular neighborhood of $R_-(\gamma) \cup \Gamma_{01}(f, v)$. Since $\partial N_{01} = R_-(\gamma) \cup \Sigma_{01} \cup (\gamma \cap N_{01})$, where $\gamma \cap N_{01}$ is a disjoint union of annuli,

$$\chi(R_-(\gamma)) + \chi(\Sigma_{01}) = \chi(\partial N_{01}) = 2\chi(N_{01}).$$

As N_{01} deformation retracts onto $R_-(\gamma) \cup \Gamma_{01}(f, v)$, we have

$$\chi(N_{01}) = \chi(R_-(\gamma)) + c_0 - c_1 + c_2,$$

where c_i is the number of i-cells in $\Gamma_{01}(f, v)$. By construction, each point $p \in C_{01}(f, v)$ contributes $i(p)/2$ to $c_0 - c_1 + c_2$. Indeed, for $p \in C_0(f)$ or p an index 0-1-0 birth-death-birth, $W^s(p) = \{p\}$, so p contributes a single 0-cell. If p is an index 0-1 birth-death, then it contributes a 0-cell and a 1-cell, while an index 1-2 birth-death contributes a 0-cell, two 1-cells, and a 2-cell. For $p \in C_1(f)$ or p an index 1-0-1 or 1-2-1 birth-death-birth, $W^s(p)$ is an arc, and p contributes a 0-cell and two 1-cells. □

COROLLARY 6.10. *If (f_0, v_0) and (f_1, v_1) lie in the same path-component of $\mathcal{FV}_0(M,\gamma)$ or $\mathcal{FV}_1(M,\gamma)$, then $\chi(f_0, v_0) = \chi(f_1, v_1)$.*

Note that this corollary is false for $\mathcal{FV}_{\leq 1}(M,\gamma)$.

PROOF. Take a path $\{\,(f_t, v_t) \in \mathcal{FV}_i(M,\gamma) : t \in I\,\}$ connecting (f_0, v_0) and (f_1, v_1). In this family, the types of critical points in $C_{01}(f_t, v_t)$ remain unchanged; in particular, the local contributions $i(p_t)$ for $p_t \in C_{01}(f_t, v_t)$ are also constant. By Lemma 6.9, we obtain that $\chi(f_0, v_0) = \chi(f_1, v_1)$. □

We denote by $B_k^m(M,\gamma)$ and $E_k^m(M,\gamma)$ the space of those $(f,v) \in B_k(M,\gamma)$ and $(f,v,\Sigma) \in E_k(M,\gamma)$ for which f is Morse. By slight abuse of notation, we also denote the projection $(f,v,\Sigma) \mapsto (f,v)$ by π_k.

PROPOSITION 6.11. *Let (M,γ) be a connected sutured manifold and $k \in \mathbb{Z}$. Then the map*

$$\pi_k \colon E_k^m(M,\gamma) \to B_k^m(M,\gamma)$$

is a fiber bundle with fibre $C^\infty(\Sigma,\mathbb{R})$ for a compact, connected, orientable surface Σ with $|\partial\Sigma| = |s(\gamma)|$ and $\chi(\Sigma) = k$.

PROOF. Given $(f,v,\Sigma) \in E_k^m(M,\gamma)$, there is a neighborhood U of $\pi_k(f,v,\Sigma) = (f,v)$ in $B_k^m(M,\gamma)$ such that, for every $(f',v') \in U$, we have $\Sigma \in \Sigma(f',v')$. To see this, note that the surface Σ separates (M,γ) into two sutured compression bodies (M_+,γ_+) and (M_-,γ_-), one of which contains $\Gamma_{01}(f,v) \cup R_-(\gamma)$, and the other one contains $\Gamma_{23}(f,v) \cup R_+(\gamma)$. If (f',v') is sufficiently close to (f,v), then Σ is transverse to v'. Furthermore, $\Gamma_{01}(f',v') \cup R_-(\gamma) \subset M_-$ and $\Gamma_{23}(f',v') \cup R_+(\gamma) \subset M_+$. Indeed, no critical point can pass through Σ as long as Σ is transverse to the vector field, and every critical point of f is stable.

We now construct a local trivialization $\phi\colon U \times C^\infty(\Sigma,\mathbb{R}) \to \pi_k^{-1}(U)$. Choose a smooth function $h\colon M \to I$ such that $h^{-1}(0) = R(\gamma)$. Then ϕ is defined by the formula $\phi((f',v'),s) = \Sigma + s$, where $(f',v') \in U$ and $s \in C^\infty(\Sigma,\mathbb{R})$, and we view $\Sigma(f',v')$ as an affine space over $C^\infty(\Sigma,\mathbb{R})$ via flowing along hv'. The trivializations ϕ define the topology on $E(M,\gamma)$ that makes $\pi_k\colon E_k^m(M,\gamma) \to B_k^m(M,\gamma)$ into a fiber bundle, and is compatible with the topology on each fiber $\Sigma(f,v)$. □

As a corollary, $\pi_k\colon E_k^m(M,\gamma) \to B_k^m(M,\gamma)$ is a Serre fibration. In particular, it satisfies the path-lifting property. For example, together with Corollary 6.10, for any family of Morse–Smale vector fields $\{\,(f_\lambda, v_\lambda) : \lambda \in \Lambda\,\}$, there is a corresponding family of surfaces $\{\,\Sigma_\lambda : \lambda \in \Lambda\,\}$ such that $\Sigma_\lambda \in \Sigma(f_\lambda, v_\lambda)$ for every $\lambda \in \Lambda$. As the fiber $C^\infty(\Sigma,\mathbb{R})$ is contractible, we can even extend a family of splitting surfaces defined over a closed subset of Λ.

6.2. Codimension-0

In the previous section, we described how to obtain a contractible space of Heegaard splittings of the sutured manifold (M,γ) from a separable pair $(f,v) \in \mathcal{FV}_{\leq 2}(M,\gamma)$. If we also assume that (f,v) is Morse–Smale, and we make an additional discrete choice, then we can enhance these splittings to sutured diagrams. In the opposite direction, we also show that every diagram of (M,γ) with $\boldsymbol{\alpha} \pitchfork \boldsymbol{\beta}$ arises from a particularly simple Morse–Smale pair (f,v), and the space of such pairs is connected.

By Proposition 6.7, every $\Sigma \in \Sigma(f,v)$ is a Heegaard surface of (M,γ). If (f,v) is Morse–Smale, then, for every $p \in C_1(f)$ and $q \in C_2(f)$, the intersections $W^u(p) \cap \Sigma$ and $W^s(q) \cap \Sigma$ are embedded circles, and $W^u(p) \cap \Sigma$ is transverse to $W^s(q) \cap \Sigma$.

Definition 6.12. Suppose that $(f,v) \in \mathcal{FV}_0(M,\gamma)$, and let $\Sigma \in \Sigma(f,v)$. Then the triple $H(f,v,\Sigma) = (\Sigma, \boldsymbol{\alpha}, \boldsymbol{\beta})$ is defined by taking the α-curves to be $W^u(p) \cap \Sigma$ for $p \in C_1(f)$, and the β-curves to be $W^s(q) \cap \Sigma$ for $q \in C_2(f)$.

In general, $H(f,v,\Sigma)$ is not a diagram of (M,γ) as $\boldsymbol{\alpha}$ and $\boldsymbol{\beta}$ might have too many components. We will refer to such diagrams as *overcomplete*, as we can remove some components of $\boldsymbol{\alpha}$ and $\boldsymbol{\beta}$ to get a sutured diagram of (M,γ).

Definition 6.13. Let (M,γ) be a sutured manifold. We say that $(\Sigma, \boldsymbol{\alpha}, \boldsymbol{\beta})$ is an *overcomplete diagram* of (M,γ) if the following hold:

(1) $\Sigma \subset M$ is an oriented surface with $\partial\Sigma = s(\gamma)$ as oriented 1-manifolds.
(2) The components of the 1-manifold $\boldsymbol{\alpha} \subset \Sigma$ bound disjoint disks to the negative side of Σ, and the components of the 1-manifold $\boldsymbol{\beta} \subset \Sigma$ bound disjoint disks to the positive side of Σ.
(3) If we compress Σ along $\boldsymbol{\alpha}$, we get a surface isotopic to $R_-(\gamma)$ relative to γ, plus some 2-spheres that bound disjoint balls in M.
(4) If we compress Σ along $\boldsymbol{\beta}$, we get a surface isotopic to $R_+(\gamma)$ relative to γ, plus some 2-spheres that bound disjoint balls in M.

Overcomplete diagrams specify handle decompositions of (M,γ) that also include 0- and 3-handles. Note that $\boldsymbol{\alpha}$ and $\boldsymbol{\beta}$ might fail to be attaching sets because $\Sigma \setminus \boldsymbol{\alpha}$ and $\Sigma \setminus \boldsymbol{\beta}$ can have some components disjoint from $\partial\Sigma$; however, all such components are planar.

To actually make the overcomplete diagram $H(f,v,\Sigma)$ into a usual Heegaard diagram of (M,γ), in addition to assuming that (f,v) is Morse–Smale, we also need to make a discrete choice. The Morse–Smale condition rules out flows between two index i critical points for $i \in \{1,2\}$, hence every point of $C_1(f) \cup C_2(f)$ has valence 2 in the graph $\Gamma(f,v)$.

Definition 6.14. Let $\Gamma_-(f,v)$ be the graph obtained from $\Gamma_{01}(f,v)$ by identifying all vertices lying in $R_-(\gamma)$, and deleting the vertices at $C_1(f)$ (and merging the two adjacent edges into one edge). So the vertices of $\Gamma_-(f,v)$ are the points of $C_0(f)$, plus at most one vertex for $R_-(\gamma)$, and its edges correspond to $W^s(p)$ for $p \in C_1(f)$. In other words, $\Gamma_-(f,v)$ is obtained from the relative CW complex $(\Gamma_{01}(f,v) \cup R_-(\gamma), R_-(\gamma))$ by taking the factor CW complex $(\Gamma_{01}(f,v) \cup R_-(\gamma))/R_-(\gamma)$, and removing the vertices at $C_1(f)$. Similarly, the graph $\Gamma_+(f,v)$ is obtained by collapsing $\Gamma_{23}(f,v) \cap R_+(\gamma)$ to a single point, and deleting the vertices at $C_2(f)$. The edges of $\Gamma_+(f,v)$ correspond to $W^u(q)$ for $q \in C_2(f)$.

Definition 6.15. Suppose that $(f,v) \in \mathcal{FV}_0(M,\gamma)$. Let $T_\pm$ be a spanning tree of $\Gamma_\pm(f,v)$, and choose a splitting surface $\Sigma \in \Sigma(f,v)$. Then the sutured diagram $H(f,v,\Sigma,T_-,T_+)$ is defined by taking the α-curves to be $W^u(p) \cap \Sigma$, where $p \in C_1(f)$ and $W^s(p)$ is not an edge of T_-. Similarly, the β-curves are the intersections $W^s(q) \cap \Sigma$, where $q \in C_2(f)$ and $W^u(q)$ is not an edge of T_+.

For brevity, we will often write $H(f,v,\Sigma,T_\pm)$ for $H(f,v,\Sigma,T_+,T_-)$. Of course, a different choice of T_- gives a diagram that is α-equivalent to the original, while changing T_+ gives a diagram that is β-equivalent.

In the opposite direction, given a Heegaard surface Σ for (M, γ), we will show that one can find a particularly nice pair $(f, v) \in \mathcal{FV}_0(M, \gamma)$ such that $\Sigma \in \Sigma(f, v)$.

DEFINITION 6.16. We say that a Morse function f on (M, γ) is *simple* if

(1) $C_i(f) = \emptyset$ for $i \in \{0, 3\}$,
(2) $f(p) < 0$ for every $p \in C_1(f)$,
(3) $f(q) > 0$ for every $q \in C_2(f)$.

We call a pair $(f, v) \in V(M, \gamma)$ *simple* if f is simple and, in addition,

(4) for every $p \in C_1(f)$, there is a local coordinate system (x_1, x_2, x_3) around p in which $f = -x_1^2 + x_2^2 + x_3^2 + f(p)$ and v has coordinates $(-2x_1, 2x_2, 2x_3)$,
(5) for every $q \in C_2(f)$, there is a local coordinate system (x_1, x_2, x_3) around q in which $f = -x_1^2 - x_2^2 + x_3^2 + f(q)$ and v has coordinates $(-2x_1, -2x_2, 2x_3)$.

Let f be a simple Morse function. Then observe that, for any gradient-like vector field v for f, the pair (f, v) is separable. Indeed, $f(p) < f(q)$ for every $p \in C_1(f)$ and $q \in C_2(f)$, so there is no flow-line of v from q to p. Furthermore, the surface $\Sigma = f^{-1}(0)$ is a Heegaard surface that separates (f, v). If, in addition, (f, v) is Morse–Smale, then we can uniquely complete this to a diagram $H(f, v) = (\Sigma, \boldsymbol{\alpha}, \boldsymbol{\beta})$ of (M, γ) using Definition 6.15. Indeed, since $\Gamma_\pm(f, v)$ is a wedge of circles, it has a unique spanning tree $T_\pm$ consisting of a single vertex. The diagram $H(f, v)$ is then $H(f, v, \Sigma, T_-, T_+)$. More explicitly, the α-curves are $W^u(p) \cap \Sigma$ for $p \in C_1(f)$, and the β-curves are $W^s(q) \cap \Sigma$ for $q \in C_2(f)$.

The following result is basically standard Morse theory, but since it provides the crucial link between sutured Heegaard diagrams and Morse functions, we give a complete proof.

PROPOSITION 6.17. *Let $(\Sigma, \boldsymbol{\alpha}, \boldsymbol{\beta})$ be a diagram of the sutured manifold (M, γ), and suppose that $\boldsymbol{\alpha} \pitchfork \boldsymbol{\beta}$. Then there exists a simple pair $(f, v) \in \mathcal{FV}_0(M, \gamma)$ such that $H(f, v) = (\Sigma, \boldsymbol{\alpha}, \boldsymbol{\beta})$.*

PROOF. Given an arbitrary attaching set $\boldsymbol{\delta} \subset \Sigma$, recall that $C(\boldsymbol{\delta})$ is the sutured compression body obtained by attaching 2-handles to $\Sigma \times I$ along $\boldsymbol{\delta} \times \{1\}$. We are going to construct a Morse function $f_\delta \colon C(\boldsymbol{\delta}) \to I$ with only index 2 critical points, and a gradient-like vector field v_δ for f_δ.

Consider the Morse function

$$h(x) = -x_1^2 - x_2^2 + x_3^2 + \frac{1}{2},$$

and its gradient $v(x) = (-2x_1, -2x_2, 2x_3)$ on the unit disk D^3. Our model 2-handle will be $Z = h^{-1}(I)$; this is a 3-manifold with boundary and corners. Let $Z^- = h^{-1}(0)$ and $Z^+ = h^{-1}(1)$. The boundary of Z is $Z^- \cup Z^+ \cup A$, where Z^- is a connected hyperboloid, hence topologically an annulus. The surface Z^+ is the disjoint union of two disks, while $A = Z \cap S^2$ is the disjoint union of two annuli. The attaching circle of Z is the curve $a = Z^- \cap \{x_3 = 0\}$. We isotope v near A such that it stays gradient-like for h, and becomes tangent to A. The function h has a single non-degenerate critical point of index 2 at the origin, with stable manifold $W^s(0) = Z \cap \{x_3 = 0\}$.

Pick an open regular neighborhood N of $\boldsymbol{\delta}$, and let $R = (\Sigma \setminus N) \times I$. We define f_δ on R to be the projection t onto the I-factor, and v_δ on R is simply $\partial/\partial t$. If the components of $\boldsymbol{\delta}$ are $\delta_1, \dots, \delta_d$, then we denote by N_i the component of N containing δ_i. For each $i \in \{1, \dots, d\}$, take a copy Z_i of Z, together with the function h_i and the vector field v_i constructed above. We glue Z_i to R using a diffeomorphism $A_i \to \partial N_i \times I$ that maps the circles $h^{-1}(t) \cap A$ to $\partial N_i \times \{t\}$ for every $t \in I$. So we can extend f_δ to Z_i with h_i and v_δ with v_i. After gluing $Z_1, \dots, Z_d$ to R, we get a compression body diffeomorphic to $C(\boldsymbol{\delta})$, together with the required pair (f_δ, v_δ). Note that here we identify $(\Sigma \setminus N) \times \{0\}$ with $\Sigma \setminus N$ and Z_i^- with $\overline{N}_i$, so that $C_-(\boldsymbol{\delta}) = \Sigma$. In addition, the attaching circle a_i of Z_i is identified with α_i. Let p_i be the center of the 2-handle Z_i. By construction, the stable manifold $W^s(p_i) \cap \Sigma = \alpha_i$.

The surface Σ cuts (M, γ) into two sutured compression bodies. Call these C_- and C_+, such that $R_\pm(\gamma) \subset C_\pm$. There are diffeomorphisms $d_- \colon C_- \to C(\boldsymbol{\alpha})$ and $d_+ \colon C_+ \to C(\boldsymbol{\beta})$ that are the identity on Σ. Then we define the Morse function f on (M, γ) by taking $-f_\alpha \circ d_-$ on C_- and $f_\beta \circ d_+$ on C_+, and smoothing along Σ. The vector field v that agrees with $-(d_-)_*^{-1} \circ v_\alpha \circ d_-$ on C_- and with $(d_+)_*^{-1} \circ v_\beta \circ d_+$ on C_+ is gradient-like for f, and is Morse–Smale since $\boldsymbol{\alpha} \pitchfork \boldsymbol{\beta}$. It follows from the construction that $H(f, v) = (\Sigma, \boldsymbol{\alpha}, \boldsymbol{\beta})$. □

Notice that the above construction is almost completely canonical, in the sense that the various choices can be easily deformed into each other. In fact, we have the following stronger statement.

PROPOSITION 6.18. *Let $(\Sigma, \boldsymbol{\alpha}, \boldsymbol{\beta})$ be a diagram of the sutured manifold (M, γ) such that $\boldsymbol{\alpha} \pitchfork \boldsymbol{\beta}$, and recall that $\mathcal{FV}_0(M, \gamma)$ is endowed with the C^∞-topology. Then the subspace of simple pairs $(f, v) \in \mathcal{FV}_0(M, \gamma)$ for which $H(f, v) = (\Sigma, \boldsymbol{\alpha}, \boldsymbol{\beta})$ is connected.*

PROOF. Suppose that the simple pairs (f, v) and (g, w) in $\mathcal{FV}_0(M, \gamma)$ satisfy $H(f, v) = (\Sigma, \boldsymbol{\alpha}, \boldsymbol{\beta})$ and $H(g, w) = (\Sigma, \boldsymbol{\alpha}, \boldsymbol{\beta})$.

As in Proposition 6.17, the surface Σ cuts (M, γ) into two compression bodies C_- and C_+, such that $R_\pm(\gamma) \subset C_\pm$. We will describe how to connect $(f, v)|_{C_+}$ and $(g, w)|_{C_+}$; the deformation on C_- is analogous. Since Σ remains the zero level set and v, w stay transverse to Σ throughout, it is easy to glue the deformations on C_+ and C_- together. We construct this deformation in several steps.

First, some terminology. Let $\{\, d_t : t \in I \,\}$ be an isotopy of C_+ such that $d_0 = \mathrm{Id}_{C_+}$. Then we call the family

$$(f_t, v_t) = (f \circ d_t^{-1}, (d_t)_* \circ v \circ d_t^{-1})$$

the isotopy of (f, v) along d_t. Note that v_t is a gradient-like vector field for f_t.

Step 1. In this step, we move the stable manifolds of (f, v) until they coincide with the stable manifolds of (g, w). We denote the components of $\boldsymbol{\beta}$ by $\beta_1, \dots, \beta_k$. In addition, let $C_2(f) = \{\, q_1, \dots, q_k \,\}$ and $C_2(g) = \{\, q_1', \dots, q_k' \,\}$, enumerated such that

$$W^s(q_i, v) \cap \Sigma = W^s(q_i', w) \cap \Sigma = \beta_i.$$

By Lemma 2.9, we have $\pi_2(C_+) = 0$. Hence, using cut-and-paste techniques, the disks $W^s(q_i, v)$ and $W^s(q_i', w)$ are isotopic relative to their boundary. So there is an isotopy $\{\, d_t : t \in I \,\}$ of C_+ fixing ∂C_+ such that $d_0 = \mathrm{Id}_{C_+}$ and $d_1(W^s(q_i, v)) = W^s(q_i', w)$; furthermore, $d_1(q_i) = q_i'$. Isotoping (f, v) along d_t, we get a path of simple pairs (f_t, v_t) in $\mathcal{FV}_0(M, \gamma)$, all compatible with the diagram $(\Sigma, \boldsymbol{\alpha}, \boldsymbol{\beta})$. Replacing (f, v) with (f_1, v_1), we can assume that $W^s(q_i, v) = W^s(q_i', w)$ and $q_i = q_i'$ for every $i \in \{\, 1 \dots, k \,\}$. The further deformation of (f, v) will preserve these properties, so from now on we will write $W^s(q_i)$ for every $q_i \in C_2(f) = C_2(g)$.

Step 2. We now isotope (f,v) until it coincides with (g,w) in a neighborhood of the critical points, without ruining what we have already achieved in Step 1. Let $i \in \{1\dots,k\}$. Since both (f,v) and (g,w) are simple, there are balls N_1 and N_2 centered at q_i and coordinate systems $x\colon N_1 \to \mathbb{R}^3$ and $y\colon N_2 \to \mathbb{R}^3$ such that

$$f = -x_1^2 - x_2^2 + x_3^2 + f(q_i),$$

and v has coordinates $(-2x_1, -2x_2, 2x_3)$ in N_1, while

$$g = -y_1^2 - y_2^2 + y_3^2 + g(q_i),$$

and w has coordinates $(-2y_1, -2y_2, 2y_3)$ in N_2. Choose an $\varepsilon > 0$ so small that the disks $D_1 = \{\,|x| \le \varepsilon\,\}$ and $D_2 = \{\,|y| \le \varepsilon\,\}$ both lie in $N_1 \cap N_2$. Consider the diffeomorphism $d\colon D_1 \to D_2$ given by the formula $y^{-1} \circ x$. Then $d(D_1 \cap W^s(q_i)) = D_2 \cap W^s(q_i)$, as $D_1 \cap W^s(q_i)$ is given by the equation $x_3 = 0$, while $D_2 \cap W^s(q_i)$ by $y_3 = 0$. We can choose an isotopy $e_t\colon D_1 \to N_1 \cap N_2$ such that $e_0 = \mathrm{Id}_{D_1}$ and $e_1 = d$; furthermore,

$$e_t(D_1 \cap W^s(q_i)) \subset W^s(q_i),$$

and $e_t(q_i) = q_i$ for every $t \in I$. This can be extended to an isotopy $d_t\colon C_+ \to C_+$ such that $d_t|_{D_1} = e_t$, the diffeomorphism d_t is the identity outside $N_1 \cap N_2$, and $d_t(W^s(q_i)) = W^s(q_i)$. If we isotope (f,v) along d_t, we get a pair (f_1, v_1) that agrees with (g,w) in D_2. Repeating this process for every q_i, we can assume that (f,v) and (g,w) agree in a neighborhood N of all the critical points $q_1, \dots, q_k$ (where N is the union of the disks D_2 for each q_i).

Step 3. In this step, we arrange that

$$v|_{W^s(q_i)} = w|_{W^s(q_i)}$$

for every $i \in \{1,\dots,k\}$. By Step 2, we know that v and w coincide on the disk $B = N \cap W^s(q_i)$, and that they are transverse to ∂B. Let A be the annulus $W^s(q_i) \setminus B$. Take a regular neighborhood of $W^s(q_i)$ of the form $W^s(q_i) \times [-1,1]$, where $W^s(q_i)$ is identified with $W^s(q_i) \times \{0\}$. We are going to construct an isotopy of C_+ that is supported in $A \times [-1,1]$ that takes v to w. For every $p \in \partial B$, we denote by $\nu_v(p,t)$ the flow-line of $-v$ starting at p and ending at $\partial W^s(q_i)$. Here t lies in some interval $[0, T(v,p)]$. Similarly, $\nu_w(p,t)$ denotes the flow-line of $-w$ starting at p and defined for $t \in [0, T(w,p)]$. After smoothly rescaling v inside $A \times [-1,1]$ such that it is unchanged in a neighborhood of $\partial(A \times [-1,1])$, we can assume that $T(v,p) = T(w,p)$ for every $p \in \partial B$. Let $a\colon A \to A$ be the diffeomorphism defined by the formula

$$a(\nu_v(p,t)) = \nu_w(p,t)$$

for $t \in [0, T(v,p)]$. This has the property that $a_* \circ v \circ a^{-1} = w$. There is an isotopy $\{\,a_t : t \in I\,\}$ of A that is fixed on ∂B, and such that $a_0 = \mathrm{Id}_A$ and $a_1 = a$. We extend this to $A \times [-1,1]$ by the formula $a_t(x,s) = (a_{r(s)t}(x), s)$, where $r\colon \mathbb{R} \to I$ is a bump function that is zero outside $[-1,1]$, and such that $r(0) = 1$. Finally, we extend a_t to the whole of C_+ as the identity. Then, isotoping (f,v) along a_t, we get a family $\{\,(f_t, v_t) : t \in I\,\}$ such that $v_1|_{W^s(q_i)} = w|_{W^s(q_i)}$. Note that $W^s(q_i)$ is invariant under a_t. Furthermore, even though a_t is not the identity on Σ, the field v_t stays transverse to Σ throughout. In fact, v_t can be made invariant under a_t if we first make v and w agree in a neighborhood of ∂A; so we can glue the deformation with the one on C_-.

Note that we do not claim that $f = g$ anywhere outside a neighborhood of the critical points. We will return to this in the last step.

Step 4. We now make v and w agree on a product neighborhood $W^s(q_i) \times [-1,1]$ of every stable manifold $W^s(q_i)$. Fix $i \in \{1, \dots, k\}$, and let the ball B and the annulus A be as in Step 3. We already know that (f,v) and (g,w) agree on a neighborhood N of B. Since v and w agree on A and have no zeroes there, there is a thin product neighborhood of $W^s(q_i)$ diffeomorphic to $W^s(q_i) \times [-2,2]$ such that, in this neighborhood, the linear homotopy $(1-t)v + tw$ from v to w stays gradient-like for f throughout. In addition, we choose this neighborhood so thin that $B \times [-2,2] \subset N$. Let $\vartheta\colon \mathbb{R} \to I$ be a smooth function that is zero outside $[-2,2]$, and is identically one in $[-1,1]$. Then we define the isotopy v_t of v to be the identity outside $W^s(q_i) \times [-2,2]$, and

$$v_t(x,s) = (1 - \vartheta(s)\,t)\,v + (\vartheta(s)\,t)\,w$$

for every $(x,s) \in W^s(q_i) \times [-2,2]$. Then v_t is gradient-like for f for every $t \in I$, and v_1 agrees with w on $W^s(q_i) \times [-1,1]$.

Step 5. In this step, we homotope v to w on the rest of C_+. Let P be the manifold obtained from

$$C_+ \setminus \bigcup_{i=1}^{k} (W^s(q_i) \times [-\varepsilon, \varepsilon])$$

by rounding the corners. Here, we choose ε so small that (after possibly a small perturbation) v and w point into P along $P_- = \partial P \setminus \mathrm{Int}(\gamma \cup R_+(\gamma))$. Notice that P_- consists of $\Sigma \setminus \bigcup_{i=1}^k (\alpha_i \times [-\varepsilon,\varepsilon])$ and $W^s(q_i) \times \{-\varepsilon, \varepsilon\}$ for $i \in \{1,\dots,k\}$. By construction, the vector fields v and w point into P along Σ and $B \times \{-\varepsilon, \varepsilon\}$ for every disk B of the form $N \cap W^s(q_i)$. As v and w are tangent to the annuli A, such an ε and perturbation clearly exist. Note that v and w coincide along P_-.

The sutured manifold $(P, \gamma \cap P)$ is diffeomorphic to the product sutured manifold $(R_+(\gamma) \times I, \partial R_+(\gamma) \times I)$. For every $x \in P_-$, let $\phi_v(x,t)$ be the flow-line of v starting at x and defined for $t \in [0, T(v,x)]$. Similarly, let $\phi_w(x,t)$ be the flow-line of w starting at x and defined for $t \in [0, T(w,x)]$. After smoothly rescaling v inside P such that it is unchanged in a neighborhood of ∂P, we can assume that $T(v,x) = T(w,x)$ for every $x \in P_-$. As in Step 3, we define a diffeomorphism $d\colon P \to P$ by the formula

$$d(\phi_v(x,t)) = \phi_w(x,t)$$

for $t \in [0, T(v,x)]$. This satisfies $d_* \circ v \circ d^{-1} = w$, and d is the identity on P_- and $\gamma \cap P$. Since any diffeomorphism of a product sutured manifold $(S \times I, \partial S \times I)$ that fixes $S \times \{0\}$ and $\partial S \times I$ is isotopic to the identity through such diffeomorphisms, there is an isotopy d_t of P that fixes P_- and $\gamma \cap P$, and $d_0 = \mathrm{Id}_P$. Since v and w agree on each $W^s(q_i) \times [-1,1]$, we can extend d_t to every $W^s(q_i) \times [-\varepsilon, \varepsilon]$ as the identity. If we isotope (f,v) along d_t, we get a path of pairs (f_t, v_t) such that $v_1 = w$.

Step 6. In this final step, we achieve $f = g$. For this, the linear homotopy $\{ f_t = (1-t)f + tg : t \in I \}$ works. Indeed, as f and g coincide in a neighborhood N of the critical points, $f_t|_N = f|_N$ for every $t \in I$. In addition,

$$v(f_t) = (1-t)v(f) + tv(g) = (1-t)v(f) + tw(g) > 0$$

away from $\{q_1, \dots, q_k\}$, the common critical set of f and g. So f_t has the same index 2 non-degenerate critical points as f, and is hence Morse. Furthermore, v is a gradient-like vector field for f_t, and the pair (f_t, v) is simple for every $t \in I$. $\square$

6.3. Codimension-1: Overcomplete diagrams

We start this section by proving a type of isotopy extension lemma for families of Heegaard diagrams.

LEMMA 6.19. *Suppose that*

$$\{ (\Sigma_t, \boldsymbol{\alpha}_t, \boldsymbol{\beta}_t) : t \in I \}$$

is a smooth 1-parameter family of possibly overcomplete Heegaard diagrams in (M, γ) *such that* $\boldsymbol{\alpha}_t \pitchfork \boldsymbol{\beta}_t$ *for every* $t \in I$. *Then there is an isotopy* $D \colon M \times I \to M$ *such that*

$$d_t(\Sigma_0, \boldsymbol{\alpha}_0, \boldsymbol{\beta}_0) = (\Sigma_t, \boldsymbol{\alpha}_t, \boldsymbol{\beta}_t)$$

for every $t \in I$, *and* d_t *fixes* ∂M *pointwise (where* $d_t = D(\cdot, t)$*). In particular,* $d_1|_{\Sigma_0} \colon \Sigma_0 \to \Sigma_1$ *is isotopic to the identity in* M. *The space of such isotopies is contractible, so the space of diffeomorphisms that arise as* d_1 *for such an isotopy* D *is path-connected.*

An analogous statement holds if $\{ \Sigma_t : t \in I \}$ *is a 1-parameter family of Heegaard surfaces of* (M, γ). *In particular, there is an induced diffeomorphism* $d_1 \colon \Sigma_0 \to \Sigma_1$, *well-defined up to isotopy.*

PROOF. In $M \times I$, consider the submanifold

$$\Sigma_* = \bigcup_{t \in I} \Sigma_t \times \{t\},$$

which, in turn, contains the submanifolds

$$\boldsymbol{\alpha}_* = \bigcup_{t \in I} \boldsymbol{\alpha}_t \times \{t\} \text{ and } \boldsymbol{\beta}_* = \bigcup_{t \in I} \boldsymbol{\beta}_t \times \{t\}.$$

(The fact that these are smooth submanifolds is in fact our definition that the family of Heegaard diagrams is smooth.) Let $\mathcal{F}$ be the horizontal foliation of $M \times I$ by leaves $M \times \{t\}$, and we coorient $\mathcal{F}$ by $\partial/\partial t$. The condition $\boldsymbol{\alpha}_t \pitchfork \boldsymbol{\beta}_t$ implies that $\boldsymbol{\alpha}_* \cap \boldsymbol{\beta}_*$ is a collection of arcs transverse to $\mathcal{F}$. Furthermore, $\boldsymbol{\alpha}_*$, $\boldsymbol{\beta}_*$, and Σ_* are also transverse to $\mathcal{F}$.

Pick a smooth vector field ν in the tangent bundle $T(\boldsymbol{\alpha}_* \cap \boldsymbol{\beta}_*)$ positively transverse to $\mathcal{F}$. This can be extended to first a vector field in $T\boldsymbol{\alpha}_*$ and $T\boldsymbol{\beta}_*$ positively transverse to $\mathcal{F}$, then to a field in $T\Sigma_*$ positively transverse to $\mathcal{F}$ such that $\nu_{s(\gamma) \times I} = \partial/\partial t$. Finally, extend the vector field to $M \times I$ positively transverse to $\mathcal{F}$ such that $\nu|_{\partial M} = \partial/\partial t$. We also denote this extension by ν. After normalizing ν such that $\nu(t) = 1$, we can assume that its flow preserves the foliation $\mathcal{F}$. Then the diffeomorphism $d_t \colon M \to M$ is defined by flowing along ν from $M \times \{0\}$ to $M \times \{t\}$. By construction, $d_t(\Sigma_0, \boldsymbol{\alpha}_0, \boldsymbol{\beta}_0) = (\Sigma_t, \boldsymbol{\alpha}_t, \boldsymbol{\beta}_t)$. Note that the embedding $\iota_0 \colon \Sigma_0 \hookrightarrow M$ is ambient isotopic to $\iota_1 \circ d_1 \colon \Sigma_0 \hookrightarrow M$ relative to $s(\gamma)$, where ι_1 is the embedding of Σ_1 in M. So d_1 is indeed isotopic to the identity in M.

On the other hand, every isotopy D arises from the above construction. Indeed, given D, take ν to be the velocity vector fields of the curves $t \mapsto (d_t(x), t)$ for $x \in M$. The space of such ν is convex, hence contractible, so the space of such isotopies D is also contractible.

The proof of the last statement about families of Heegaard surfaces is completely analogous, but simpler as our isotopies now do not have to preserve sets of attaching curves $\boldsymbol{\alpha}_t$ and $\boldsymbol{\beta}_t$. □

LEMMA 6.20. *Let* $\{\mathcal{H}_t : t \in I\}$ *and* $\{\mathcal{H}'_t : t \in I\}$ *be 1-parameter families of possibly overcomplete Heegaard diagrams of* (M, γ)*, both connecting* $\mathcal{H}_0$ *and* $\mathcal{H}_1$*. If the two families are homotopic relative to their endpoints, then the induced diffeomorphisms* d_1*,* $d'_1 \colon M \to M$ *(in the sense of Lemma 6.19) are isotopic through diffeomorphisms mapping* $\mathcal{H}_0$ *to* $\mathcal{H}_1$*. An analogous statement holds for homotopic families of Heegaard surfaces* $\{\Sigma_t : t \in I\}$ *and* $\{\Sigma'_t : t \in I\}$*; i.e., the induced diffeomorphisms* d_1*,* $d'_1 \colon \Sigma_0 \to \Sigma_1$ *are isotopic.*

PROOF. Let

$$\mathcal{H}_{t,u} = (\Sigma_{t,u}, \boldsymbol{\alpha}_{t,u}, \boldsymbol{\beta}_{t,u})$$

for $(t,u) \in I \times I$ be the homotopy between $\{\mathcal{H}_t\}$ and $\{\mathcal{H}'_t\}$; i.e., $\mathcal{H}_{t,0} = \mathcal{H}_t$ and $\mathcal{H}_{t,1} = \mathcal{H}'_t$ for $t \in I$, while $\mathcal{H}_{i,u} = \mathcal{H}_i$ for $i \in \{0,1\}$ and $u \in I$. As in the proof of Lemma 6.19, we can construct a vector field ν on $M \times I \times I$ such that $\nu(u) = 0$, $\nu(t) = 1$ (in particular, it is transverse to the foliation of $M \times I \times I$ with leaves $M \times \{t\} \times I$), and which is tangent to the submanifolds $\bigcup_{t,u \in I} \Sigma_{t,u}$, $\bigcup_{t,u\in I} \boldsymbol{\alpha}_{t,u}$, and $\bigcup_{t,u \in I} \boldsymbol{\beta}_{t,u}$. Then the flow of ν defines a diffeomorphism

$$g_u \colon M \times \{0\} \times \{u\} \to M \times \{1\} \times \{u\}$$

that maps $\mathcal{H}_{0,u} = \mathcal{H}_0$ to $\mathcal{H}_{1,u} = \mathcal{H}_1$ for every $u \in I$. Notice that $g_0 = d_1$ and $g_1 = d'_1$ (up to isotopy). Hence $\{g_u \colon u \in I\}$ provides the required isotopy between the diffeomorphisms d_1 and d'_1. □

LEMMA 6.21. *Suppose that*

$$\{(f_t, v_t) \in \mathcal{FV}_0(M,\gamma) : t \in I\}$$

is a 1-parameter family of gradient-like vector fields, and let $\Sigma_i \in \Sigma(f_i, v_i)$ *be a Heegaard surface of* (M, γ) *for* $i \in \{0,1\}$*. Choose a spanning tree* $T^0_\pm$ *of* $\Gamma_\pm(f_0, v_0)$*. The isotopy* $\Gamma(f_t, v_t)$ *takes* $T^0_\pm$ *to a spanning tree* $T^1_\pm$ *of* $\Gamma_\pm(f_1, v_1)$*, and consider the diagrams* $(\Sigma_0, \boldsymbol{\alpha}_0, \boldsymbol{\beta}_0) = H(f_0, v_0, \Sigma_0, T^0_\pm)$ *and* $(\Sigma_1, \boldsymbol{\alpha}_1, \boldsymbol{\beta}_1) = H(f_1, v_1, \Sigma_1, T^1_\pm)$*. Then there is a (non-unique) induced diffeomorphism*

$$d \colon (\Sigma_0, \boldsymbol{\alpha}_0, \boldsymbol{\beta}_0) \to (\Sigma_1, \boldsymbol{\alpha}_1, \boldsymbol{\beta}_1)$$

isotopic to the identity in M*, and the space of such diffeomorphisms is path-connected. If we do not pick spanning trees, we obtain a similar statement for overcomplete diagrams.*

REMARK 6.22. If $T^0_\pm$ and $T^1_\pm$ are not related as above, then one can get from $(\Sigma_0, \boldsymbol{\alpha}_0, \boldsymbol{\beta}_0)$ to $(\Sigma_1, \boldsymbol{\alpha}_1, \boldsymbol{\beta}_1)$ via a diffeomorphism isotopic to the identity in M, an α-equivalence, and a β-equivalence.

PROOF. Note that, if $\Sigma \in \Sigma(f_t, v_t)$ for some $t \in I$, then Σ also separates (f_s, v_s) for every s sufficiently close to t. Indeed, each (f_s, v_s) is Morse–Smale, hence separable, so Σ stays separating as long as v_s is transverse to Σ, which is an open condition. By the compactness of I, there is a sequence

$$0 = t_0 < t_1 < \cdots < t_n = 1,$$

and surfaces $\Sigma_{t_i} \in \Sigma(f_{t_i}, v_{t_i})$ for every $i \in \{0, \dots, n-1\}$ such that $\Sigma_{t_i} \in \Sigma(f_s, v_s)$ for every $s \in [t_i, t_{i+1}]$.

By the discussion preceding Proposition 6.7, we can view $\Sigma(f_t, v_t)$ as an affine space over $C^\infty(\Sigma)$ for any $\Sigma \in \Sigma(f_t, v_t)$. For this, choose a smooth function $h \colon M \to I$ such that $h^{-1}(0) = R(\gamma)$, and let $i \in \{2, \dots, n\}$. Since both $\Sigma_{t_{i-1}}$, $\Sigma_{t_i} \in$

$\Sigma(f_{t_i}, v_{t_i})$, we can talk about the difference $d_{\Sigma_{t_i},\Sigma_{t_{i-1}}} \in C^\infty(\Sigma_{t_{i-1}})$, obtained by flowing along hv_{t_i}. Let $\varphi_i\colon \mathbb{R} \to I$ be a smooth function such that $\varphi_i(t) = 0$ for $t \le t_{i-1}$ and $\varphi_i(t) = 1$ for $t \ge t_i$. For $t \in [t_{i-1}, t_i]$, let

$$\Sigma_t = \Sigma_{t_{i-1}} + \varphi_i(t)\, d_{\Sigma_{t_i},\Sigma_{t_{i-1}}},$$

where the sum is taken using the flow of hv_t. Then Σ_t is a smooth 1-parameter family of surfaces connecting Σ_0 to Σ_1 such that $\Sigma_t \in \Sigma(f_t, v_t)$ for every $t \in I$. (Note that this path-lifting also follows from Proposition 6.11, which claims that $E_k^m(M,\gamma) \to B_k^m(M,\gamma)$ is a fibre bundle with connected fiber $C^\infty(\Sigma, \mathbb{R})$.)

The isotopy $\{\,\Gamma(f_s, v_s) : 0 \le s \le t\,\}$ takes $T_\pm^0$ to a spanning tree $T_\pm^t$ of $\Gamma_\pm(f_t, v_t)$. Then

$$(\Sigma_t, \boldsymbol{\alpha}_t, \boldsymbol{\beta}_t) = H(f_t, v_t, \Sigma_t, T_\pm^t)$$

provides a smooth 1-parameter family of diagrams connecting $(\Sigma_0, \boldsymbol{\alpha}_0, \boldsymbol{\beta}_0)$ and $(\Sigma_1, \boldsymbol{\alpha}_1, \boldsymbol{\beta}_1)$. Since (f_t, v_t) is Morse–Smale for every $t \in I$, we have $\boldsymbol{\alpha}_t \pitchfork \boldsymbol{\beta}_t$, so we can apply Lemma 6.19 to obtain an isotopy $D\colon M \times I \to M$ such that $d_t(\Sigma_0, \boldsymbol{\alpha}_0, \boldsymbol{\beta}_0) = (\Sigma_t, \boldsymbol{\alpha}_t, \boldsymbol{\beta}_t)$ for every $t \in I$. If we take d to be d_1, then d is isotopic to the identity in M.

Also by Lemma 6.19, the diffeomorphism d_1 is unique up to isotopy in the space of diffeomorphisms mapping $(\Sigma_0, \boldsymbol{\alpha}_0, \boldsymbol{\beta}_0)$ to $(\Sigma_1, \boldsymbol{\alpha}_1, \boldsymbol{\beta}_1)$ once we fix the family of surfaces Σ_t. For a different family of surfaces $\Sigma_t' \in \Sigma(f_t, v_t)$ connecting Σ_0 and Σ_1, consider the homotopy $\Sigma_{t,u} = \Sigma_t + u d_{\Sigma_t',\Sigma_t}$ for t, $u \in I$ (where the sum means flowing along hv_t). Then $\Sigma_{t,0} = \Sigma_t$ and $\Sigma_{t,1} = \Sigma_t'$ for every $t \in I$. Applying Lemma 6.20 to the homotopy

$$\mathcal{H}_{t,u} = H(f_t, v_t, \Sigma_{t,u}, T_\pm^t),$$

we obtain that d_1 is also unique up to isotopy if we are allowed to vary the path $t \mapsto \Sigma_t$. □

Summarizing the proof of Lemma 6.21, the diffeomorphism induced by the family $\{\,(f_t, v_t) : t \in I\,\}$ is obtained by first picking an arbitrary family of surfaces $\Sigma_t \in \Sigma(f_t, v_t)$, and then applying Lemma 6.19 to the diagrams $H(f_t, v_t, \Sigma_t, T_\pm^t)$. We have the following analogue of Lemma 6.21 for 1-parameter families in $\mathcal{FV}_1(M,\gamma)$, which is somewhat weaker as an element of $\mathcal{FV}_1(M,\gamma)$ does not induce a Heegaard diagram.

LEMMA 6.23. *Let*

$$\{\,(f_t, v_t) \in \mathcal{FV}_1(M,\gamma) : t \in I\,\}$$

be a 1-parameter family, and let $\Sigma_i \in \Sigma(f_i, v_i)$ be a Heegaard surface of (M,γ) for $i \in \{0,1\}$. This family induces a diffeomorphism $d\colon \Sigma_0 \to \Sigma_1$ that is well-defined up to isotopy. Furthermore, there is an isotopy $d_t\colon \Sigma_0 \to \Sigma_t$ for $t \in I$ connecting Id_{Σ_0} and d such that $\Sigma_t \in \Sigma(f_t, v_t)$ for every $t \in I$.

More precisely, if the (f_t, v_t) have an index 1-2 birth-death critical point, then we can choose either $\Sigma_0 \in \Sigma_-(f_0, v_0)$ and $\Sigma_1 \in \Sigma_-(f_1, v_1)$, or $\Sigma_0 \in \Sigma_+(f_0, v_0)$ and $\Sigma_1 \in \Sigma_+(f_1, v_1)$.

PROOF. Just like in the proof of Lemma 6.21, there exists a smooth family of Heegaard surfaces $\{\,\Sigma_t : t \in I\,\}$ such that $\Sigma_t \in \Sigma(f_t, v_t)$ for every $t \in I$. If we apply the second part of Lemma 6.19 to $\{\,\Sigma_t : t \in I\,\}$, we obtain a family of diffeomorphisms $d_t\colon \Sigma_0 \to \Sigma_t$ for $t \in I$, and d_1 is unique up to isotopy. Independence of d_1

from the choice of family $\{ \Sigma_t : t \in I \}$ (up to isotopy) is obtained just like in the proof of Lemma 6.21, except now we apply the second part of Lemma 6.20. □

LEMMA 6.24. *Let $\Lambda\colon I \to \mathcal{FV}_0(M,\gamma)$ be a loop of gradient-like vector fields. Furthermore, let $\Sigma \in \Sigma(f_0, v_0)$ be a Heegaard surface, pick a spanning tree $T_\pm$ of $\Gamma_\pm(f_0,v_0)$, and set $\mathcal{H} = H(f_0, v_0, \Sigma, T_\pm)$. By Lemma 6.21, the loop Λ induces a diffeomorphism $d\colon \mathcal{H} \to \mathcal{H}$. If Λ is null-homotopic in $\mathcal{FV}_0(M,\gamma)$, then d is isotopic to $Id_{\mathcal{H}}$ in the space of diffeomorphisms from $\mathcal{H}$ to itself. If we do not pick a tree $T_\pm$, we obtain an analogous statement for overcomplete diagrams.*

PROOF. Let $L\colon I \times I \to \mathcal{FV}_0(M,\gamma)$ be the null-homotopy; i.e., $L(t,0) = (f_t, v_t)$ for every $t \in I$, and $L(t,u) = (f_0, v_0)$ for $t \in \{0,1\}$ or $u = 1$. By Proposition 6.11, there is a smooth 2-parameter family of Heegaard surfaces $\{ \Sigma_{t,u} : t,\, u \in I \}$ such that $\Sigma_{t,u} \in \Sigma(L(t,u))$ for every $t,\, u \in I$, and $\Sigma_{t,u} = \Sigma$ whenever $t \in \{0,1\}$ or $u = 1$. Furthermore, $T_\pm$ naturally induces a spanning tree $T_\pm^{t,u}$ of $\Gamma_\pm(L(t,u))$ such that $T_\pm^{t,u} = T_\pm$ for $t \in \{0,1\}$ or $u = 1$. So we have a smooth 2-parameter family of diagrams

$$\mathcal{H}_{t,u} = H\left(L(t,u), \Sigma_{t,u}, T_\pm^{t,u}\right)$$

such that $\mathcal{H}_{t,u} = \mathcal{H}$ for $t \in \{0,1\}$ or $u = 1$. Now Lemma 6.20 provides the required isotopy between d and $\mathrm{Id}_{\mathcal{H}}$. □

COROLLARY 6.25. *Let $(f_i, v_i) \in \mathcal{FV}_0(M,\gamma)$ for $i \in \{0,1\}$, and let*

$$\Gamma_0,\, \Gamma_1 \colon I \to \mathcal{FV}_0(M,\gamma)$$

be paths such that $\Gamma_j(i) = (f_i, v_i)$ for $i,\, j \in \{0,1\}$. Given surfaces $\Sigma_i \in \Sigma(f_i, v_i)$ for $i \in \{0,1\}$, consider the (overcomplete) diagrams $\mathcal{H}_i = H(f_i, v_i, \Sigma_i)$. According to Lemma 6.21, the path Γ_i induces a diffeomorphism $d_i\colon \mathcal{H}_0 \to \mathcal{H}_1$. Suppose the paths Γ_0 and Γ_1 are homotopic in $\mathcal{FV}_0(M,\gamma)$ fixing their endpoints. Then d_0 and d_1 are isotopic through diffeomorphisms from $\mathcal{H}_0$ to $\mathcal{H}_1$.

DEFINITION 6.26. The sutured diagram $(\Sigma', \boldsymbol{\alpha}', \boldsymbol{\beta}')$ is obtained from $(\Sigma, \boldsymbol{\alpha}, \boldsymbol{\beta})$ by a *(k,l)-stabilization* if there is a disk $D \subset \Sigma$ and a punctured torus $T \subset \Sigma'$, and there are curves $\alpha \in \boldsymbol{\alpha}'$ and $\beta \in \boldsymbol{\beta}'$ such that

- $\Sigma \setminus D = \Sigma' \setminus T$,
- $\boldsymbol{\alpha} \setminus D = \boldsymbol{\alpha}' \setminus T$ and $\boldsymbol{\beta} \setminus D = \boldsymbol{\beta}' \setminus T$,
- $\boldsymbol{\alpha} \cap D$ and $\boldsymbol{\beta} \cap D$ consist of l and k arcs, respectively, and each component of $\boldsymbol{\alpha} \cap D$ intersects each component of $\boldsymbol{\beta} \cap D$ transversely in a single point,
- $\alpha,\, \beta \subset T$, and they intersect each other transversely in a single point,
- $(\boldsymbol{\alpha}' \setminus \alpha) \cap T$ consists of l parallel arcs, each of which intersects β transversely in a single point,
- $(\boldsymbol{\beta}' \setminus \beta) \cap T$ consists of k parallel arcs, each of which intersects α transversely in a single point,
- for each component of $\boldsymbol{\alpha} \cap D$, there is a corresponding component of $\boldsymbol{\alpha}' \cap T$ with the same endpoints, and similarly for the β-curves.

In the above case, we also say that $(\Sigma, \boldsymbol{\alpha}, \boldsymbol{\beta})$ is obtained from $(\Sigma', \boldsymbol{\beta}', \boldsymbol{\alpha}')$ by a (k,l)-destabilization. Notice that a $(0,0)$-(de)stabilization agrees with the "simple" (de)stabilization of Definition 2.18. The two diagrams in the bottom of Figure 18 are related by a $(3,3)$-stabilization.

DEFINITION 6.27. The sutured diagram $(\Sigma, \boldsymbol{\alpha}', \boldsymbol{\beta})$ is obtained from $(\Sigma, \boldsymbol{\alpha}, \boldsymbol{\beta})$ by a *generalized α-handleslide of type* (m, n) if there are curves $\alpha_1, \alpha_2 \in \boldsymbol{\alpha}$, a curve $\alpha_1' \in \boldsymbol{\alpha}'$, and an embedded arc $a \subset \Sigma$ such that

- $\boldsymbol{\alpha}' \setminus \alpha_1' = \boldsymbol{\alpha} \setminus \alpha_1$,
- $\partial a \subset \alpha_2$ and the interior of a is disjoint from $\boldsymbol{\alpha}$,
- there is a regular neighborhood N of $\alpha_2 \cup a$ such that $\partial N = \alpha_1 \cup \alpha_1' \cup c$, where c is a curve parallel to α_2, and the interior of N is disjoint from $\boldsymbol{\alpha} \cup \boldsymbol{\alpha}' \setminus \{\alpha_2\}$, and
- if $\alpha_2 \setminus \partial a = \alpha_2^0 \cup \alpha_2^1$, where $\alpha_2^0 \cup a$ is parallel to α_1 and $\alpha_2^1 \cup a$ is parallel to α_1', then $|\alpha_2^0 \cap \boldsymbol{\beta}| = m$ and $|\alpha_2^1 \cap \boldsymbol{\beta}| = n$.

See Figure 37. Generalized β-handleslides are defined similarly.

In particular, an "ordinary" handleslide is a generalized handleslide of type $(0, n)$, where the endpoints of the arc a lie very close to each other.

The bifurcations that appear in generic 1-parameter families of gradient vector fields were given in Section 5.2.1. We now translate these to moves on sutured diagrams. For clarity, we state what happens to overcomplete diagrams.

PROPOSITION 6.28. *Suppose that*

$$\{\, (f_t, v_t) : t \in [-1, 1] \,\}$$

is a generic 1-parameter family of sutured functions and gradient-like vector fields on (M, γ) that has a bifurcation at $t = 0$. Since $(f_0, v_0) \in \mathcal{FV}_1(M, \gamma)$, it is separable; pick a separating surface $\Sigma \in \Sigma(f_0, v_0)$. Then there exists an $\epsilon = \epsilon(\Sigma) > 0$ such that $\Sigma \pitchfork v_t$ for every $t \in (-\epsilon, \epsilon)$. Furthermore, for every $x \in (-\epsilon, 0)$ and $y \in (0, \epsilon)$, the following hold.

If the bifurcation is not an index 1-2 birth-death, then $\Sigma \in \Sigma(f_x, v_x) \cap \Sigma(f_y, v_y)$. Furthermore, the (overcomplete) diagrams

$$(\Sigma, \boldsymbol{\alpha}, \boldsymbol{\beta}) = H(f_x, v_x, \Sigma) \quad \text{and} \quad (\Sigma, \boldsymbol{\alpha}', \boldsymbol{\beta}') = H(f_y, v_y, \Sigma),$$

possibly after a small isotopy of the immersed submanifold $\boldsymbol{\alpha} \cup \boldsymbol{\beta}$, are related in one of the following ways:

1. *If the bifurcation is an index 0-1 or 2-3 birth-death, adding or removing a redundant α- or β-curve, not necessarily disjoint from curves of the opposite type. "Redundant" means this α- or β-curve is null-homotopic in Σ compressed along the remaining α- or β-curves, or equivalently that it bounds a planar region together with the other α- or β-curves.*
2. *If the bifurcation is a tangency of $W^u(p)$ and $W^s(q)$ for $p \in C_1(f_0)$ and $q \in C_2(f_0)$, an isotopy of the α- and β-curves cancelling or creating a pair of intersection points.*
3. *If the bifurcation is a tangency between $W^u(p)$ and $W^s(q)$ for p, $q \in C_1(f_0)$ or p, $q \in C_2(f_0)$, a generalized α- or β-handleslide. Specifically, the α-curve corresponding to p slides over the α-curve corresponding to q if p, $q \in C_1(f_0)$, while the β-curve corresponding to q slides over the β-curve corresponding to p if p, $q \in C_2(f_0)$.*

If the bifurcation is an index 1-2 birth, then $\Sigma \in \Sigma(f_x, v_x)$. Furthermore, there exists a surface $\Sigma' \in \Sigma(f_y, v_y)$ such that the (overcomplete) diagrams $(\Sigma, \boldsymbol{\alpha}, \boldsymbol{\beta}) = H(f_x, v_x, \Sigma)$ and $(\Sigma', \boldsymbol{\alpha}', \boldsymbol{\beta}') = H(f_y, v_y, \Sigma')$, possibly after a small isotopy of the immersed submanifold $\boldsymbol{\alpha} \cup \boldsymbol{\beta}$, are related by a (k, l)-stabilization if there are l flows *from index 1 critical points into the degenerate singularity and k flows from the degenerate singularity to index 2 critical points. For an index 1-2 death, the same statements hold, but with x and y reversed.*

PROOF. Since the family is generic, $(f_t, v_t) \in \mathcal{FV}_0(M,\gamma)$ for every $t \in [-1,1] \setminus \{0\}$, and $(f_0, v_0) \in \mathcal{FV}_1(M,\gamma)$. By Proposition 6.7, the surface Σ divides (M,γ) into two sutured compression bodies (M_-, γ_-) and (M_+, γ_+). Let $\epsilon > 0$ be so small that for every $t \in (-\epsilon, \epsilon)$, the surface Σ is transverse to v_t.

First, suppose we are in case (1). Without loss of generality, we can assume that the bifurcation is an index 0-1 birth. The function f_0 has a degenerate critical point at $p_0 \in M$, which splits into an index 0 critical point $p_t^0 \in C_0(f_t)$ and an index 1 critical point $p_t^1 \in C_1(f_t)$ for $t > 0$. Recall that the stable manifold $W^s(p_0)$ is a 1-manifold with boundary at p_0, while the unstable manifold $W^u(p_0)$ is locally diffeomorphic to $\mathbb{R}^3_+$, with boundary the strong unstable manifold $W^{uu}(p_0)$; see Figure 12. The critical points p_0 at $t = 0$ and p_t^0 for $t > 0$ both have valence $k+1$ in $\Gamma(f_t, v_t)$, where k is the number of flow lines from p_0 to index 1 critical points within $W^u(p_0)$.

Recall that $p_0 \in C_{01}(f_0, v_0) \subset M_-$. Since v_t is transverse to Σ for every $t \in (-\epsilon, \epsilon)$, both p_t^0 and p_t^1 lie in M_-, hence $C_{01}(f_t, v_t) \subset M_-$ for every $t \in (-\epsilon, \epsilon)$. This implies that $\Sigma \in \Sigma(f_t, v_t)$ for every $t \in (-\epsilon, \epsilon)$.

The attaching sets $\boldsymbol{\beta}$ and $\boldsymbol{\beta}'$ are just small isotopic translates of each other. The isotopy is provided by

$$\bigcup_{q_t \in C_2(f_t)} W^s(q_t) \cap \Sigma$$

for $t \in [x, y]$. The same holds for $\boldsymbol{\alpha}$ and $\boldsymbol{\alpha}'$, except that $\boldsymbol{\alpha}'$ has one new component due to the appearance of the new index 1 critical point p_y^1. The new α-circle $W^u(p_y^1) \cap \Sigma$ is a small translate of $W^{uu}(p_0) \cap \Sigma$. For every index 2 critical point $q \in C_2(f_0)$ for which $W^s(q)$ intersects $W^{uu}(p_0)$, the corresponding β-circle $W^s(q) \cap \Sigma$ intersects the new, redundant, α-circle. (This does happen generically in 1-parameter families.)

Now we look at case (2). Consider the family of diagrams $(\Sigma, \boldsymbol{\alpha}_t, \boldsymbol{\beta}_t) = H(f_t, v_t, \Sigma)$ for $t \in [x, y]$. Then we have $(\Sigma, \boldsymbol{\alpha}, \boldsymbol{\beta}) = (\Sigma, \boldsymbol{\alpha}_x, \boldsymbol{\beta}_x)$ and $(\Sigma, \boldsymbol{\alpha}', \boldsymbol{\beta}') = (\Sigma, \boldsymbol{\alpha}_y, \boldsymbol{\beta}_y)$. The 1-manifolds $\boldsymbol{\alpha}_t$ and $\boldsymbol{\beta}_t$ remain transverse, except for $t = 0$, when there is a generic tangency between $W^u(p) \cap \Sigma \in \boldsymbol{\alpha}_0$ and $W^s(q) \cap \Sigma \in \boldsymbol{\beta}_0$.

Next, assume we are in case (3). Without loss of generality, we can suppose that p, $q \in C_1(f_0)$. Then we show that the α-curve corresponding to p slides over the α-curve corresponding to q. Since Σ is transverse to v_t for every $t \in (-\epsilon, \epsilon)$, we have $C_0(f_t) \cup C_1(f_t) \cup R_-(\gamma) \subset M_-$ and $C_2(f_t) \cup C_3(f_t) \cup R_+(\gamma) \subset M_+$ for every $t \in (-\epsilon, \epsilon)$. Hence $\Sigma \in \Sigma(f_t, v_t)$ for every $t \in (-\epsilon, \epsilon)$.

Let $\tau = W^u(p) \cap W^s(q)$ be the flow-line of v_0 from p to q. Recall from Section 5.2.1 that, generically, inside the 2-dimensional unstable manifold $W^u(q)$, there is a 1-dimensional strong unstable manifold $W^{uu}(q)$. Furthermore, for every $r \in \tau$, there is a 1-dimensional subspace $E_r < T_r W^u(p)$ complementary to $\langle v_0(r) \rangle = T_r \tau$ that limits to $T_q W^{uu}(q)$ under the flow of v_0; see Figure 17. It follows that the curve $\alpha_p^0 = W^u(p) \cap \Sigma$ is diffeomorphic to $\mathbb{R}$, with ends limiting to the two points of $W^{uu}(q) \cap \Sigma$. Consider the circle $\alpha_q^0 = W^u(q) \cap \Sigma$, and take a thin regular neighborhood P of $\alpha_p^0 \cup \alpha_q^0$. Notice that P is a pair-of-pants, and one component α_q' of ∂P is a small isotopic translate of α_q^0. For $t \in (-\epsilon, \epsilon)$, let p_t and q_t be the

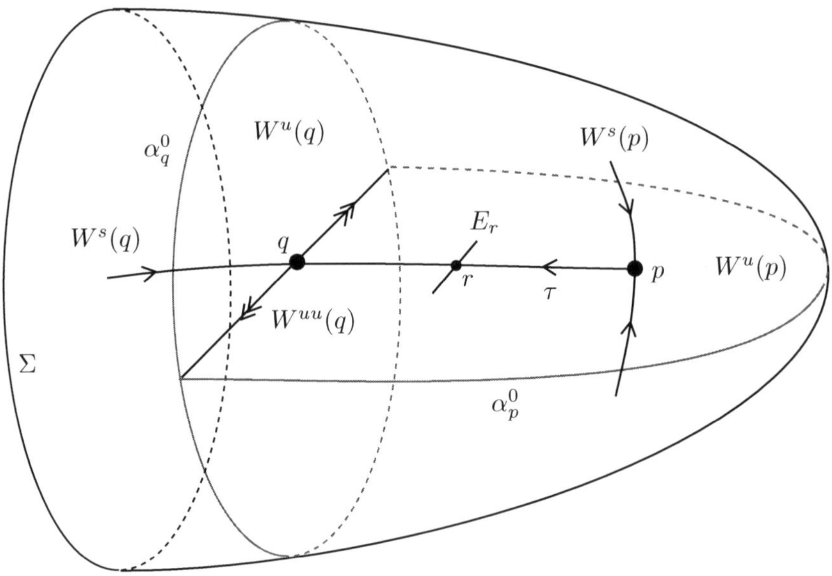

FIGURE 17. The situation in case (3), a tangency between $W^u(p)$ and $W^s(q)$, leading to a generalized handleslide.

points of $C_1(f_t)$ corresponding to $p = p_0$ and $q = q_0$, and let $\alpha_p^t = W^u(p_t) \cap \Sigma$ and $\alpha_q^t = W^u(q_t) \cap \Sigma$. Then α_q^x and α_q^y are small isotopic translates of α_q^0, while α_p^x and α_p^y are small isotopic translates of the other two components of ∂P. Hence $\alpha_p^y \in \boldsymbol{\alpha}'$ is obtained (up to a small isotopy) by a generalized handleslide of $\alpha_p^x \in \boldsymbol{\alpha}$ over $\alpha_q^x \in \boldsymbol{\alpha}$ using the arc $a = \text{cl}\left(\alpha_p^0\right)$, and every other component of $\boldsymbol{\alpha}'$ is a small translate of a component of $\boldsymbol{\alpha}$. The type (m,n) of the generalized handleslide is given by the number of flow lines from q to index 2 critical points that intersect the two components of $W^u(q) \setminus W^{uu}(q)$.

Finally, consider the case of an index 1-2 birth-death, as illustrated in Figure 18. Without loss of generality, we can assume that a pair of index 1 and 2 critical points are born. So f_0 has a degenerate singularity at p_0 that splits into $p_t^1 \in C_1(f_t)$ and $p_t^2 \in C_2(f_t)$ for $t > 0$. Recall that we can either include p_0 in $C_{01}(f_0, v_0)$ or in $C_{23}(f_0, v_0)$. For now, we assume that $p_0 \in C_{23}(f_0, v_0)$, but the other choice works as well.

Observe that $\Sigma \in \Sigma(f_t, v_t)$ for every $t \in (-\epsilon, 0)$, as $\Sigma \pitchfork v_t$, $C_{01}(f_t, v_t) \subset M_-$, and $C_{23}(f_t, v_t) \subset M_+$. We have $p_0 \in C_{23}(f_0, v_0) \subset M_+$, thus p_t^1, $p_t^2 \in M_+$ for every $t \in (0, \epsilon)$. Indeed, neither of the points p_t^1 and p_t^2 can pass through Σ as v_t remains transverse to Σ throughout.

Since v_y is generic, $W^s(p_y^1)$ has both ends in $C_0(f_y) \cup R_-(\gamma) \subset M_-$. This implies that $W^s(p_y^1)$ intersects Σ transversely in two points. On the other hand, as $p_y^2 \in M_+$ and both ends of $W^u(p_y^2)$ lie in $C_3(f_y) \cup R_+(\gamma) \subset M_+$, we have $W^u(p_y^2) \cap \Sigma = \emptyset$. We obtain Σ' by smoothing the corners of

$$\partial\left(M_- \cup N\left(W^s(p_y^1)\right)\right) \setminus \partial M,$$

where $N\left(W^s(p_y^1)\right)$ is a thin tubular neighborhood of $W^s(p_y^1)$ whose boundary in M_+ is transverse to v_y. It is apparent that Σ' is transverse to v_y, and Σ' cuts (M, γ)

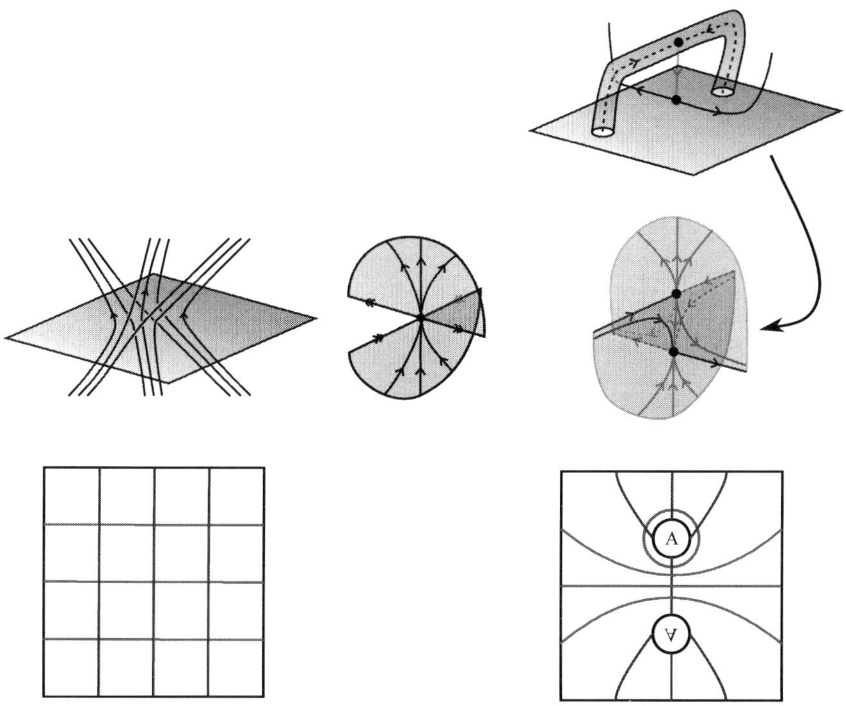

FIGURE 18. **An index 1-2 birth death.** Here, we see the two sides of a codimension-1 index 1-2 birth-death singularity, which is included in $C_{23}(f_0, v_0)$. There may be flows from index 1 critical points to this singularity, or from this singularity to index 2 critical points; in the example, there are three of each. The top row shows locally the gradient flow, together with the Heegaard surface (drawn in grey). In the bottom row, we illustrate the corresponding Heegaard diagrams. As usual, we identify the two circles labeled "A". On the side of the singularity where the two critical points die (on the left), we see a grid of flows between these critical points.

into two sutured compression bodies, one of which contains $C_{01}(f_y, v_y) \cup R_-(\gamma)$, while the other one contains $C_{23}(f_y, v_y) \cup R_+(\gamma)$. Hence $\Sigma' \in \Sigma(f_y, v_y)$. Notice that $\Sigma' \setminus \Sigma$ is an annulus A, and $\Sigma \setminus \Sigma'$ is a disjoint union of two disks D_1 and D_2.

We now describe the attaching sets $\boldsymbol{\alpha}'$ and $\boldsymbol{\beta}'$. Observe that $\alpha = W^u(p_y^1) \cap \Sigma' \in \boldsymbol{\alpha}'$ is a homologically non-trivial curve in A. The curve $\beta = W^s(p_y^2) \cap \Sigma' \in \boldsymbol{\beta}'$ intersects α transversely in a single point, and $\beta \cap A$ is an arc connecting ∂D_1 and ∂D_2. Let T be a thin regular neighborhood of $A \cup \beta$. Then T is homeomorphic to a punctured torus. In addition, let

$$D = (T \setminus A) \cup D_1 \cup D_2;$$

this is diffeomorphic to a disk. Observe that $\boldsymbol{\alpha}' \cap (\Sigma \setminus T)$ is a small isotopic translate of $\boldsymbol{\alpha} \cap (\Sigma \setminus D)$, and similarly, $\boldsymbol{\beta}' \cap (\Sigma \setminus T)$ is a small isotopic translate of $\boldsymbol{\beta} \cap (\Sigma \setminus D)$.

If $q_0 \in C_1(f_0)$ is a non-degenerate critical point, and $q_y \in C_1(f_y)$ is the corresponding critical point, then $W^u(q_y) \cap \Sigma' \in \boldsymbol{\alpha}'$ intersects β in precisely $|W^u(q_0) \cap$

$W^s(p_0)|$ points. In addition, $W^u(q_y) \cap T$ consists of parallel arcs that do not enter A. Similarly, for every $r_0 \in C_2(f_0)$ with corresponding $r_y \in C_2(f_y)$, the β-curve $W^s(r_y) \cap \Sigma' \in \boldsymbol{\beta}'$ intersects α in $|W^s(r_0) \cap W^u(p_0)|$ points, and A in the same number of parallel arcs. Hence $(\Sigma', \boldsymbol{\alpha}', \boldsymbol{\beta}')$ is indeed obtained from $(\Sigma, \boldsymbol{\alpha}, \boldsymbol{\beta})$ by a (k, l)-stabilization, as stated.

Note that, if we include p_0 in $C_{01}(f_0, v_0)$, then an analogous argument applies, with the difference that inside the stabilization tube A we have a β-curve and an α-arc. □

REMARK 6.29. In case of an index 1-2 stabilization, the stabilized surface $\Sigma' \in \Sigma(f_y, v_y)$ does not separate (f_t, v_t) if $t \in (0, y)$ is sufficiently small. Indeed, generically, the saddle-node $p_0 \notin \Sigma'$, and consequently the index 1 and 2 critical points p_1^t and p_2^t will both lie on the same side of Σ' for $t > 0$ sufficiently small.

Essentially the same argument gives the following analogue of Proposition 6.28 for 2-parameter families.

PROPOSITION 6.30. *Suppose that*

$$\{ (f_\mu, v_\mu) \in \mathcal{FV}(M, \gamma) : \mu \in \mathbb{R}^2 \}$$

is a generic 2-parameter family that has a codimension-1 bifurcation at $\mu = 0$. Let S be the stratum of the bifurcation set passing through the origin (S is a non-singular curve near 0). Since $(f_0, v_0) \in \mathcal{FV}_1(M, \gamma)$, it is separable; pick a separating surface $\Sigma \in \Sigma(f_0, v_0)$. Then there exists an $\epsilon = \epsilon(\Sigma) > 0$ such that $D^2_\epsilon \setminus S$ consists of two components C_1 and C_2, and, for every $x \in C_1$ and $y \in C_2$, the same conclusion holds as in Proposition 6.28.

Recall that, in Definition 2.29, we introduced the notion of distinguished rectangles of Heegaard moves.

DEFINITION 6.31. A *generalized distinguished rectangle* is defined just like in Definition 2.29, except we replace the word "stabilization" with "(k, l)-stabilization," and allow overcomplete diagrams.

The following result relates isotopies with codimension-1 moves.

PROPOSITION 6.32. *Suppose that*

$$\{ (f_\mu, v_\mu) \in \mathcal{FV}_{\le 1}(M, \gamma) : \mu \in \mathbb{R}^2 \}$$

is a generic 2-parameter family, and let

$$V_1 = \{ \mu \in \mathbb{R}^2 : (f_\mu, v_\mu) \in \mathcal{FV}_1(M, \gamma) \}$$

be the codimension-1 bifurcation set. Let $a \subset V_1$ be an arc with endpoints μ_0 and μ_1, and suppose we are given surfaces $\Sigma_i \in \Sigma(f_{\mu_i}, v_{\mu_i})$ for $i \in \{0, 1\}$. Let b_0 and b_1 be arcs transverse to V_1 such that the only bifurcation value in b_i is μ_i, and $\Sigma_i \pitchfork v_\mu$ for every $\mu \in b_i$. Orient b_0 and b_1 such that they have the same intersection sign with V_1, and let $\partial b_0 = y_0 - x_0$ and $\partial b_1 = y_1 - x_1$; see Figure 19.

After possibly flipping the orientation of b_0 and b_1, we can assume that $\Sigma_i \in \Sigma(f_{x_i}, v_{x_i})$ for $i \in \{0, 1\}$. Furthermore, suppose that we are given surfaces $\Sigma_i' \in \Sigma(f_{y_i}, v_{y_i})$ for $i \in \{0, 1\}$ such that Σ_i' is obtained from Σ_i by a stabilization if (f_{μ_i}, v_{μ_i}) is an index 1-2 birth, and $\Sigma_i' = \Sigma_i$ otherwise. (Such surfaces always exist by applying Proposition 6.28 to the 1-parameter family parametrized by b_i.) Then the isotopy diagrams $H_1 = [H(f_{x_0}, v_{x_0}, \Sigma_0)]$, $H_2 = [H(f_{y_0}, v_{y_0}, \Sigma_0')]$, $H_3 =$

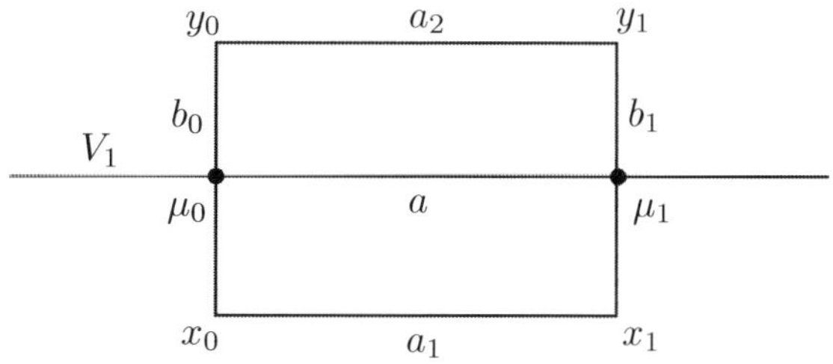

FIGURE 19. A rectangle about a the codimension-1 bifurcation set V_1.

$[H(f_{x_1}, v_{x_1}, \Sigma_1)]$, and $H_4 = [H(f_{y_1}, v_{y_1}, \Sigma_1')]$ fit into a generalized distinguished rectangle

$$\begin{array}{ccc} H_1 & \xrightarrow{e} & H_2 \\ \Big\downarrow f & & \Big\downarrow g \\ H_3 & \xrightarrow{h} & H_4 \end{array}$$

of type (3) if (f_{μ_i}, v_{μ_i}) is a handleslide, or type (5) if (f_{μ_i}, v_{μ_i}) is an index 1-2 birth-death; see Definition 2.29. For other types of bifurcations, we have a rectangle with e and h the identity or adding/removing a redundant α- or β-curve, and $f = g$ a diffeomorphism. If we pick any curves a_1 and a_2 outside the bifurcation set parallel to a with $\partial a_1 = x_1 - x_0$ and $\partial a_2 = y_1 - y_0$ and apply Lemma 6.21, then a_1 will induce a diffeomorphism isotopic to f, and a_2 will induce a diffeomorphism isotopic to g. In particular, $f \in \mathcal{G}^0_{\text{diff}}(H_1, H_3)$ and $g \in \mathcal{G}^0_{\text{diff}}(H_2, H_4)$. The arrows e and h are given by Proposition 6.28.

Note that, in case of an index 1-2 birth-death singularity, we mean $\Sigma_0 \in \Sigma_\pm(f_{\mu_0}, v_{\mu_0})$ and $\Sigma_1 \in \Sigma_\pm(f_{\mu_1}, v_{\mu_1})$, and we allow all four combinations of signs.

PROOF. For now, assume that, in case of an index 1-2 birth death, we have either $\Sigma_0 \in \Sigma_+(f_{\mu_0}, v_{\mu_0})$ and $\Sigma_1 \in \Sigma_+(f_{\mu_1}, v_{\mu_1})$, or $\Sigma_0 \in \Sigma_-(f_{\mu_0}, v_{\mu_0})$ and $\Sigma_1 \in \Sigma_-(f_{\mu_1}, v_{\mu_1})$.

Choose an arbitrary parametrization $a(t)$ of the arc a, then apply Lemma 6.23 to the 1-parameter family $\{\, (f_{a(t)}, v_{a(t)}) : t \in I \,\}$ inside $\mathcal{FV}_1(M, \gamma)$. We obtain a family of diffeomorphisms $d_t \colon \Sigma_0 \to \Sigma_t$ such that

$$\Sigma_t = d_t(\Sigma_0) \in \Sigma(f_{a(t)}, v_{a(t)})$$

for every $t \in I$. There exists an $\epsilon > 0$ such that, for every $t \in I$ and $\mu \in \mathbb{R}^2$ with $|a(t) - \mu| < \epsilon$, we have $\Sigma_t \pitchfork v_\mu$. Indeed, as transversality is an open relation,

$$U = \{\, (t, \mu) \in I \times \mathbb{R}^2 : \Sigma_t \pitchfork v_\mu \,\}$$

is an open neighborhood of the graph $\overline{a} = \{\, (t, a(t)) : t \in I \,\}$ in $I \times \mathbb{R}^2$. In particular, we can take ϵ to be the distance of $\overline{a}$ and $(I \times R^2) \setminus U$. Furthermore, we take ϵ so small that, for every $\mu \in N_\epsilon(a)$, the pair (f_μ, v_μ) is Morse–Smale unless μ lies in the component of V_1 containing a. We denote by C_1 and C_2 the components of $N_\epsilon(a) \setminus V_1$, labeled such that b_0 and b_1 are both oriented from C_1 to C_2.

First, suppose that (f_{μ_i}, v_{μ_i}) is not an index 1-2 birth-death. Then, as explained in the proof of Proposition 6.28, for every $t \in I$ and $\mu \in \mathbb{R}^2$ with $|a(t) - \mu| < \epsilon$, we even have $\Sigma_t \in \Sigma(f_\mu, v_\mu)$. Furthermore, $\Sigma_0 = \Sigma_0'$ and $\Sigma_1 = \Sigma_1'$, and the corresponding isotopy diagrams H_1 and H_2, and similarly, H_3 and H_4, are related by a handleslide, adding or removing a redundant α- or β-curve, or they are the same (for an orbit of tangency between an index 1 and an index 2 critical point). In this case, we take $f = g = d_1 \colon \Sigma_0 \to \Sigma_1$. What we need to show is that $d_1(H_1) = H_3$ and $d_1(H_2) = H_4$. Pick points $x_0' \in C_1 \cap b_0$, $x_1' \in C_1 \cap b_1$, $y_0' \in C_2 \cap b_0$, and $y_1' \in C_2 \cap b_1$, then choose arcs $c_1 \subset C_1$ and $c_2 \subset C_2$ with $\partial c_1 = x_1' - x_0'$ and $\partial c_2 = y_1' - y_0'$. These can be parametrized such that $|a(t) - c_j(t)| < \epsilon$ for every $t \in I$ and $j \in \{1,2\}$. Since $\Sigma_t \in \Sigma(f_{c_j(t)}, v_{c_j(t)})$ and $(f_{c_j(t)}, v_{c_j(t)})$ is Morse–Smale, there is an induced overcomplete diagram

$$\mathcal{H}_t^j = H(f_{c_j(t)}, v_{c_j(t)}, \Sigma_t).$$

If we apply the first part of Lemma 6.19 to the family of diagrams $\{\, \mathcal{H}_t^j : t \in I \,\}$, we obtain an induced diffeomorphism $d_1^j \colon \Sigma_0 \to \Sigma_1$ such that $d_1^1(\mathcal{H}_0^1) = \mathcal{H}_1^1$ and $d_1^2(\mathcal{H}_0^2) = \mathcal{H}_1^2$. Since $\Sigma_i \in \Sigma(f_\mu, v_\mu)$ for every $\mu \in b_i$ and $i \in \{0,1\}$, the isotopy diagrams $[\mathcal{H}_0^1] = H_1$, $[\mathcal{H}_1^1] = H_3$, $[\mathcal{H}_0^2] = H_2$, and $[\mathcal{H}_1^2] = H_4$. Hence $d_1^1(H_1) = H_3$ and $d_1^2(H_2) = H_4$. The second part of Lemma 6.19 implies that d_1^j is isotopic to d_1 for $j \in \{1,2\}$, so indeed $d_1(H_1) = H_3$ and $d_1(H_2) = H_4$.

Let $a_1' \subset \mathbb{R}^2$ be the path obtained by going from x_0 to x_0' along b_0, then from x_0' to x_1' along c_1, finally, from x_1' to x_1 along b_1. We define the path $a_2' \subset \mathbb{R}^2$ from y_0 to y_1 in an analogous manner. Since $\Sigma_i \in \Sigma(f_\mu, v_\mu)$ for every $\mu \in b_i$ and $i \in \{0,1\}$, the path a_j' induces a diffeomorphism $\delta_1^j \colon H_j \to H_{j+2}$ isotopic to d_1^j for $j \in \{1,2\}$. If a_1 is an arbitrary path from x_0 to x_1 in the complement of the bifurcation set and parallel to a, then a_1 is homotopic to a_1' relative to their boundary. So, by Corollary 6.25, the path a_1 induces a diffeomorphism $f \colon H_1 \to H_3$ isotopic to δ_1^1, hence also to d_1. Similarly, a path a_2 from y_0 to y_1 avoiding the bifurcation set and parallel to a induces a diffeomorphism $g \colon H_2 \to H_4$ isotopic to δ_1^2, hence also to d_1.

Suppose that (f_{μ_i}, v_{μ_i}) is an index 1-2 birth; furthermore, $\Sigma_0 \in \Sigma_+(f_{\mu_0}, v_{\mu_0})$ and $\Sigma_1 \in \Sigma_+(f_{\mu_1}, v_{\mu_1})$. The case when $\Sigma_0 \in \Sigma_-(f_{\mu_0}, v_{\mu_0})$ and $\Sigma_1 \in \Sigma_-(f_{\mu_1}, v_{\mu_1})$ is completely analogous. Then, by Proposition 6.30, the diagram H_2 is obtained from H_1 by a stabilization, and similarly, H_4 is obtained from H_3 by a stabilization. Pick arcs $c_1 \colon I \to C_1$ and $c_2 \colon I \to C_2$ as above, and extend them to arcs $a_1' \colon [-1,2] \to \mathbb{R}^2$ connecting x_0 and x_1 and $a_2' \colon [-1,2] \to \mathbb{R}^2$ connecting y_0 and y_1 in a similar manner. I.e., $a_j'([-1,0]) \subset b_0$, $a_j'\big|_I = c_j$, and $a_j'([1,2]) \subset b_1$ for $j \in \{1,2\}$. For every μ on the "birth" side C_2 of V_1, let the index 1 and 2 critical points born be $p^1(\mu)$ and $p^2(\mu)$, respectively. The surface Σ_t divides M into two sutured compression bodies $M_-(t)$ and $M_+(t)$. Let $\Sigma_t = \Sigma_0$ for $t \in [-1,0]$ and $\Sigma_t = \Sigma_1$ for $t \in [1,2]$. Furthermore, we write $p^j(t)$ for $p^j(a_2'(t))$, where $j \in \{1,2\}$ and $t \in [-1,2]$. For each $t \in [-1,2]$, we construct a surface

$$\Sigma_t^* \in \Sigma(f_{a_2'(t)}, v_{a_2'(t)})$$

from Σ_t by adding a tube around $W^s(p^1(t))$ as in the proof of Proposition 6.28, but now in a way that the construction depends smoothly on t. For this, simply pick a thin regular neighborhood N of

$$\bigcup_{t \in [-1,2]} W^s(p^1(t)) \times \{t\} \subset M \times [-1,2],$$

let $N_t = N \cap (M \times \{t\})$, then take

$$\Sigma_t^* = \partial(M_-(t) \cup N_t) \setminus \partial M.$$

Finally, let $A_t = \overline{\Sigma_t^* \setminus \Sigma_t}$ be the added tube (i.e., annulus). We can do this in such a way that $\Sigma_{-1}^* = \Sigma_0'$ and $\Sigma_2^* = \Sigma_1'$.

We take f to be d_1, defined at the beginning of this proof. Furthermore, we define g to be d_1 outside the extra tube A_{-1}, and extend it to A_{-1} using the family Σ_t^* (this follows from a straightforward relative version of Lemma 6.19). Similarly to the other cases, take

$$\mathcal{H}_t^1 = H(f_{a_1'(t)}, v_{a_1'(t)}, \Sigma_t) \text{ and } \mathcal{H}_t^2 = H(f_{a_2'(t)}, v_{a_2'(t)}, \Sigma_t^*) = (\Sigma_t^*, \boldsymbol{\alpha}_t^2, \boldsymbol{\beta}_t^2),$$

then apply Lemma 6.19 to these families of diagrams to obtain diffeomorphisms d_1^1 and d_1^2, respectively, such that $d_1^1(H_1) = H_3$ and $d_1^2(H_2) = H_4$. As above, d_1^1 is isotopic to d_1, hence $d_1(H_1) = H_3$. Similarly, d_1^2 agrees with d_1 up to isotopy outside A_{-1}, and inside A_{-1} it has to map the curve $\boldsymbol{\alpha}_{-1}^2 \cap A_{-1}$ to $\boldsymbol{\alpha}_2^2 \cap A_2$ up to isotopy, hence $d_1^2(H_2) = H_4$. So we indeed have a generalized distinguished rectangle of type (5). The fact that any curve a_i homotopic to a_i' relative to their boundary induces an isotopic diffeomorphism follows from Corollary 6.25.

Finally, we consider the case of an index 1-2 birth-death singularity with $\Sigma_0 \in \Sigma_+(f_{\mu_0}, v_{\mu_0})$ and $\Sigma_1 \in \Sigma_-(f_{\mu_1}, v_{\mu_1})$, or $\Sigma_0 \in \Sigma_-(f_{\mu_0}, v_{\mu_0})$ and $\Sigma_1 \in \Sigma_+(f_{\mu_1}, v_{\mu_1})$. We will only discuss the former possibility, as the latter is completely analogous. We first assume that a is a constant path mapping to the point $\overline{\mu} \in V_1$, and $b_0 = b_1$. We denote the arc $b_0 = b_1$ by b, and write $\partial b = y - x$. Let b_x and b_y be the components of $b \setminus \{\overline{\mu}\}$ containing x and y, respectively. Then Σ_0, $\Sigma_1 \in \Sigma(f_x, v_x)$, while Σ_0', $\Sigma_1' \in \Sigma(f_y, v_y)$. If a_1 is the constant path at x, then it induces the diffeomorphism $f\colon \Sigma_0 \to \Sigma_1$ obtained by flowing along v_x. Similarly, if a_2 is the constant path at y, then it induces the diffeomorphism $g\colon \Sigma_0' \to \Sigma_1'$ obtained by flowing along v_y. Then $f(H_1) = H_3$ and $g(H_2) = H_4$.

All we need to show is that g is isotopic to the stabilization of f. Let $p \in M$ be the degenerate critical point of $f_{\overline{\mu}}$, and let $p^1(\mu) \in C_1(f_\mu)$ and $p^2(\mu) \in C_2(f_\mu)$ be the corresponding critical points of f_μ for $\mu \in b_y$. Let $N \subset M$ be a $(v_{\overline{\mu}})$-saturated regular neighborhood of $W^s(p) \cup W^u(p)$. Then the disk $D_0 = \Sigma_0 \cap N$ is a regular neighborhood of the arc $W^s(p) \cap \Sigma_0$, and the disk $D_1 = \Sigma_1 \cap N$ is a regular neighborhood of the arc $W^u(p) \cap \Sigma_1$. Furthermore, for $i \in \{0, 1\}$, let

$$A_i = \overline{\Sigma_i' \setminus \Sigma_i}$$

be the stabilization tubes, and $\alpha_i = \Sigma_i' \cap W^u(p^1(y))$ and $\beta_i = \Sigma_i' \cap W^s(p^2(y))$ the new α- and β-curves. Note that $\alpha_0 \subset A_0$ and $\beta_0 \cap A_0$ is an arc, whereas $\beta_1 \subset A_1$ and $\alpha_1 \cap A_1$ is an arc. Recall that $B_i = \Sigma_i \setminus \Sigma_i'$ is a pair of open disks; we choose D_i such that $B_i \subset D_i$. Then

$$T_i = (D_i \setminus B_i) \cup A_i$$

is a punctured torus that is a regular neighborhood of $\alpha_i \cup \beta_i$ for $i \in \{0, 1\}$. By construction, $\Sigma_i' = (\Sigma_i \setminus D_i) \cup T_i$. The flow of $v_{\overline{\mu}}$ induces a diffeomorphism

$$d\colon \Sigma_0 \setminus D_0 \to \Sigma_1 \setminus D_1.$$

If $i\colon \Sigma_1 \setminus D_1 \hookrightarrow \Sigma_1$ is the embedding, then both

$$f|_{\Sigma_0 \setminus D_0}\colon \Sigma_0 \setminus D_0 \to \Sigma_1 \text{ and}$$
$$g|_{\Sigma_0' \setminus T_0}\colon \Sigma_0' \setminus T_0 = \Sigma_0 \setminus D_0 \to \Sigma_1'$$

are isotopic to $i \circ d$. Hence $f|_{\Sigma_0 \setminus D_0}$ is isotopic to $g|_{\Sigma_0' \setminus T_0}$. Since $g(\alpha_0) = \alpha_1$ and $g(\beta_0) = \beta_1$, we can isotope g such that it maps the regular neighborhood T_0 of

$\alpha_0 \cup \beta_0$ to the regular neighborhood T_1 of $\alpha_1 \cup \beta_1$. So, up to isotopy of f and g, the diagram

$$\begin{array}{ccc} [H(f_x, v_x, \Sigma_0)] & \xrightarrow{e} & [H(f_y, v_y, \Sigma_0')] \\ \downarrow{\scriptstyle f} & & \downarrow{\scriptstyle g} \\ [H(f_x, v_x, \Sigma_1)] & \xrightarrow{h} & [H(f_y, v_y, \Sigma_1')] \end{array}$$

is a distinguished rectangle of type (5).

We are now ready to prove the general case, when $a \subset V_1$ is an arbitrary arc, and we have an index 1-2 birth-death singularity with $\Sigma_0 \in \Sigma_+(f_{\mu_0}, v_{\mu_0})$ and $\Sigma_1 \in \Sigma_-(f_{\mu_1}, v_{\mu_1})$. Choose a surface $\Sigma \in \Sigma_+(f_{\mu_1}, v_{\mu_1})$. There exists an $\epsilon = \epsilon(\Sigma) > 0$ such that $\Sigma \pitchfork v_\mu$ for every $\mu \in D^2_\epsilon(\mu_1)$, and let $b \subset D^2_\epsilon(\mu_1)$ be a sub-arc of b_1 such that $\mu_1 \in \mathrm{Int}(b)$. Suppose that $\partial b = y - x$, and denote by b_{x,x_1} the sub-arc of b_1 between x and x_1, and by b_{y,y_1} the sub-arc of b_1 between y and y_1. If we apply Proposition 6.28 to the 1-parameter family b, then we see that $\Sigma \in \Sigma(f_x, v_x)$, and we obtain a surface $\Sigma' \in \Sigma(f_y, v_y)$ stabilizing $\Sigma \in \Sigma(f_x, v_x)$. We write $H = [H(f_x, v_x, \Sigma)]$ and $H' = [H(f_y, v_y, \Sigma')]$.

Let $a_1 \subset \mathbb{R}^2$ be the path obtained by going from x_0 to x along an arc a_1' parallel to a, then from x to x_1 along b_{x,x_1}. The path $a_2 \subset \mathbb{R}^2$ is obtained by going from y_0 to y along an arc a_2' parallel to a, then from y to y_1 along b_{y,y_1}. We also assume that a_1' and a_2' are disjoint from the bifurcation set. Then a_1' induces a diffeomorphism

$$f' \colon H_1 = [H(f_{x_0}, v_{x_0}, \Sigma_0)] \to H = [H(f_x, v_x, \Sigma)],$$

and the arc a_2' induces a diffeomorphism

$$g' \colon H_2 = [H(f_{y_0}, v_{y_0}, \Sigma_0')] \to H' = [H(f_y, v_y, \Sigma')].$$

Furthermore, the constant x path induces a diffeomorphism

$$f'' \colon H = [H(f_x, v_x, \Sigma)] \to [H(f_x, v_x, \Sigma_1)],$$

and the constant y path induces a diffeomorphism

$$g'' \colon H' = [H(f_y, v_y, \Sigma')] \to [H(f_y, v_y, \Sigma_1')].$$

Since $\Sigma \in \Sigma(f_\mu, v_\mu)$ for every $\mu \in b_{x,x_1}$, both $H(f_x, v_x, \Sigma_1)$ and $H(f_{x_1}, v_{x_1}, \Sigma_1)$ define the same isotopy diagram H_3. Similarly, as $\Sigma' \in \Sigma(f_\mu, v_\mu)$ for every $\mu \in b_{y,y_1}$, both $H(f_y, v_y, \Sigma_1')$ and $H(f_{y_1}, v_{y_1}, \Sigma_1')$ define the same isotopy diagram H_4. Furthermore, the arc b_{x,x_1} induces a diffeomorphism isotopic to f'', and the path b_{y,y_1} induces a diffeomorphism isotopic to g''. Hence, the path a_1 induces a diffeomorphism $f \colon H_1 \to H_3$ isotopic to $f'' \circ f'$, and the path a_2 induces a diffeomorphism $g \colon H_2 \to H_4$ isotopic to $g'' \circ g'$. Let s denote the stabilization from H to H', and consider the following subgraph of $\mathcal{G}$:

$$\begin{array}{ccc} H_1 & \xrightarrow{e} & H_2 \\ \downarrow{\scriptstyle f'} & & \downarrow{\scriptstyle g'} \\ H & \xrightarrow{s} & H' \\ \downarrow{\scriptstyle f''} & & \downarrow{\scriptstyle g''} \\ H_3 & \xrightarrow{h} & H_4. \end{array}$$

The top rectangle is distinguished of type (5), as we already know the result for $\Sigma_0 \in \Sigma_+(f_{\mu_0}, v_{\mu_0})$ and $\Sigma \in \Sigma_+(f_{\mu_1}, v_{\mu_1})$, together with the arc $a \subset V_1$ and transverse arcs b_0 and b. The bottom rectangle is also distinguished of type (5), since we have proved the proposition for the special case $\Sigma \in \Sigma_+(f_{\mu_1}, v_{\mu_1})$ and $\Sigma_1 \in \Sigma_-(f_{\mu_1}, v_{\mu_1})$, the constant μ_1 path, and the transverse arc b. It follows that the large rectangle is also distinguished of type (5), and by the above discussion, it agrees with the rectangle in the statement. □

6.4. Codimension-1: Ordinary diagrams

In this section, we will show how to choose spanning trees appropriately in Propositions 6.28 and 6.30 to pass from overcomplete to actual Heegaard diagrams, without altering the relationship of the diagrams before and after the bifurcation in an essential way. We are going to write Γ for $\Gamma(f_x, v_x)$ and Γ' for $\Gamma(f_y, v_y)$. Similarly, we use the shorthand $\Gamma_\pm$ for $\Gamma_\pm(f_x, v_x)$ and $\Gamma'_\pm$ for $\Gamma_\pm(f_y, v_y)$, where $\Gamma_\pm(f, v)$ is defined in Definition 6.14. By abuse of notation, if p is a non-degenerate critical point of f_0, then we also write p for the corresponding critical points of f_x and f_y. Furthermore, if p is index 1 or 2, we also view p as the midpoint of the appropriate edge of $\Gamma_+ \cup \Gamma_-$. Even though this graph is not strictly speaking a subset of M, it is obtained from Γ – which does contain p – by identifying its vertices lying in $R_+(\gamma)$ and its vertices lying in $R_-(\gamma)$. Similarly, we can view p as the midpoint of the appropriate edge of $\Gamma'_+ \cup \Gamma'_-$.

Suppose we are in case (1) of Proposition 6.28 (0-1 or 2-3 birth-death), and without loss of generality, consider the case of the birth of the critical points $p \in C_0(f_y)$ and $q \in C_1(f_y)$. Then Γ is obtained from Γ' by a small isotopy, deleting the vertex q of valence two, along with its two adjacent edges, and merging the two vertices in Γ' it was connected to (one of which is p). There is a map b from spanning trees of $\Gamma_\pm$ to spanning trees of $\Gamma'_\pm$, given by small isotopy and adding the edge whose midpoint is q; then $H(f_x, v_x, \Sigma, T_\pm)$ and $H(f_y, v_y, \Sigma, b(T_\pm)$ are the same isotopy diagram.

Similarly, in case (2) (1-2 tangency), the graphs Γ and Γ' are the same, except for a small isotopy. This induces a bijection b of spanning trees of $\Gamma_\pm$ and $\Gamma'_\pm$ such that $H(f_x, v_x, \Sigma, T_\pm)$ and $H(f_y, v_y, \Sigma, b(T_\pm))$ represent the same isotopy diagram. So bifurcations (1) and (2) have no effect on isotopy diagrams if we choose the spanning trees consistently.

Now consider the case of an index 1-2 birth. Then Γ'_- is obtained from Γ_- by adding an edge corresponding to the new index 1 critical point, and similarly, Γ'_+ is obtained from Γ_+ by adding an edge corresponding to the new index 2 critical point. Furthermore, Γ_- and Γ_+ are both connected. So spanning trees $T_\pm$ of $\Gamma_\pm$ remain spanning trees $T'_\pm$ of $\Gamma'_\pm$. The diagram $H(f_y, v_y, \Sigma', T'_\pm)$ is obtained from $H(f_x, v_x, \Sigma, T_\pm)$ by a (k', l')-stabilization, where l' is the number of flows from index 1 critical points of f_0 not in T_- to the saddle-node singular point, and k' is the number of flows from the saddle-node to index 2 critical points not in T_+. Note that, in this case, not all spanning trees of $\Gamma'_\pm$ come from spanning trees of $\Gamma_\pm$.

Finally, consider case (3) (same-index tangency). Without loss of generality, assume that the curve α_p slides over α_q, yielding α'_p, where p, $q \in C_1(f_x)$, the curve $\alpha_p = W^u(p) \cap \Sigma$, and $\alpha_q = W^u(q) \cap \Sigma$. Then the graph Γ'_- is obtained from Γ_- by sliding the edge $e_q \in E(\Gamma_-)$ containing q over the edge $e_p \in E(\Gamma_-)$ containing p, yielding the edge e'_q (note the change of roles as we pass to the spanning trees).

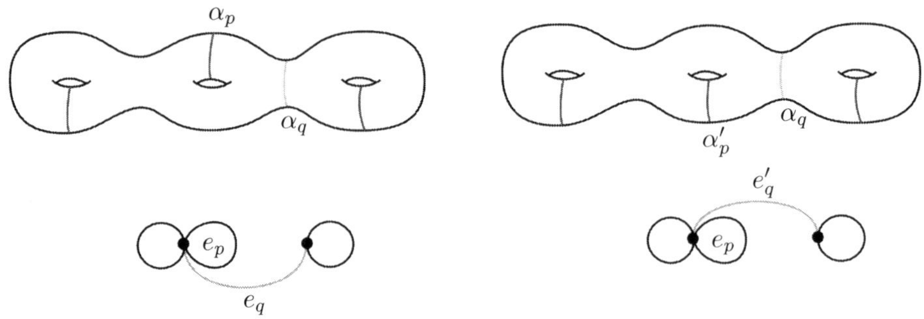

FIGURE 20. A handleslide in an overcomplete diagram. In this example, $T_- = \{e_q\}$ is the only spanning tree of Γ_-, and $T'_- = \{e'_q\}$ is the unique spanning tree of Γ'_-. However, the diagram $\mathcal{H}'$ cannot be obtained from $\mathcal{H}$ by a single handleslide. The α-curves and graph edges corresponding to T_- and T'_- are drawn in a lighter color. The tree T_- is not adapted to the handleslide.

Issues arise when $e_q \in T_-$ and $e_p \notin T_-$, since then the curve α_p is sliding over the "invisible" curve α_q. In fact, there are situations where, for any spanning tree T_- of Γ_- and any spanning tree T'_- of Γ'_-, the corresponding Heegaard diagrams do not differ by a single handleslide. For such a situation, see Figure 20. This motivates the following definition.

DEFINITION 6.33. Suppose we have a handleslide of α_p over α_q as above. Then we say that the spanning tree $T_\pm$ of $\Gamma_\pm$ is *adapted to the handleslide* if either

- $e_q \notin T_\pm$, or
- both $e_p, e_q \in T_\pm$.

We denote by $A_{\alpha_p/\alpha_q}(\Gamma_\pm)$ the set of spanning trees of $\Gamma_\pm$ adapted to sliding the curve α_p over α_q.

LEMMA 6.34. *Given a handleslide as above, $A_{\alpha_p/\alpha_q}(\Gamma_\pm) \neq \emptyset$ if either e_p is not a loop or e_q is not a cut-edge. Furthermore, there is a bijection*

$$b\colon A_{\alpha_p/\alpha_q}(\Gamma_\pm) \to A_{\alpha'_p/\alpha_q}(\Gamma'_\pm)$$

such that, for every spanning tree $T_\pm \in A_{\alpha_p/\alpha_q}(\Gamma_\pm)$, the sutured diagrams

$$\mathcal{H} = H(f_{-\epsilon}, v_{-\epsilon}, \Sigma, T_\pm) \text{ and } \mathcal{H}' = H(f_\epsilon, v_\epsilon, \Sigma, b(T_\pm))$$

are related by sliding α_p over α_q if $e_p, e_q \notin T_\pm$, and represent the same isotopy diagram otherwise.

PROOF. If e_p is not a loop, then there is a spanning tree $T_\pm$ of $\Gamma_\pm$ that contains e_p. Alternatively, if e_q is not a cut-edge, then there is a spanning tree $T_\pm$ of $\Gamma_\pm$ such that $e_q \notin T_\pm$. In either case $T_\pm \in A_{\alpha_p/\alpha_q}(\Gamma_\pm)$ and $A_{\alpha_p/\alpha_q}(\Gamma_\pm) \neq \emptyset$.

We now define the map b. If $e_q \notin T_\pm$, then $b(T_\pm) = T_\pm$. In this case, the diagram $\mathcal{H}'$ is obtained from $\mathcal{H}$ by sliding α_p over α_q if $e_p \notin T_\pm$, and $\mathcal{H}'$ represents the same isotopy diagram as $\mathcal{H}$ otherwise. If $e_p, e_q \in T_\pm$, then $b(T_\pm)$ is $T_\pm \setminus \{e_q\} \cup \{e'_q\}$, where e'_q is obtained by sliding e_q across e_p. Now $\mathcal{H}$ and $\mathcal{H}'$ represent the same isotopy diagram. $\square$

Even if $A_{\alpha_p/\alpha_q}(\Gamma_\pm) = \emptyset$ (with p, q of index 1), since $\Gamma_{23}(f_x, v_x)$ and $\Gamma_{23}(f_y, v_y)$ are small isotopic translates of each other, there is a natural bijection b between spanning trees of Γ_+ and Γ'_+. If $T_\pm$ and $T'_\pm$ are spanning trees of $\Gamma_\pm$ and $\Gamma'_\pm$, respectively, such that $T'_+ = b(T_+)$, then the corresponding diagrams are α-equivalent. As the example in Figure 20 shows, this is the best we can hope for, unless we are in one of the lucky situations of Lemma 6.34.

6.5. Codimension-1: Converting Heegaard moves to function moves

We now turn to the other direction: Given a move on Heegaard diagrams, can it be converted to a path of functions?

PROPOSITION 6.35. *Suppose that $\mathcal{H}_i = (\Sigma_i, \boldsymbol{\alpha}_i, \boldsymbol{\beta}_i)$ for $i \in \{0,1\}$ are diagrams of the sutured manifold (M,γ) such that $\boldsymbol{\alpha}_i \pitchfork \boldsymbol{\beta}_i$. In addition, let $(f_i, v_i) \in \mathcal{FV}_0(M,\gamma)$ for $i \in \{0,1\}$ be simple Morse–Smale pairs with $H(f_i,v_i) = \mathcal{H}_i$.*

1. *Given a diffeomorphism $d\colon \mathcal{H}_0 \to \mathcal{H}_1$ isotopic to the identity in M, there is a family*
$$\{\, (f_t, v_t) : t \in [0,1] \,\}$$
of simple Morse–Smale pairs connecting (f_0,v_0) and (f_1,v_1) that induces d in the sense of Lemma 6.21.
2. *If $\mathcal{H}_0$ and $\mathcal{H}_1$ are α- or β-equivalent, then (f_0,v_0) and (f_1,v_1) can be connected by a family of simple (but not necessarily Morse–Smale) pairs (f_t,v_t) such that*
$$\Sigma_0 = \Sigma_1 \in \Sigma(f_t, v_t)$$
for every $t \in [0,1]$. In particular, every isotopy and handleslide can be realized by such a family.
3. *If $\mathcal{H}_1$ is obtained from $\mathcal{H}_0$ by a (de)stabilization, then there is a generic family (f_t,v_t) of sutured functions connecting (f_0,v_0) and (f_1,v_1) such that for every $t \neq 1/2$, the pair (f_t,v_t) is simple and Morse–Smale, and at $t = 1/2$, there is an index 1-2 birth-death bifurcation of (f_t,v_t) realizing the stabilization.*

PROOF. We first prove claim (1). Let $\iota_i\colon \Sigma_i \hookrightarrow M$ be the embedding. The statement that d is isotopic to the identity in M means that there exists an isotopy $e_t\colon \Sigma_0 \to M$ such that $e_0 = \iota_0$ and $e_1 = \iota_1 \circ d$, while $e_t(\partial\Sigma_0) = s(\gamma)$ for every $t \in [0,1]$. This can be extended to a diffeotopy $E_t\colon M \to M$ such that $E_t|_{\Sigma_0} = e_t$ and $E_0 = \mathrm{Id}_M$. Consider the function $g_t = f_0 \circ E_t^{-1}$ and the vector field
$$w_t = dE_t \circ v_0 \circ E_t^{-1}.$$
Then (g_t,w_t) is a simple Morse–Smale pair. If $\Sigma_t = e_t(\Sigma_0)$, $\boldsymbol{\alpha}_t = e_t(\boldsymbol{\alpha}_0)$, and $\boldsymbol{\beta}_t = e_t(\boldsymbol{\beta}_0)$, then
$$(\Sigma_t, \boldsymbol{\alpha}_t, \boldsymbol{\beta}_t) \in \Sigma(g_t, w_t).$$
Clearly, $(g_0,w_0) = (f_0,v_0)$, but (g_1,w_1) and (f_1,v_1) might differ. We define (f_t,v_t) to be (g_{2t}, w_{2t}) for $t \in [0,1/2]$. By Proposition 6.18, the pairs (g_1,w_1) and (f_1,v_1) can be connected by a family $\{\, (f_t,v_t) : t \in [1/2,1] \,\}$ of simple Morse–Smale pairs, all adapted to $\mathcal{H}_1$. In the proof of Lemma 6.21, if we take d_t to be e_{2t} for $t \in [0,1/2]$, and to be e_1 for $t \in [1/2,1]$, then d_t satisfies
$$d_t(\mathcal{H}_0) = \mathcal{H}_t \in \Sigma(f_t, v_t)$$

for every $t \in [0,1]$. Hence, the family $\{ (f_t, v_t) : t \in [0,1] \}$ indeed induces the diffeomorphism $d_1 = d$, which concludes the proof of (1).

Now consider claim (2), and suppose that $\mathcal{H}_0 = (\Sigma, \boldsymbol{\alpha}_0, \boldsymbol{\beta})$ and $\mathcal{H}_1 = (\Sigma, \boldsymbol{\alpha}_1, \boldsymbol{\beta})$ are α-equivalent. Then Lemma 2.11 implies that, after applying a sequence of handleslides to $\boldsymbol{\alpha}_0$, we get an attaching set $\boldsymbol{\alpha}_1'$ that is isotopic to $\boldsymbol{\alpha}_1$. Hence, it suffices to prove the claim when $\mathcal{H}_2$ can be obtained from $\mathcal{H}_1$ by an isotopy of the α-curves, or by an α-handleslide.

First, assume that $\boldsymbol{\alpha}_0$ and $\boldsymbol{\alpha}_1$ are related by an isotopy. As described by Milnor [**24**, Section 4], there is an isotopy $\{ w_t : t \in [0,1] \}$ of v_0, supported in a collar neighborhood of Σ in the α-handlebody, such that $w_0 = v_0$, every w_t is gradient-like for f_0, and $H(f_0, w_1) = \mathcal{H}_1$. Once we have arranged that the Heegaard diagrams are equal, by Proposition 6.18, we can connect (f_0, w_1) and (f_1, v_1) through a family of simple Morse–Smale pairs, all adapted to $\mathcal{H}_1$.

Now suppose that $\boldsymbol{\alpha}_0$ and $\boldsymbol{\alpha}_1$ are related by a handleslide. In particular, the circle $\alpha_p = W^u(p) \cap \Sigma$ corresponding to $p \in C_1(f_0)$ slides over the curve $\alpha_q = W^u(q) \cap \Sigma$ corresponding to $q \in C_1(f_0)$ along some arc $a \subset \Sigma$ connecting α_p and α_q. Again, by Milnor [**24**, Section 4], there is a deformation

$$\{ (g_t, w_t) : t \in [0,1] \}$$

of (f_0, v_0) such that $(g_0, w_0) = (f_0, v_0)$, every (g_t, w_t) is a simple Morse–Smale pair, and p, q are neighboring index 1 critical points of g_1. *Neighboring* means that, if $\xi = g_1(p)$ and $\eta = g_1(q)$, then $\xi < \eta$, and the only critical points of g_1 in $M_{[\xi,\eta]} = g_1^{-1}([\xi,\eta])$ are p and q. We can also assume that $H(g_t, w_t) = \mathcal{H}_0$ for every $t \in [0,1]$, and that (g_t, w_t) coincides with (f_0, v_0) outside a small regular neighborhood of

$$W^s(p) \cup W^u(p) \cup W^s(q) \cup W^u(q).$$

Let $c = (\xi + \eta)/2$ and $M_c = g_1^{-1}(c)$. By flowing backwards along w_1, the arc a gives rise to an arc $a' \subset M_c$. Then there is an isotopy $\{ w_t : t \in [1,2] \}$ of w_1 such that

- the isotopy is supported in $M_{[\xi,\eta]}$,
- w_t is a gradient-like vector field for g_1 for every $t \in [1,2]$,
- it isotopes the circle $W^u(p) \cap M_c$ by a finger move along a' across one of the points of the 0-sphere $W^s(q) \cap M_c$,
- $W^u(r) \cap M_c$ is fixed for every $r \in C_1(g_1) \setminus \{p\}$.

The last condition can be satisfied because a' is disjoint from the circles $W^u(r) \cap M_c$. This realizes the handleslide of α_p over α_q; i.e., $H(g_2, w_2) = \mathcal{H}_1$. Again, using Proposition 6.18, the pairs (g_2, w_2) and (f_1, v_1) can be connected by a family of simple Morse–Smale pairs, all adapted to $\mathcal{H}_1$, concluding the proof of claim (2). Notice that (f_t, v_t) ceases to be Morse–Smale at values of t for which there is a tangency between an α- and a β-curve, or when there is an α-handleslide.

Finally, consider statement (3). Without loss of generality, we can suppose that $\mathcal{H}_1$ can be obtained from $\mathcal{H}_0$ by a stabilization. The case of a destabilization follows by time-reversal.

By definition, there is a disk $D \subset \Sigma_0$ and a punctured torus $T \subset \Sigma_1$ such that $\Sigma_0 \setminus D = \Sigma_1 \setminus T$. Furthermore, $\boldsymbol{\alpha}_0 = \boldsymbol{\alpha}_1 \cap (\Sigma_1 \setminus T)$, $\boldsymbol{\beta}_0 = \boldsymbol{\beta}_1 \cap (\Sigma_1 \setminus T)$, and there are circles $\alpha = \boldsymbol{\alpha}_1 \cap T$ and $\beta = \boldsymbol{\beta}_1 \cap T$ that intersect each other transversely in a single point. Let $p \in C_1(f_1)$ and $q \in C_2(f_1)$ be the critical points of f_1 for which $W^u(p) \cap \Sigma_2 = \alpha$ and $W^s(q) \cap \Sigma_2 = \beta$. Let Z_0 be the union of the flow lines of v_0 passing through D. As $D \cap (\boldsymbol{\alpha}_0 \cup \boldsymbol{\beta}_0) = \emptyset$, the manifold Z_0 is diffeomorphic to $D \times I$.

Define Z_1 to be the union of the flow lines of v_1 passing through T, together with

$$W^s(p) \cup W^u(p) \cup W^s(q) \cup W^u(q).$$

Then Z_1 is also diffeomorphic to $D \times I$, since it can be obtained from $T \times I$ by attaching 3-dimensional 2-handles along $\alpha \times \{0\}$ and $\beta \times \{1\}$.

The vertical boundary of Z_i is the annulus A_i obtained by taking the union of the flow lines of v_i passing through $\partial D = \partial T$. There is an isotopy $\{ d_t : t \in [0,1] \}$ of M such that $d_0 = \mathrm{Id}_M$, $d_1(A_0) = A_1$, and d_t fixes Σ_0 pointwise. Then consider the 1-parameter family

$$(f_0 \circ d_t^{-1}, (d_t)_* \circ v_0 \circ d_t^{-1})$$

of simple Morse–Smale pairs. This isotopes A_0 to A_1. Hence, we can assume that $A_0 = A_1$, which implies that $Z_0 = Z_1$. So we will write A for A_i and Z for Z_i. The attaching sets $\boldsymbol{\alpha}_0$ and $\boldsymbol{\beta}_0$ might move during this process via an isotopy avoiding D, but we can undo this using claim (2) without changing A anymore. So we still have $\boldsymbol{\alpha}_0 = \boldsymbol{\alpha}_1 \cap (\Sigma_1 \setminus T)$ and $\boldsymbol{\beta}_0 = \boldsymbol{\beta}_1 \cap (\Sigma_1 \setminus T)$. It is straightforward to arrange that (f_0, v_0) and (f_1, v_1) agree on a regular neighborhood of A.

Take the sutured manifold

$$(N, \nu) = \left(\overline{M \setminus Z}, \gamma \cup A\right).$$

Then

$$\mathcal{H}_0' = (\Sigma_0 \setminus D, \boldsymbol{\alpha}_0, \boldsymbol{\beta}_0) \text{ and } \mathcal{H}_1' = (\Sigma_1 \setminus T, \boldsymbol{\alpha}_1 \setminus \{\alpha\}, \boldsymbol{\beta}_1 \setminus \{\beta\})$$

are both diagrams of (N, ν). If we write $(f_i', v_i') = (f_i, v_i)|_N$ for $i \in \{0,1\}$, then $\mathcal{H}_i' = H(f_i', v_i')$. However, as $\mathcal{H}_0' = \mathcal{H}_1'$, we can apply Proposition 6.18 to get a family (f_t', v_t') of simple Morse–Smale pairs on (N, ν) connecting (f_0', v_0') and (f_1', v_1'). On the other hand, observe that $(D, \emptyset, \emptyset)$ and (T, α, β) are both diagrams of the product sutured manifold (Z, A) that are related by a stabilization, hence it now suffices to prove claim (3) for this special case. Indeed, we can simply glue the family connecting $(f_0, v_0)|_Z$ and $(f_1, v_1)|_Z$ to the family (f_t', v_t').

Consider $\mathbb{R}^3$ with the standard coordinates (x, y, z). Let

$$G_t(x, y, z) = x^3 - y^2 + z^2 + (1/2 - t)x,$$

with gradient vector field

$$W_t(x, y, z) = (3x^2 + 1/2 - t, -2y, 2z).$$

Then G_t has a bifurcation at $t = 1/2$, where a pair of index 1 and 2 critical points are born. Let

$$B_t = G_t^{-1}([-1, 1]) \cap D_2^3 \quad \text{and} \quad \eta_t = B_t \cap \partial D_2^3,$$

where D_2^3 is the unit disk in $\mathbb{R}^3$ of radius 2. Furthermore, let $g_t = G_t|_{B_t}$ and $w_t = W_t|_{B_t}$. It is straightforward to check that (B_t, η_t) is diffeomorphic to the product sutured manifold $(D^2, \partial D^2 \times I)$ for every $t \in [0,1]$. In addition, $H(g_0, w_0) = (D', \emptyset, \emptyset)$, where $D' = g_0^{-1}(0)$ is a disk, while $H(g_1, w_1) = (T', \alpha', \beta')$, where $T' = g_1^{-1}(0)$ is a punctured torus, and α' and β' are simple closed curves that intersect each other in a single point. There exists a smooth family of diffeomorphisms $h_t \colon (B_t, \eta_t) \to (Z, A)$ such that $h_0(D') = D$, $h_1(T') = T$, $h_1(\alpha') = \alpha$, and $h_1(\beta') = \beta$. Pushing (g_t, w_t) forward along h_t, we get a family on (Z, A) that we also denote by (g_t, w_t). According to Proposition 6.18, for $i \in \{0,1\}$, the pair $(f_i, v_i)|_Z$ can be connected with (g_i, w_i) via a family of simple Morse–Smale pairs. This concludes the proof of claim (3). □

6.6. Codimension-2

The singularities of gradient vector fields that appear in generic 2-parameter families were given in Section 5.2.2. This also applies to gradient-like vector fields on sutured manifolds by Proposition 5.19. In this section, we associate a loop of sutured Heegaard diagrams and Heegaard moves to each codimension-2 singularity.

Let (f_μ, v_μ) for $\mu \in \mathbb{R}^2$ be a generic 2-parameter family of sutured functions and gradient-like vector fields on the sutured manifold (M, γ) that has a codimension-2 singularity for $\mu = 0$; i.e., $(f_0, v_0) \in \mathcal{FV}_2(M, \gamma)$. Recall the notion of the bifurcation set in the parameter space from Definition 5.10; this is the set S of parameter values $\mu \in \mathbb{R}^2$ for which v_μ fails to be Morse–Smale. Then, for $\epsilon > 0$ sufficiently small, the set $(S \cap D^2_\epsilon) \setminus \{0\}$ is the disjoint union of smooth arcs (strata) $S_1, \dots, S_r$ with $0 \in \partial S_i$ and $\partial S_i \setminus \{0\} \in S^1_\epsilon$. We label the arcs S_i in a clockwise manner. The components of $D^2_\epsilon \setminus S$ are chambers $C_1, \dots, C_r$, labeled such that C_i lies between S_{i-1} and S_i for $i \in \{1, \dots, r\}$ (where $S_0 = S_r$ by definition).

In this section, the bifurcation diagrams that we draw illustrate the bifurcation set $S \subset \mathbb{R}^2$ in a neighborhood D^2_ϵ of 0, and for each chamber C_i, we indicate the relevant part of the corresponding (overcomplete) Heegaard diagram $H(f_\mu, v_\mu, \Sigma)$ for $\mu \in C_i$ near 0 and some Heegaard surface $\Sigma \in \Sigma(f_\mu, v_\mu)$. (Note that, if μ, $\mu' \in C_i$, then the vector fields v_μ and $v_{\mu'}$ are topologically equivalent, hence the corresponding diagrams are homeomorphic and close to each other.) We only show certain subsurfaces of Σ in our illustrations, and draw the boundary of these in green. Outside these subsurfaces, the diagrams are related by a small isotopy of $\boldsymbol{\alpha} \cup \boldsymbol{\beta}$. Following our previous conventions, α-circles are drawn in red, while β-circles are drawn in blue.

Consider an arc S_i, and pick a short curve $c\colon [-\nu, \nu] \to \mathbb{R}^2$ transverse to S_i at $c(0)$. This gives rise to a 1-parameter family

$$\{\, (f_{c(t)}, v_{c(t)}) : t \in [-\nu, \nu] \,\},$$

to which we can apply Proposition 6.28. If the diagrams for $(f_{c(-\nu)}, v_{c(-\nu)})$ and $(f_{c(\nu)}, v_{c(\nu)})$ are related by an α-equivalence, then we draw S_i in red; if they are related by a β-equivalence, then we draw S_i in blue; and S_i is black if they are related by a (de)stabilization.

Definition 6.36. Suppose that $\{\, (f_\mu, v_\mu) : \mu \in \mathbb{R}^2 \,\}$ is a generic 2-parameter family such that $(f_0, v_0) \in \mathcal{FV}_2(M, \gamma)$. For $\epsilon > 0$ as above, a *link of the bifurcation* at 0 is an embedded polygonal curve $P \subset D^2_\epsilon$ such that

- the bifurcation value 0 lies in the interior of P,
- $P \pitchfork S$ and $|S_i \cap P| = 1$ for every $i \in \{1, \dots, r\}$,
- each chamber C_i contains exactly one or two vertices of P.

We say that P is *minimal* if each C_i contains precisely one vertex of P. We orient the curve P in a clockwise manner.

A *surface enhanced link* of the bifurcation at 0 is a link P, together with a choice of Heegaard surface $\Sigma_\mu \in \Sigma(f_\mu, v_\mu)$ for each vertex μ of P.

We will use the following notational convention. If C_i contains one vertex of P, then we denote that by μ_i. The edge of P that intersects S_i is called a_i. If C_i contains two vertices, then they are denoted by μ_i and μ_i', ordered coherently with the orientation of the edge a_i' of P between them. So ∂a_i is either $\mu_{i+1} - \mu_i$ or $\mu_{i+1} - \mu_i'$, and $\partial a_i' = \mu_i' - \mu_i$. In particular, if P is minimal, then the vertices of P are $\mu_1, \dots, \mu_r$, and its edges are $a_1, \dots, a_r$. For simplicity, we write (f_i, v_i) for (f_{μ_i}, v_{μ_i}), (f_i', v_i') for $(f_{\mu_i'}, v_{\mu_i'})$, Σ_i for Σ_{μ_i}, and Σ_i' for $\Sigma_{\mu_i'}$. Furthermore, we write $H_i = (\Sigma_i, [\boldsymbol{\alpha}_i], [\boldsymbol{\beta}_i])$ for the (overcomplete) isotopy diagram $[H(f_i, v_i, \Sigma_i)]$, and $H_i' = (\Sigma_i', [\boldsymbol{\alpha}_i'], [\boldsymbol{\beta}_i'])$ for $[H(f_i', v_i', \Sigma_i')]$.

The cases distinguished in the following result are labeled consistently with the ones appearing in the bifurcation analysis of Section 5.2.2.

THEOREM 6.37. *Suppose that*

$$\mathcal{F} = \{ (f_\mu, v_\mu) : \mu \in \mathbb{R}^2 \}$$

is a generic 2-parameter family such that $(f_0, v_0) \in \mathcal{FV}_2(M, \gamma)$. *Using the above notation, for every* $\epsilon > 0$ *sufficiently small, there exists a surface enhanced link* $P \subset D^2_\epsilon$ *of the bifurcation at* 0 *such that the following hold:*

(1) *The polygon* P *is minimal unless the bifurcation at* 0 *is of type* (C) *or* (E1).
(2) *For* $i \in \{1, \dots, r\}$, *there is a point* $x_i \in \partial a_i$ *such that* $\Sigma_{x_i} \pitchfork v_\mu$ *for every* $\mu \in a_i$.
(3) *Consecutive isotopy diagrams* H_i *and* H_{i+1}, *or* H_i' *and* H_{i+1}, *are related by a move corresponding to the type of the stratum* S_i.
(4) *As in Lemma* 6.21, *each edge* a_i' *induces a diffeomorphism* $d_i \colon H_i \to H_i'$ *isotopic to the identity in* M.

We explicitly describe the surface enhanced link P *for each bifurcation of type* (A)–(E) *in the proof below.*

PROOF. We may ignore the strata S_i that correspond to an index 0-1 or an index 2-3 birth-death, or a tangency between the unstable manifold of an index 1 critical point and the stable manifold of an index 2 critical point, as the isotopy diagrams defined by

$$H(f_i, v_i, \Sigma_i, T^i_\pm) \text{ and } H(f_{i+1}, v_{i+1}, \Sigma_{i+1}, T^{i+1}_\pm)$$

coincide if we take $\Sigma_i = \Sigma_{i+1}$, and choose $T^i_\pm$ and $T^{i+1}_\pm$ consistently (see the discussion of trees following Proposition 6.28). If this reduces the bifurcation set to a single curve of codimension-1 bifurcations or eliminates it completely, then we do not list the bifurcation below. This simplification reduces the number of cases considerably, though no extra technical difficulty arises in the proof of Theorem 6.37 in the omitted cases. We use the notation of Section 5.2.2, with the codimension-2 bifurcation appearing at the parameter value $\overline{\mu} = 0$. Whenever we talk about handleslides, we mean generalized handleslides, as in Definition 6.27.

In all the cases where (f_0, v_0) is separable (i.e., everywhere except in case (E1)), we construct the surfaces $\Sigma_1, \dots, \Sigma_r$ (and Σ_3' in case (C)) from a common surface $\Sigma \in \Sigma(f_0, v_0)$ with the aid of Proposition 6.7. In these cases, we take ϵ so small that $\Sigma \pitchfork v_\mu$ for every $|\mu| < \epsilon$. Often, $\Sigma \in \Sigma(f_i, v_i)$ for every $i \in \{1, \dots, r\}$; for example, when f_0 is Morse (this includes all bifurcations of type (A)), or has an index 0-1 or 2-3 birth-death singularity, or an index 0-1-0, 1-0-1, 2-3-2, or 3-2-3 birth-death-birth. When f_0 has an index 1-2 birth-death, then we can construct surfaces on the two sides of the corresponding stratum as in the proof of Proposition 6.32. We only explain how to construct the surfaces $\Sigma_1, \dots, \Sigma_r$ whenever a new idea is needed.

As stated above, in cases (C) and (E1), the link P is not minimal. In the corresponding figures, if C_i contains two vertices of P, then we will draw a yellow ray in C_i emanating from 0 that separates μ_i and μ_i'. The reader should think of

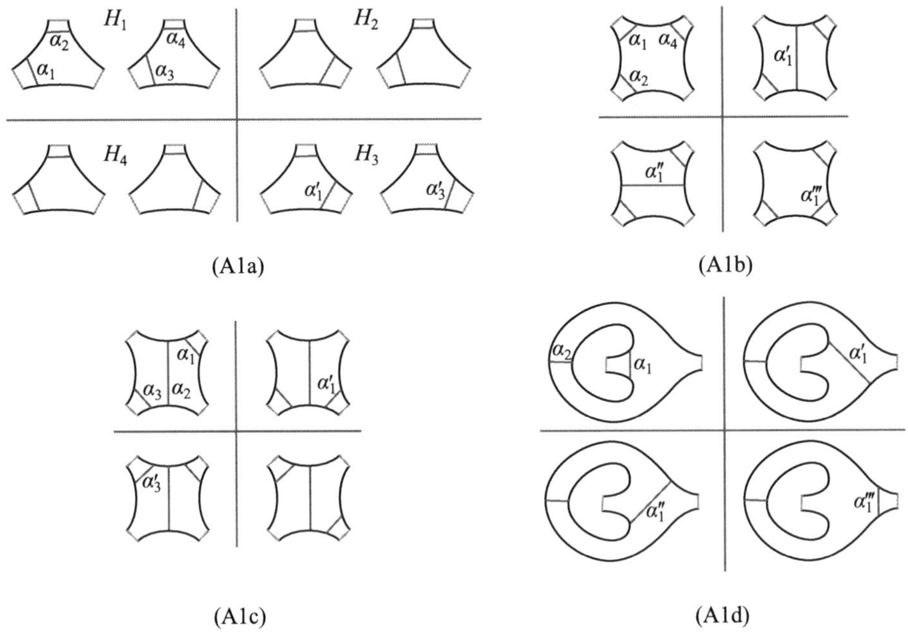

FIGURE 21. The links of bifurcations of type (A1a)–(A1d). The surfaces shown should be doubled along their black boundary arcs to obtain the relevant subsurface of the Heegaard diagram. This way, the red arcs become the α-circles, and the green arcs become the boundary components of the subsurface. We do not draw the β-circles as they remain unchanged.

this ray as a "diffeomorphism stratum" of the bifurcation set. The purpose of this will be explained in the following chapter.

We start by looking at bifurcations of type (A), which were illustrated schematically in Figure 13. For cases (B) and (C), the reader should consult Figure 14, while for cases (D) and (E1), see Figure 15.

In all subcases of case (A), we can take an arbitrary minimal link $P \subset D^2_\epsilon$, and $\Sigma_i = \Sigma$ for $i \in \{1, \dots, r\}$. First, we describe the possibilities in case (A1); see Figure 21. In each case, $r = 4$, and the bifurcation set S is the union of two smooth curves that intersect transversely at 0.

(A1a) Suppose $\mathcal{F}$ is of type (A1), and using the notation of Section 5.2.2, all p_i^0 are distinct, $\mathcal{I}(p_1^0) = \mathcal{I}(p_2^0) \in \{1,2\}$, and $\mathcal{I}(p_3^0) = \mathcal{I}(p_4^0) \in \{1,2\}$. We describe what happens to the diagrams H_i when all the p_i^0 have index 1; the other cases are analogous. Then $\boldsymbol{\beta}_i = \boldsymbol{\beta}_1$ for $i \in \{2,3,4\}$. Furthermore, the attaching set $\boldsymbol{\alpha}_1$ contains four distinct curves $\alpha_1, \dots, \alpha_4$ (corresponding to $p_1^0, \dots, p_4^0$, respectively), and $\boldsymbol{\alpha}_3$ contains two distinct curves α_1' and α_3' such that α_1' is obtained by sliding α_1 over α_2, and α_3' is obtained by sliding α_3 over α_4. In addition, $\boldsymbol{\alpha}_2 = (\boldsymbol{\alpha}_1 \setminus \alpha_1) \cup \alpha_1'$, $\boldsymbol{\alpha}_4 = (\boldsymbol{\alpha}_1 \setminus \alpha_3) \cup \alpha_3'$, and $\boldsymbol{\alpha}_3 = (\boldsymbol{\alpha}_1 \setminus (\alpha_1 \cup \alpha_3)) \cup (\alpha_1' \cup \alpha_3')$.

(A1b) Suppose $\mathcal{F}$ is of type (A1) and $p_1^0 = p_3^0$. The points p_1^0, p_2^0, and p_4^0 all have index 1, or they all have index 2. We discuss the case when they are all index 1. Then there are curves $\alpha_1, \alpha_2, \alpha_4 \in \boldsymbol{\alpha}_1$, and curves $\alpha_1' \in H_2$, $\alpha_1'' \in H_4$, and $\alpha_1''' \in H_3$ such that α_1' is obtained by sliding α_1 over α_2, the curve α_1'' is obtained by sliding α_1 over α_4, while α_1''' can be obtained by either sliding α_1' over α_4, or α_1'' over α_2. Furthermore, $\boldsymbol{\alpha}_2 = (\boldsymbol{\alpha}_1 \setminus \alpha_1) \cup \alpha_1'$, $\boldsymbol{\alpha}_3 = (\boldsymbol{\alpha}_1 \setminus \alpha_1) \cup \alpha_1'''$, and $\boldsymbol{\alpha}_4 = (\boldsymbol{\alpha}_1 \setminus \alpha_1) \cup \alpha_1''$. In other words, H_2 is obtained from H_1 by sliding α_1 over α_2, the diagram H_4 is obtained from H_1 by sliding α_1 over α_4, and H_3 is obtained from H_1 by sliding α_1 over α_2, and then sliding the resulting curve over α_4.

(A1c) Suppose $\mathcal{F}$ is of type (A1) and $p_2^0 = p_4^0$. The set $\boldsymbol{\alpha}_1$ contains three distinct curves α_1, α_2, and α_3; furthermore, there is an arc a_1 with $\partial a_1 \subset \alpha_2$ and an arc a_3 with $\partial a_3 \subset \alpha_2$ such that a_1 and a_3 reach α_2 from opposite sides, and H_2 is obtained from H_1 by sliding α_1 over α_2 using a_1 (resulting in a curve α_1'), H_4 is obtained from H_1 by sliding α_3 over α_2 using a_3 (resulting in the curve α_3'), while H_3 differs from H_1 by removing α_1 and α_3 and adding α_1' and α_3'.

(A1d) Suppose $\mathcal{F}$ is of type (A1), $p_1^0 = p_3^0$, and $p_2^0 = p_4^0$. Both p_1^0 and p_2^0 have index 1, or they both have index 2. This is similar to case (A1b), except that α_1 is sliding over the same curve α_2 in two different ways from opposite sides.

(A2) In this case, we have

$$\mathcal{I}(p_1^0) = \mathcal{I}(p_2^0) = \mathcal{I}(p_3^0) \in \{1, 2\}$$

and $r = 5$. We consider the case when all the p_i^0 are index 1. The $\boldsymbol{\beta}_i$ coincide up to a small isotopy. The attaching set $\boldsymbol{\alpha}_1$ contains three distinct curves α_1, α_2, and α_3 corresponding to p_1^0, p_2^0, and p_3^0, respectively. Then the pentagon is formed by α_1 sliding over α_2, which is itself sliding over α_3. More precisely, let α_1' be the curve obtained from α_1 by sliding it over α_2, let α_2' be the curve obtained from α_2 by sliding it over α_3, and finally let α_1'' be the curve obtained from α_1 by sliding it over α_2'. Then $\boldsymbol{\alpha}_2 = (\boldsymbol{\alpha}_1 \setminus \alpha_2) \cup \alpha_2'$, $\boldsymbol{\alpha}_3 = (\boldsymbol{\alpha}_2 \setminus \alpha_1) \cup \alpha_1''$, $\boldsymbol{\alpha}_4 = (\boldsymbol{\alpha}_3 \setminus \alpha_2') \cup \alpha_2$, and $\boldsymbol{\alpha}_5 = (\boldsymbol{\alpha}_4 \setminus \alpha_1'') \cup \alpha_1'$. In particular, this implies that $\boldsymbol{\alpha}_1 = (\boldsymbol{\alpha}_5 \setminus \alpha_1') \cup \alpha_1$. For a schematic illustration, see Figure 22.

We now look at bifurcations of type (B); i.e., codimension-2 singularities that include a single stabilization. See Figure 14 for schematic drawings. The link P and the surfaces Σ_i are obtained as follows. We label the strata such that S_1 and S_3 are the stabilizations, and there is a single stratum S_2 on the stabilized side. We choose $\Sigma \in \Sigma(f_0, v_0)$ and ϵ as above. For $i \notin \{2, 3\}$, the vertex μ_i of P is an arbitrary point of C_i, and $\Sigma_i = \Sigma$. Pick a parameter value $\nu \in S_2$ with $|\nu| < \epsilon$. Let $p^0 \in C(f_0)$ be the index 1-2 birth-death singularity; it breaks into the critical points $p_\nu^1 \in C_1(f_\nu)$ and $p_\nu^2 \in C_2(f_\nu)$. The surface

$$\Sigma_\nu \in \Sigma(f_\nu, v_\nu)$$

is obtained from Σ by attaching a tube around $W^u(p_\nu^2)$ if $p^0 \in C_{01}(f_0, v_0)$, or a tube around $W^s(p_\nu^1)$ if $p^0 \in C_{23}(f_0, v_0)$. The side a_2 of P is chosen short enough such that $\Sigma_\nu \pitchfork v_\mu$ for every $\mu \in a_2$. The endpoints of a_2 are μ_2 and μ_3. Both Σ_2 and Σ_3 are defined to be Σ_ν. Every side a_i of P for $i \neq 2$ can be an arbitrary arc connecting μ_i and μ_{i+1} that intersects S_i in a single point.

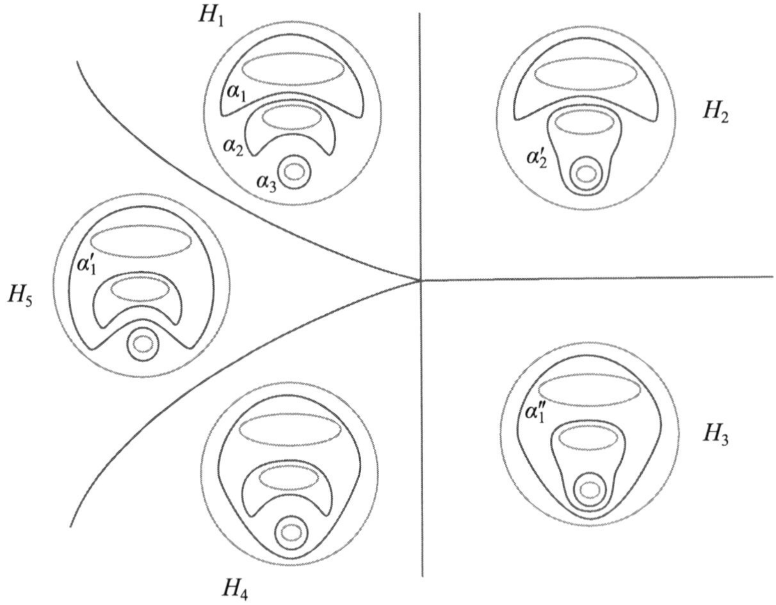

FIGURE 22. The link of a bifurcation of type (A2).

(B1) For definiteness, suppose that p^0 is an index 1-2 birth, while p_1^0 and p_2^0 are index 1 critical points. Then $r = 4$, and the strata S_1 and S_3 are stabilizations, while S_2 and S_4 are handleslides. Recall from Definition 6.2 that, in this case, $p^0 \in C_{01}(f_0, v_0)$. The type of the stabilizations $H_1 \to H_2$ and $H_4 \to H_3$ depend on the number of flows from p_1^μ to p^μ for $\mu \in S_1$ and $\mu \in S_3$, respectively. Recall from (ND-2) that $W^{uu}(p_2^0)$ is 1-dimensional, and it is transverse to $W^s(p^0)$ by (NH); i.e., disjoint from it. Hence, the flows from p_2^0 to p^0 are split into two parts by $W^{uu}(p_2^0)$. For $\mu \in S_1$, flows in one part will be glued to the orbit of tangency from p_1^0 to p_2^0, while for $\mu \in S_3$ flows in the other part will be glued to the orbit of tangency. If there are k_1 and k_2 flows from p_2^0 to p^0 in the two parts, and l flows from p^0 to index 2 critical points, and m flows to p^0 from index 1 critical points, then the two stabilizations $H_1 \to H_2$ and $H_4 \to H_3$ are of types $(l, m + k_1)$ and $(l, m + k_2)$, respectively.

Figure 23 shows an example with $k_1 = k_2 = 1$. In general, the α-curve corresponding to p_2^λ intersects the green disk in $k_1 + k_2$ horizontal segments, k_1 of which lie on one side of $W^{uu}(p_2^\lambda)$, and k_2 on the other side. So the α-curve corresponding to p_1^λ intersects the green disk in k_1 arcs on one side of the handleslide stratum, and in k_2 arcs on the other side. When p_1^0 and p_2^0 are index 2, then we obtain a similar picture, but with red and blue reversed. (This is ensured by the convention of Definition 6.1 that now $p^0 \in C_{23}(f_0, v_0)$.)

(B2) An orbit of tangency from an index 1-2 birth-death point p^0 to an index 1 critical point $\overline{p}^0$ (in which case $p^0 \in C_{01}(f_0, v_0)$), or an orbit of tangency from an index 2 critical point to an index 1-2 birth-death point (in which case $p^0 \in C_{23}(f_0, v_0)$). For definiteness, we consider the first case. The

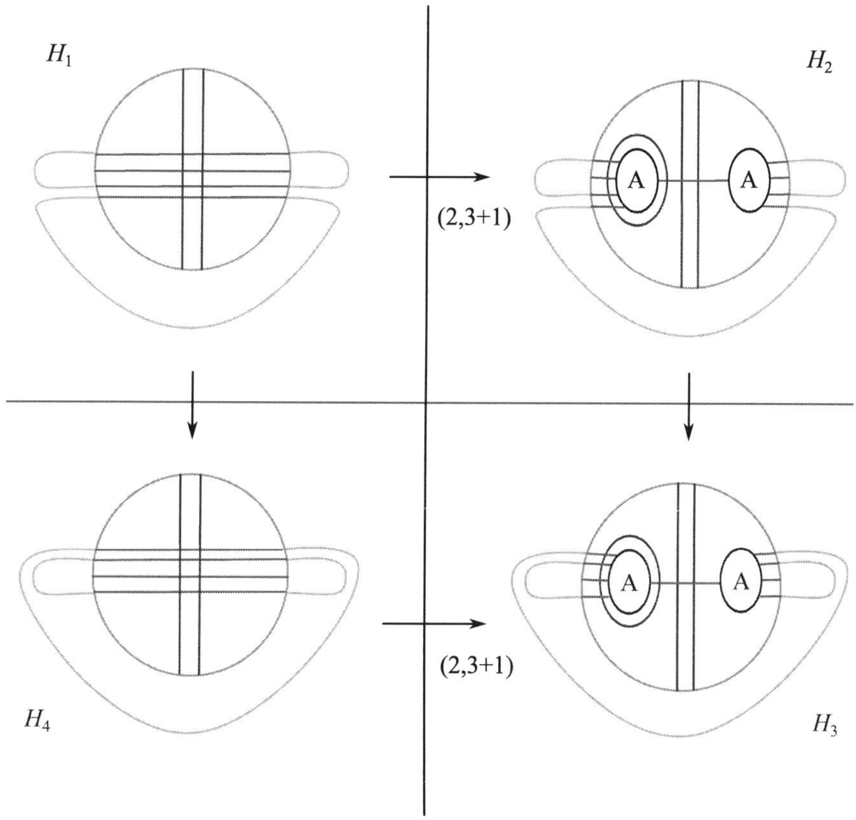

FIGURE 23. The link of a singularity of type (B1). This example has $l = 2$, $m = 3$, and $k_1 = k_2 = 1$.

bifurcation diagram has at least $r \geq 3$ strata, where S_1 and S_3 are stabilizations, and the other S_i for $i \notin \{1, 3\}$ are α-handleslides. Indeed, for any flow from an index 1 critical point p_*^0 to p^0, we can perturb the neighborhood of p^0 on the "death" side of $S_1 \cup S_3$ such that there is a flow from p_*^μ to $\overline{p}^\mu$. The number of flows from index 1 critical points to p^0 is equal to $r - 3$.

For the types of the stabilizations, suppose that there are k flows from index 1 critical points to p^0, and l flows from p^0 to index 2 critical points. Furthermore, the flows from $\overline{p}^0$ to index 2 critical points are divided into two parts by $W^{uu}(\overline{p}^0)$; let these two parts have m_1 and m_2 flows, respectively. Then the two stabilizations $H_1 \to H_2$ and $H_4 \to H_3$ have types $(l + m_1, k)$ and $(l + m_2, k)$, respectively (where $H_4 = H_1$ if $r = 3$). The pair (m_1, m_2) can be seen as the type of the generalized handleslide $H_2 \to H_3$. Figure 24 shows an example. When $\overline{p}$ is index 2, we obtain a similar picture, but with red and blue reversed.

(B3) An orbit of tangency between the strong stable manifold of an index 1-2 birth-death point p and the unstable manifold of an index 1 critical point $\overline{p}$, or between the strong unstable manifold of an index 1-2 birth-death point and the stable manifold of an index 2 critical point. Without

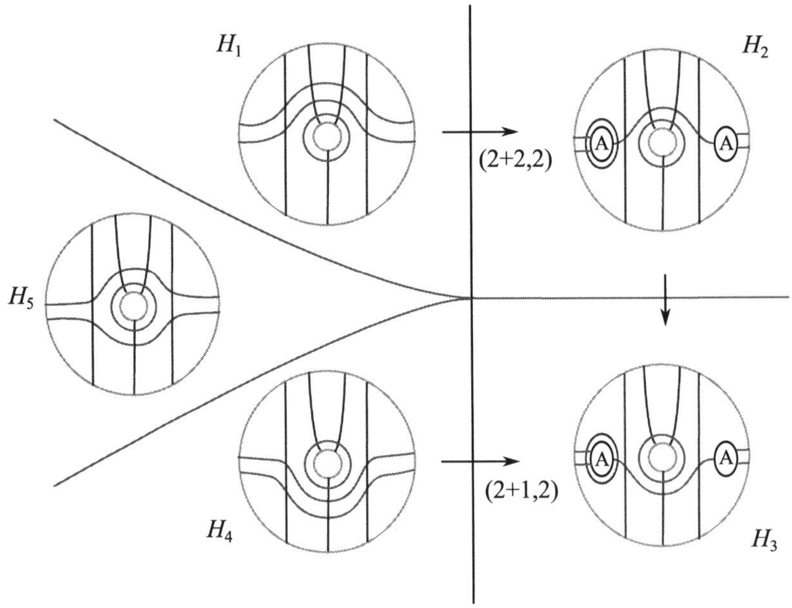

FIGURE 24. The link of a singularity of type (B2). This example has $k = 2$, $l = 2$, $m_1 = 2$, and $m_2 = 1$.

loss of generality, suppose we are in the former case. Then $r = 3$, the strata S_1 and S_3 are stabilizations, while the stratum S_2 is a handleslide. Recall that we chose $p \in C_{01}(f_0, v_0)$. If there are k flow lines from index 1 critical point to p (not counting the flow from $\overline{p}$ in $W^{ss}(p)$) and l flows from p to index 2 critical points, then the stabilizations $H_1 \to H_2$ and $H_1 \to H_3$ are of types $(k, l+1)$ and (k, l), respectively. For $\mu \in S_2$, the 2-dimensional unstable manifold $W^u(\overline{p})$ has a tangency with the 1-dimensional stable manifold of the index 1 critical point born from p; see Figure 18. Hence, the α-curve $W^u(\overline{p}) \cap \Sigma$ slides over the α-curve appearing in the stabilization as we move from H_3 to H_2. Figure 25 shows an example. As before, Definition 6.1 ensures that we obtain a picture with colors reversed when $\overline{p}$ is index 2.

(C) Two simultaneous index 1-2 birth-death critical points at p_1 and p_2, labeled such that $f(p_1) < f(p_2)$. Recall that, in this case, (f_0, v_0) is separable with $p_1 \in C_{01}(f_0, v_0)$ and $p_2 \in C_{23}(f_0, v_0)$. Let the number of flows from p_1 to p_2 be t, and let the number of flows from p_1 to index 2 critical points, and from index 1 critical points to p_1 be m and n, respectively. Similarly, let the number of flows to index 2 critical points from p_2, and from index 1 critical points to p_2 be k and l, respectively. Then $r = 4$, and each stratum S_i is a (de)stabilization. The strata S_1 and S_3 correspond to the birth-death at p_1, while S_2 and S_4 are the birth-death strata for p_2. The type of the stabilization $H_1 \to H_2$ is $(kt + m, n)$, for $H_2 \to H_3$ it is $(k, l + t)$, for $H_1 \to H_4$ it is $(k, nt + l)$, and finally, for $H_3 \to H_4$ it is $(m + t, n)$. Figure 26 shows an example with $t = 1$, and Figure 27 shows an example with $t = 2$.

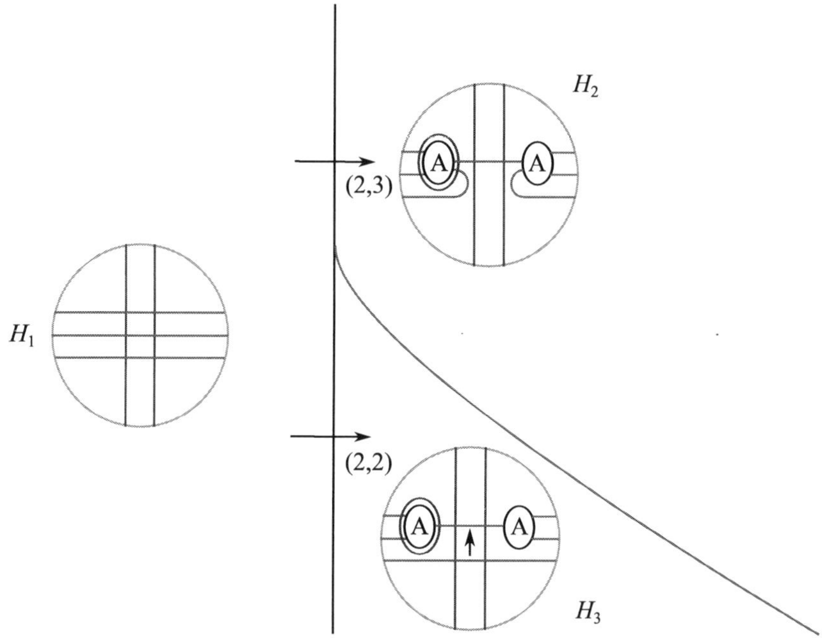

FIGURE 25. The link of a singularity of type (B3). This example has $k = 2$ and $l = 2$.

The vertices μ_1, μ_2, μ_3, μ_3', and μ_4 of P, and the Heegaard surfaces $\Sigma_i \in \Sigma(f_i, v_i)$ for $i \in \{1, \dots, 4\}$ and $\Sigma_3' \in \Sigma(f_3', v_3')$ are obtained as follows. As before, pick a surface $\Sigma \in \Sigma(f_0, v_0)$, and let $\epsilon > 0$ be so small that $\Sigma \pitchfork v_\mu$ for every $|\mu| < \epsilon$. Then $\mu_1 \in C_1$ is arbitrary and $\Sigma_1 = \Sigma$. For $\nu \in C_2 \cup S_2 \cup C_3$, let $q_1(\nu) \in C_2(f_\nu)$ be the index 2 critical point born from p_1. Similarly, for $\eta \in C_3 \cup S_3 \cup C_4$, let $q_2(\eta) \in C_1(f_\eta)$ be the index 1 critical point born from p_2. By taking ϵ to be sufficiently small, we can assume that

$$W^u(q_1(\nu)) \cap W^s(q_2(\eta)) = \emptyset$$

for every ν, $\eta \in S_2 \cup C_3 \cup S_3$. Choose points $\nu \in S_2$ and $\eta \in S_3$. The surface Σ_ν is obtained from Σ by attaching a tube around $W^u(q_1(\nu))$ such that $\Sigma_\nu \in \Sigma(f_\nu, v_\nu)$. Similarly, Σ_η is obtained from Σ by attaching a tube around $W^s(q_2(\eta))$ such that $\Sigma_\eta \in \Sigma(f_\eta, v_\eta)$. Pick short arcs a_2 and a_3 transverse to S_2 and S_3 at ν and η, respectively, such that $\Sigma_\nu \pitchfork v_\mu$ for every $\mu \in a_2$ and $\Sigma_\eta \pitchfork v_\mu$ for every $\mu \in a_3$. We take $\mu_2 = \partial a_2 \cap C_2$, $\mu_3 = \partial a_2 \cap C_3$, $\mu_3' = \partial a_3 \cap C_3$, and $\mu_4 = \partial a_3 \cap C_4$. Furthermore, $\Sigma_2 = \Sigma_\nu$ and $\Sigma_4 = \Sigma_\eta$. To obtain $\Sigma_3 \in \Sigma(f_3, v_3)$, add a tube to Σ_2 around $W^s(q_2(\mu_3))$. Similarly, $\Sigma_3' \in \Sigma(f_3', v_3')$ is obtained from Σ_4 by adding a tube around $W^u(q_1(\mu_3'))$. The edges a_1, a_3', and a_4 of P are chosen arbitrarily (subject to $|a_i \cap S_i| = 1$ for $i \in \{1, 4\}$ and $a_3' \subset C_3$).

The regions of Σ shown in Figures 26 and 27 are obtained by taking a regular neighborhood N of

$$(W^u(p_1) \cup W^s(p_2)) \cap \Sigma;$$

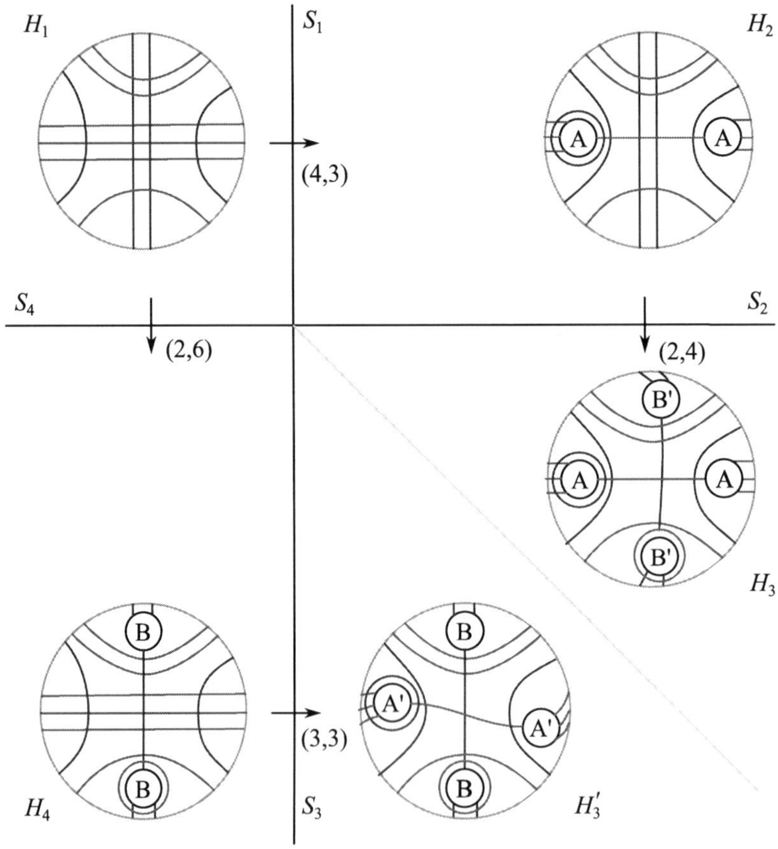

FIGURE 26. The link of a singularity of type (C). This example has $t = 1$, $(k, l) = (2, 3)$, and $(m, n) = (2, 3)$.

so the green curves are the components of ∂N. Recall that both $W^u(p_1) \cap \Sigma$ and $W^s(p_2) \cap \Sigma$ are arcs, which intersect each other in t points $x_1, \dots, x_t$. For $i \in \{1, \dots, t-1\}$, let c_i be a properly embedded arc in N that intersects $W^s(p_2) \cap \Sigma$ transversely in a single point between x_i and x_{i+1}. Cutting N along $c_1, \dots, c_{t-1}$, we obtain a disk with distinguished pairs of arcs in its boundary. This is the disk that we draw in our figures, with the c_i shown in black.

REMARK 6.38. To avoid the "diffeomorphism stratum" in case (C), one would need a quantitative result that the arcs a_2 and a_3 can be chosen to be so long that they intersect, in which case we could take $\mu_3 = a_2 \cap a_3$ and drop μ_3'. This does not seem possible for an arbitrary 2-parameter family.

Note that the diffeomorphism $d_3 \colon H_3 \to H_3'$ induced by a_3' can be destabilized to a diffeomorphism $d_3' \colon \Sigma \to \Sigma$. This follows from the fact that $\Sigma \pitchfork v_\mu$ for every $\mu \in a_3'$, and both Σ_3 and Σ_3' are obtained by attaching tubes to Σ. Indeed, consider the family of surfaces $\Sigma_\mu \in \Sigma(f_\mu, v_\mu)$ for $\mu \in a_3'$ obtained by adding tubes around $W^u(q_1(\mu))$ and $W^s(q_2(\mu))$. Then one can apply Lemma 6.19 to lift this family of surfaces to an isotopy that preserves the "tubes."

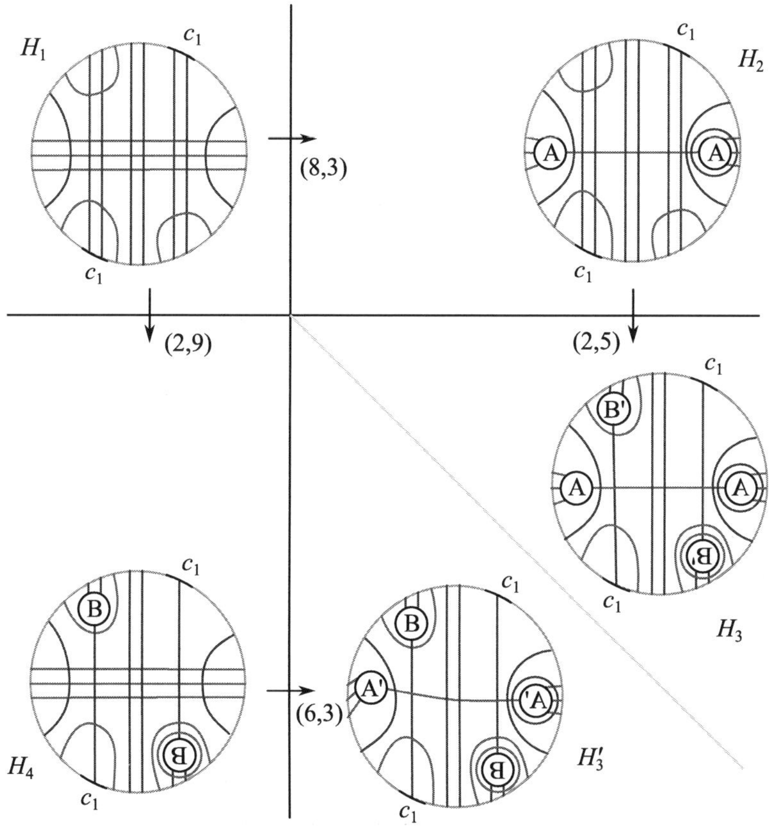

FIGURE 27. The link of a more complicated singularity of type (C). This example has $t = 2$, $(k, l) = (2, 3)$, and $(m, n) = (4, 3)$. The black arcs labeled with c_1 in the boundary of the green circle are identified.

(D) An index 1-2-1 (A_3^+) or 2-1-2 (A_3^-) degenerate critical point: a birth-death-birth singularity. See Figure 28 for an example in the index 2-1-2 case, which we will discuss. In this case, $r = 2$, and on the stabilized side C_2, we have three critical points, p_1, p_2, and p_3, with p_1 and p_3 of index 2 and p_2 of index 1. In the birth-death strata S_1 and S_2, the critical points cancel each other in two different ways: p_2 cancels against either p_1 or p_3. For both cancellations to be possible, there is necessarily a unique flow from p_2 to both p_1 and p_3, and no other flows from p_2 to index 2 critical points. To see there are no other flows from p_2, recall that the local form of an A_3^- singularity p is $-x_1^2 + x_2^2 - x_3^4$, hence it has a 1-dimensional unstable manifold, which is generically disjoint from the stable manifolds of all index 2 critical points. So, after a sufficiently small deformation of f_0, these stable manifolds will still avoid a neighborhood of p. The parameters are the numbers k and l of flows from index 1 critical points to p_1 and p_3, respectively, not counting the flows from p_2. The two stabilizations corresponding to passing S_1 and S_2 are of types $(1, k)$ and

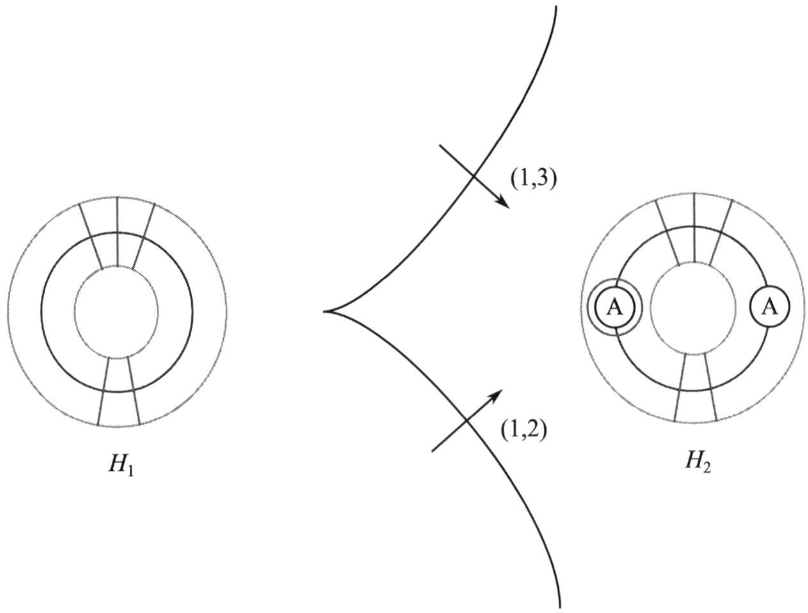

FIGURE 28. The link of a birth-death-birth singularity, type (D). In this example, $k = 3$ and $l = 2$.

$(1, l)$, respectively. Note that, on the common destabilized diagram H_1, there is a single β-circle meeting $k + l$ of the α-strands.

The link P is an arbitrary bigon around 0 inside D^2_ϵ. The Heegaard surface $\Sigma_1 = \Sigma$. The part shown in Figure 28 is a neighborhood of $W^s(p) \cap \Sigma$, where p is the degenerate critical point of f_0. The surface Σ divides M into two pieces, M_- and M_+, such that $p \in M_+$. To obtain Σ_2, we add a tube around $W^s(p_2)$ to Σ so thin that it separates p_2 from p_1 and p_3. Recall that $W^s(p)$ is a 2-disk, while $W^{ss}(p)$ is a curve inside it. The numbers k and l are in fact the number of flow lines from index 1 critical points to p on the two sides of $W^{ss}(p)$ in $W^s(p)$.

(E1) A flow from an index 2 critical point p_1 to an index 1 critical point p_2. Suppose there are k flows from p_2 to index 2 critical points, and l flows from index 1 critical points to p_1. Then $r = k + l$, and k of the strata S_i are β-handleslides, while l of them are α-handleslides. Indeed, as we move the parameter value μ in a circle around 0, for each flow from another index 1 critical point q to p_1, we pass a stratum S_i where we see an orbit of tangency in $W^u(q) \cap W^s(p_2)$, which translates to an α-handleslide. Similarly, for each flow from p_2 to an index 2 critical point r, for some value $\mu \in S_i$, we see an orbit of tangency in $W^u(p_1) \cap W^s(r)$, which translates to a β-handleslide. In each case, another curve slides over the α-circle $W^u(p_2) \cap \Sigma$ or the β-circle $W^s(p_1) \cap \Sigma$.

We now explain how to choose ϵ, the link P – which is a $2r$-gon – and the corresponding diagrams $H_1, H_1', \dots, H_r, H_r'$. Since (f_0, v_0) is not separable, we describe the construction in detail. For an illustration, see

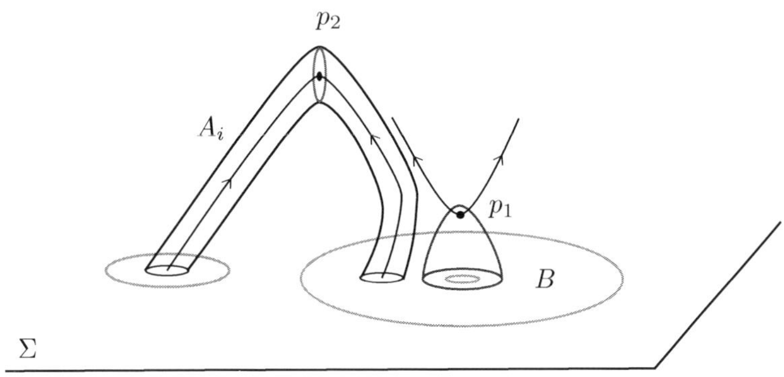

FIGURE 29. Construction of the Heegaard surface in case (E1).

Figure 29. Take N_0 to be a thin regular neighborhood of

$$\bigcup \{\, W^s(p) : p \in C_0(f_0) \cup C_1(f_0) \setminus \{p_2^0\} \,\} \cup R_-(\gamma),$$

and let $\Sigma = \overline{\partial N_0 \setminus \partial M}$. Generically, Σ is transverse to v_0, and if we choose ϵ small enough, Σ is transverse to v_μ for every $\mu \in D^2_\epsilon$. Consider the circle $\beta = W^s(p_1) \cap \Sigma$, and let B be an annular neighborhood of β in Σ so small that it is disjoint from the stable flow line into p_2 not starting at p_1. Pick values $\nu_i \in S_i$, and let A_i be a thin tube $\partial N(W^s(p_2^{\nu_i})) \setminus N_0$ disjoint from $W^u(p_1^{\nu_i}) \cup W^s(p_1^{\nu_i})$, and let $D_i \cup D_i'$ be the pair of disks $N(W^s(p_1^{\nu_i})) \cap \Sigma$ (the feet of the tube A_i). If ν_i lies sufficiently close to 0, then we can assume that $D_i \subset B$. Define

$$\Sigma_i := (\Sigma \setminus (D_i \cup D_i')) \cup A_i;$$

this is a separating surface for the Morse–Smale pair (f_i, v_i). For every $i \in \{\, 1, \dots, r \,\}$, pick an arc a_i transverse to S_i at ν_i so short that $\Sigma_i \in \Sigma(f_\mu, v_\mu)$ and $D_i \cap W^s(p_1^\mu) = \emptyset$ for every $\mu \in a_i$. According to our conventions, $\partial a_i \cap C_i = \mu_i'$ and $\partial a_i \cap C_{i+1} = \mu_{i+1}$. Note that A_i is an annular neighborhood of the circle $\alpha^i = W^u(p_2^{\nu_i}) \cap \Sigma_i$. We also pick a small disk $D \subset \Sigma$ around the point $W^s(p_2) \cap (\Sigma \setminus B)$. Again, taking A_i sufficiently thin, all the D_i' will lie in D. The side a_i' of the link P is an arbitrary curve in C_i connecting μ_i and μ_i'. To obtain the surface enhanced link, we take $\Sigma_i' = \Sigma_{i+1}$.

The diagrams $H_i = H(f_i, v_i, \Sigma_i)$ and $H_i' = H(f_i', v_i', \Sigma_i')$ all agree outside the subsurfaces $T_i = (D \setminus D_i') \cup A_i \cup (B \setminus D_i)$, up to a small isotopy of $\boldsymbol{\alpha} \cup \boldsymbol{\beta}$. What happens inside the twice punctured disks T_i is depicted in Figure 30. There, ∂T_i are the green circles, and the two components of ∂A_i are labeled by A. (We have omitted the tubes A_i for clarity.) The only difference between Σ_i and Σ_{i+1} is that the foot of the tube A_i lying in B is moved around by an isotopy. The diagrams H_i' and H_{i+1} are related by a handleslide, while a_i' induces a diffeomorphism $\varphi_i \colon H_i \to H_i'$ isotopic to the identity in M (and well-defined up to isotopy). The composition $\varphi_n \circ \cdots \circ \varphi_1 \colon \Sigma_1 \to \Sigma_1$ is a diffeomorphism that is the product of Dehn twists about the components of ∂T_1; see Definition 7.9.

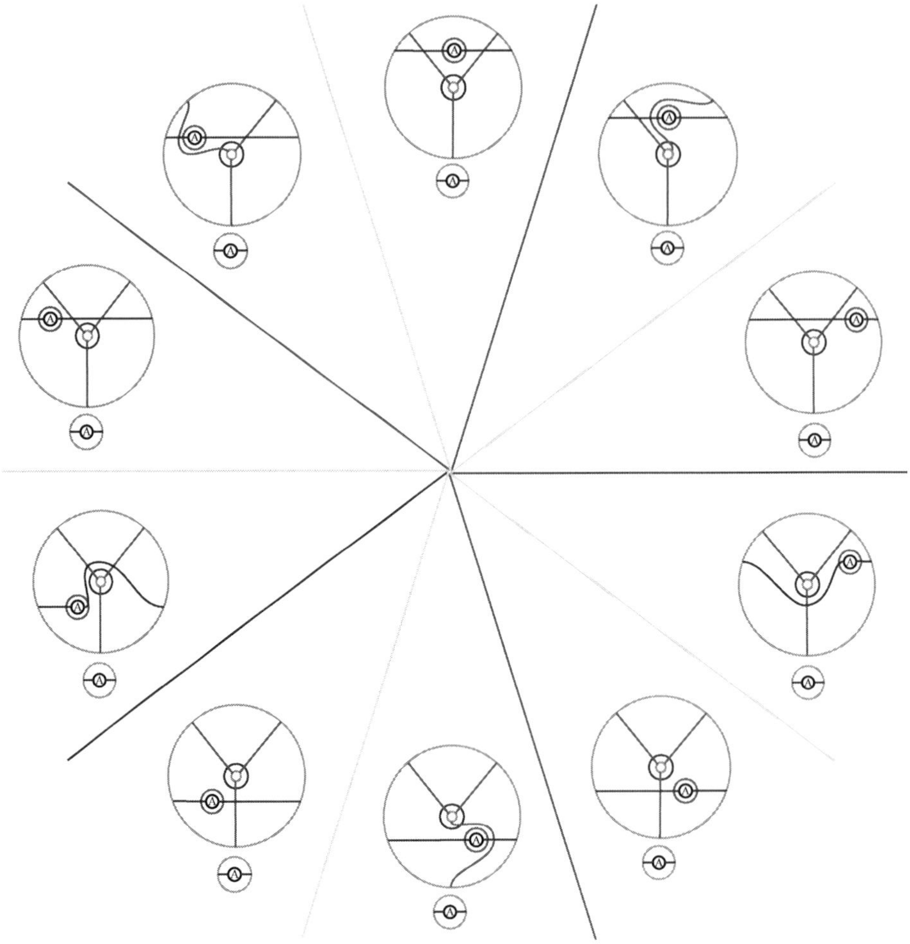

FIGURE 30. The bifurcation diagram for singularity (E1), a flow from an index 1 critical point to an index 2 critical point. In this example, $(k, l) = (2, 3)$. Outside the green circles, all five diagrams are small isotopic translates of each other. The Heegaard surface inside the green circles is not constant; there is a tube that moves around joining the two black boundary circles labeled by A.

In cases (E2)–(E4), the same splitting surface Σ can be chosen for every (f_i, v_i), and the curves $\boldsymbol{\alpha}_i$ (respectively $\boldsymbol{\beta}_i$) are all isotopic to each other for an appropriate choice of spanning trees. Since we are going to pass to isotopy diagrams, this description suffices for our purposes. This concludes the proof of Theorem 6.37. $\square$

CHAPTER 7

Simplifying Moves on Heegaard Diagrams

In this chapter, we break down (k, l)-stabilizations, generalized handleslides, and the loops of diagrams of type (A)–(E) appearing in Theorem 6.37 into the simpler moves and loops that come up in the definition of strong Heegaard invariants, Definition 2.32. During the simplification procedure, we work with overcomplete diagrams, and will only later choose spanning trees to pass to actual sutured diagrams.

We now overview the relevant combinatorial structures we are going to use. A generic 2-parameter family of gradient-like vector fields

$$\mathcal{F}\colon D^2 \to \mathcal{FV}(M, \gamma)$$

gives rise to a bordered stratification of D^2; see Definition 7.2. We subdivide this stratification into a certain CW decomposition of D^2, and consider a dual CW decomposition. We label the vertices of this dual decomposition with overcomplete diagrams such that neighboring diagrams are related by (generalized) Heegaard moves. The simplification procedure consists of modifying the CW decompositions in a combinatorial way, as opposed to modifying the 2-parameter family of gradients. Our goal is to obtain a CW decomposition of D^2 where the boundary of every 2-cell corresponds to one of the elementary loops in the definition of strong Heegaard invariants (Definition 2.32). We now give the relevant definitions.

DEFINITION 7.1. A *polyhedral decomposition* of D^2 is a regular CW decomposition of D^2 (i.e., the attaching map of every cell is an embedding) such that every closed 1-cell is smoothly embedded in D^2.

A *bordered polyhedral decomposition* of D^2 is a partition of D^2 that arises as follows: Choose a polyhedral decomposition of D^2 such that every 0-cell in the boundary S^1 has valence 3, every closed 1-cell not contained in S^1 intersects S^1 in at most one 0-cell, and every 2-cell intersects S^1 in at most one 1-cell. Then take the union of each open i-cell in S^1 with the neighboring open $(i+1)$-cell in $\operatorname{Int}(D^2)$ for $i \in \{0, 1\}$ (where an open 0-cell is just a 0-cell). We call these *bordered* $(i+1)$*-cells*.

A polyhedral decomposition $\mathcal{P}$ and a bordered polyhedral decomposition $\mathcal{R}$ are *dual* if $\operatorname{sk}_0(\mathcal{P}) \cap \operatorname{sk}_1(\mathcal{R}) = \emptyset$ and $\operatorname{sk}_0(\mathcal{R}) \cap \operatorname{sk}_1(\mathcal{P}) = \emptyset$, in each 2-cell of $\mathcal{P}$ there is a unique vertex of $\mathcal{R}$, and in each (bordered) 2-cell of $\mathcal{R}$ there is a unique vertex of $\mathcal{P}$. Furthermore, for each 1-cell e of $\mathcal{P}$, we have $|e \cap \operatorname{sk}_1(\mathcal{R})| = 1$, and for each (bordered) 1-cell e^* of $\mathcal{R}$, we have $|e^* \cap \operatorname{sk}_1(\mathcal{P})| = 1$. For an example, see Figure 31.

DEFINITION 7.2. We say that the partition

$$\mathfrak{S} = V_0 \sqcup V_1 \sqcup V_2$$

is a *bordered stratification* of the disk D^2 if the following hold:

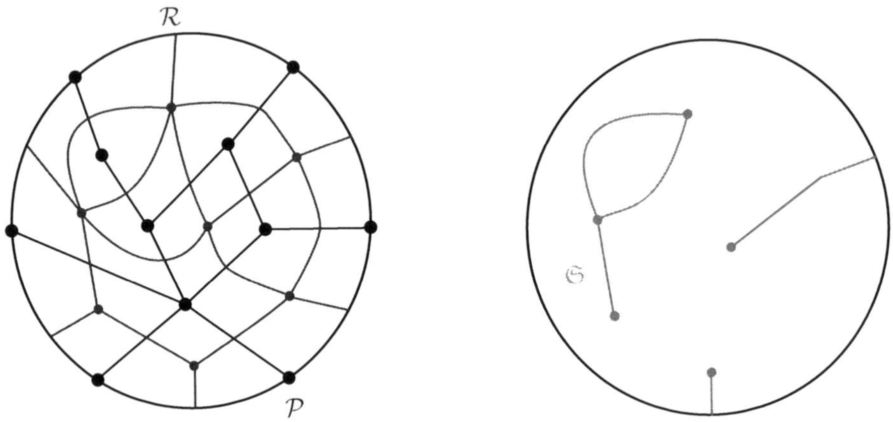

FIGURE 31. On the left, a polyhedral decomposition $\mathcal{P}$ of D^2 is shown in black. It is dual to the blue bordered polyhedral decomposition $\mathcal{R}$ that refines the green bordered stratification $\mathfrak{S}$ of D^2 on the right.

(1) V_0 is a finite set of points in the interior of D^2,
(2) V_1 is a properly embedded 1-dimensional submanifold-with-boundary of $D^2 \setminus V_0$, and
(3) each point $x \in V_0$ has a neighborhood N_x such that the pair $(N_x, V \cap N_x)$ is homeomorphic to a cone $(D^2, I \cdot H)$ for some finite set $H \subset S^1$, where $V = V_0 \cup V_1$, and $I \cdot H$ is the union of the line segments connecting the origin with each point of H.

Note that we can view every bordered polyhedral decomposition as a bordered stratification if we set V_0 to be the union of the 0-cells, and V_1 to be the union of the (bordered) 1-cells. A bordered polyhedral decomposition $\mathcal{R}$ of D^2 is a *refinement of* $\mathfrak{S}$ if $\mathrm{sk}_i(\mathcal{R}) \supset V_i$ for $i \in \{0, 1\}$ (in particular, every open 1-cell of $\mathcal{R}$ is either contained in V_1, or is disjoint from it). We say that a polyhedral decomposition of D^2 is *dual to* $\mathfrak{S}$ if it is dual to some bordered polyhedral decomposition $\mathcal{R}$ refining $\mathfrak{S}$. For an example, see Figure 31.

A generic 2-parameter family of gradient-like vector fields $\mathcal{F}\colon D^2 \to \mathcal{FV}(M, \gamma)$ gives rise to a bordered stratification $\mathfrak{S}(\mathcal{F})$ of D^2 by taking

$$V_i = \{\, \mu \in D^2 : \mathcal{F}(\mu) \in \mathcal{FV}_{2-i}(M, \gamma) \,\}$$

for $i \in \{\, 0, 1, 2 \,\}$.

DEFINITION 7.3. A polyhedral decomposition $\mathcal{P}$ of D^2 is *adapted to the family* $\mathcal{F}$ if

(1) $\mathcal{P}$ is dual to $\mathfrak{S}(\mathcal{F})$,
(2) each edge intersecting V_1 is so short that Proposition 6.28 applies to it,
(3) if $\overline{\mu} \in V_0$ and σ is the 2-cell of $\mathcal{P}$ containing $\overline{\mu}$, then $\partial\sigma$ is a link of $\overline{\mu}$ as in Theorem 6.37,
(4) every 2-cell σ of $\mathcal{P}$ that intersects V_1 but is disjoint from V_0 is a quadrilateral, and $\sigma \cap V_1$ is an arc connecting opposite sides of σ,

(5) any two closed 2-cells of $\mathcal{P}$ containing two different points of V_0 are disjoint, and any two closed 1-cells of $\mathcal{P}$ that intersect $V_1 \setminus S^1$ are either disjoint, or they both belong to a 2-cell containing a point of V_0.

LEMMA 7.4. *Let $\mathcal{F}\colon D^2 \to \mathcal{FV}(M,\gamma)$ be a generic 2-parameter family. Then there exists a polyhedral decomposition $\mathcal{P}$ of D^2 adapted to $\mathcal{F}$. Furthermore, given a triangulation of S^1 such that each 1-cell contains at most one bifurcation point of $\mathcal{F}$ and satisfies Condition* (2) *of Definition* 7.3, *then we can choose $\mathcal{P}$ such that it extends this triangulation.*

PROOF. First, choose the 2-cells of $\mathcal{P}$ containing the points of V_0 using Theorem 6.37, all taken to be sufficiently small, and denote by $N(V_0)$ their union. Pick short arcs transverse to $V_1 \setminus N(V_0)$ such that Proposition 6.28 applies to each, and such that there is an arc through each boundary point of V_1 lying inside S^1. These will all be 1-cells of $\mathcal{P}$. Next, as in Proposition 6.32, connect the endpoints of neighboring 1-cells intersecting V_1 such that we obtain a collection of rectangles that, together with $N(V_0)$, completely cover V_1. The rectangles are 2-cells of $\mathcal{P}$. Finally, we subdivide the remaining regions until the attaching map of each 2-cell becomes an embedding. This is possible if we choose sufficiently many 1-cells intersecting V_1. It is apparent from the construction that $\mathcal{P}$ is dual to a bordered polyhedral decomposition of D^2 refining the bordered stratification $\mathfrak{S}(\mathcal{F})$. If we are already given $\mathcal{P} \cap S^1$, then the extension to D^2 proceeds in an analogous manner. $\square$

If $\mathcal{P}$ is adapted to the generic 2-parameter family $\mathcal{F}\colon D^2 \to \mathcal{FV}(M,\gamma)$, then we label each edge of $\mathcal{P}$ with the type of move that occurs as we move along it, which is either a diffeomorphism isotopic to the identity if the edge does not cross V_1, or a generalized handleslide, a (k,l)-stabilization, or birth/death of a redundant α/β curve if the edge does cross V_1.

DEFINITION 7.5. Let $\mathcal{F}\colon D^2 \to \mathcal{FV}(M,\gamma)$ be a generic 2-parameter family, and $\mathcal{P}$ an adapted polyhedral decomposition. A choice of Heegaard surfaces

$$\{\, \Sigma_\mu \in \Sigma(\mathcal{F}(\mu)) \,:\, \mu \in \mathrm{sk}_0(\mathcal{P}) \,\}$$

is *coherent* with $\mathcal{P}$ if, for every edge e of $\mathcal{P}$ with $\partial e = \mu - \mu'$, the isotopy diagrams $[H(\mathcal{F}(\mu),\Sigma_\mu)]$ and $[H(\mathcal{F}(\mu'),\Sigma_{\mu'})]$ are related as indicated by the label of e. A *surface enhanced polyhedral decomposition of D^2 adapted to $\mathcal{F}$* is a polyhedral decomposition of D^2 adapted to $\mathcal{F}$, together with a coherent choice of Heegaard surfaces.

LEMMA 7.6. *Let $\mathcal{F}\colon D^2 \to \mathcal{FV}(M,\gamma)$ be a generic 2-parameter family, and suppose that $\mathcal{P}$ is a polyhedral decomposition of D^2 adapted to $\mathcal{F}$. If we are given Heegaard surfaces $\Sigma_\mu \in \Sigma(\mathcal{F}(\mu))$ for $\mu \in sk_0(\mathcal{P}) \cap S^1$ such that, for every edge e of $\mathcal{P}$ in S^1 with $\partial e = \mu - \mu'$, the isotopy diagrams $[H(\mathcal{F}(\mu),\Sigma_\mu)]$ and $[H(\mathcal{F}(\mu'),\Sigma_{\mu'})]$ are related as indicated by the label of e, then this can be extended to a choice of Heegaard surfaces coherent with $\mathcal{P}$.*

PROOF. For the vertices of each 2-cell containing a point of V_0, we choose the surfaces Σ_μ using Theorem 6.37. Then, for the remaining vertices of edges e that intersect $V_1 \setminus S^1$, we pick the Σ_μ using Proposition 6.28. This is possible because these edges have no vertices in common by condition (5) of Definition 7.3. For the rest of the vertices in $\mathrm{sk}_0(\mathcal{P}) \setminus S^1$, we choose Σ_μ arbitrarily. $\square$

From now on, let

$$\mathcal{F}\colon D^2 \to \mathcal{FV}(M,\gamma)$$

be a generic 2-parameter family, $\mathfrak{S} = \mathfrak{S}(\mathcal{F})$ the induced bordered stratification of D^2, and $\mathcal{P}$ an adapted surface enhanced polyhedral decomposition of D^2 with dual bordered polyhedral decomposition $\mathcal{R}$ refining $\mathfrak{S}$. In the rest of Chapter 7, we give a method for resolving $\mathcal{R}$, giving rise to a new bordered polyhedral decomposition $\mathcal{R}'$ of D^2. This consists of first replacing the strata in $V_1 \setminus N(V_0)$ corresponding to (k,l)-stabilizations by a collection of parallel strata labeled by simple stabilizations and handleslides. Then, at each point $\overline{\mu}$ of V_0, we connect these strata in a particular manner depending on the type of $\overline{\mu}$. We do not claim the existence of a family $\mathcal{F}'$ giving rise to the new decomposition $\mathcal{R}'$, though constructing such is probably straightforward but tedious. (This would be the 2-parameter analogue of Proposition 6.35.)

The role of the resolved stratification $\mathcal{R}'$ is that we can refine the polyhedral decomposition $\mathcal{P}$ adapted to $\mathcal{F}$ to obtain a decomposition $\mathcal{P}'$ dual to $\mathcal{R}'$, and we can choose (overcomplete) isotopy diagrams for the new vertices $\mathrm{sk}_0(\mathcal{P}') \setminus \mathrm{sk}_0(\mathcal{P})$ in a natural manner such that neighboring diagrams are now related by simple stabilizations, simple handleslides, or diffeomorphisms. Furthermore, along the boundary of each 2-cell of $\mathcal{P}'$, after an appropriate choice of spanning trees, each strong Heegaard invariant will commute by definition.

As in the previous chapters, we suppress the strata corresponding to index 0-1 and 2-3 saddle-nodes, since these disappear for any choice of spanning trees. For simplicity, we will often only draw the bordered polyhedral decomposition $\mathcal{R}'$, possibly the dual decomposition $\mathcal{P}'$, and the Heegaard diagrams for a few vertices μ of $\mathcal{P}'$ if the other intermediate diagrams are easy to recover. Consistently with our previous color conventions, edges of $\mathcal{R}$ and $\mathcal{R}'$ are red if the diagrams on the two sides are related by an α-equivalence, blue for β-equivalences, black for (de)stabilizations, and yellow for diffeomorphisms.

7.1. Codimension-1

Suppose that the possibly overcomplete isotopy diagram $H' = (\Sigma', [\boldsymbol{\alpha}'], [\boldsymbol{\beta}'])$ is obtained from $H = (\Sigma, [\boldsymbol{\alpha}], [\boldsymbol{\beta}])$ by a (k,l)-stabilization, as in Definition 6.26. In particular, we remove the disk $D \subset \Sigma$, and replace it with the punctured torus T to obtain Σ'. Inside T, we have two new attaching curves; namely, $\alpha \in \boldsymbol{\alpha}'$ and $\beta \in \boldsymbol{\beta}'$.

Such a (k,l)-stabilization can be replaced by a simple stabilization, k consecutive β-handleslides, and l consecutive α-handleslides. For convenience, we describe this procedure in the direction of the destabilization going from H' to H. Specifically, pick an orientation on both α and β. Let $\alpha_1, \dots, \alpha_l$ be the α-curves that intersect β, labeled in order given by the orientation of β, and possibly listing the same α-curve several times. This gives rise to a sequence of diagrams

$$H' = H_0,\ H_1, \dots, H_l,$$

where H_i is obtained from H_{i-1} by sliding α_i over α in the direction opposite to the orientation of β. Similarly, let $\beta_1, \dots, \beta_k$ be the β-curves that intersect α, labeled in order given by the orientation of α. Sliding these over β one by one in the direction opposite to the orientation of α, we obtain the diagrams $H_{l+1}, \dots, H_{l+k}$. The result is a rectangular grid between the arcs coming from $\alpha_1, \dots, \alpha_l$ and $\beta_1, \dots, \beta_k$, plus the handle T over which α and β run. We can now perform a simple destabilization

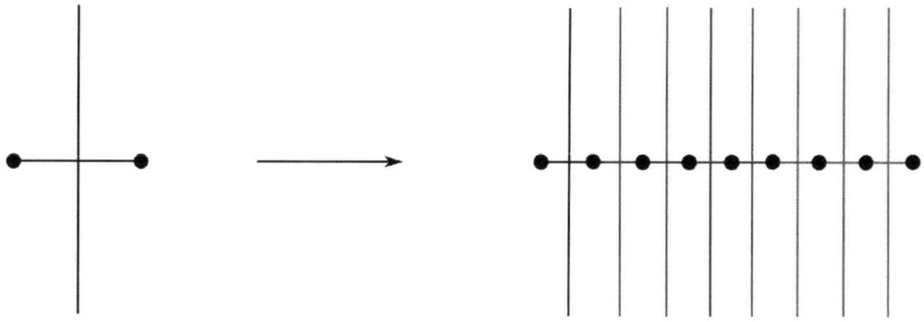

FIGURE 32. Resolving a stabilization. A stabilization of type (k, l) can be replaced by a simple stabilization, followed by k consecutive β-handleslides and l consecutive α-handleslides. When forming the resolution $\mathcal{R}'$ of the bordered polyhedral decomposition $\mathcal{R}$, we replace the (k, l)-stabilization stratum of $\mathcal{R}$ on the left with a simple stabilization stratum, k consecutive β-handleslide strata (shown in blue), and l consecutive α-handleslide strata (shown in red) of $\mathcal{R}'$ on the right.

on (T, α, β) to obtain H. See Figure 32 for the resolution of a (k, l)-stabilization stratum in a 2-parameter family.

Note that there were several choices involved in this construction, namely the orientations on α and β, and also whether to do the α-handleslides or the β-handleslides first. (In the opposite direction, going from H to H' via a (k, l)-stabilization, the choice of orientations corresponds to a choice of which quadrant around the grid of intersections to stabilize in.) It will be helpful here to introduce the notion of a stabilization slide.

DEFINITION 7.7. A *stabilization slide* is a subgraph of $\mathcal{G}$ of the form

$$\begin{array}{ccc} H_1 & \xrightarrow{e} & H_2 \\ & \searrow^{f} & \downarrow{\scriptstyle g} \\ & & H_3 \end{array}$$

such that

(1) $H_i = (\Sigma_i, [\boldsymbol{\alpha}_i], [\boldsymbol{\beta}_i])$ are (possibly overcomplete) isotopy diagrams for $i \in \{1, 2, 3\}$ such that $\Sigma_2 = \Sigma_3$,
(2) the edges e and f are stabilizations, while g is an α- or β-equivalence,
(3) there are a disk $D \subset \Sigma_1$ and a punctured torus $T \subset \Sigma_2 = \Sigma_3$ such that the restrictions $H_1|_D$, $H_2|_T$, and $H_3|_T$ are conjugate to the pictures in Figure 33 if the edge g is an α-equivalence, and to the same pictures with red and blue reversed if g is a β-equivalence, and
(4) we have

$$H_1|_{\Sigma_1 \setminus D} = H_2|_{\Sigma_2 \setminus T} = H_3|_{\Sigma_3 \setminus T}.$$

Note that, if we apply a strong Heegaard invariant to a stabilization slide, we obtain a commutative triangle. Indeed, consider the rectangle obtained from the stabilization slide triangle by taking two copies of H_1, and connecting them by

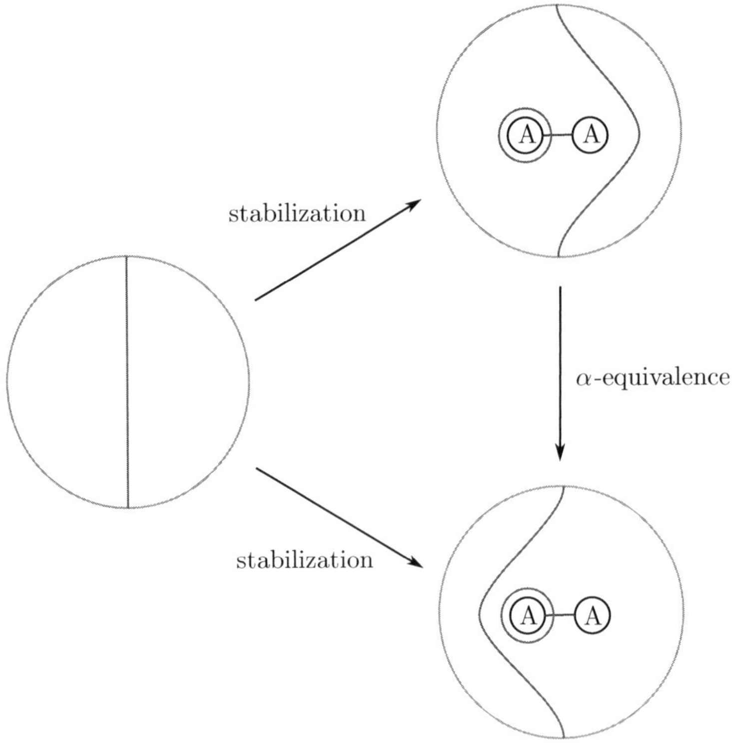

FIGURE 33. A stabilization slide. Such a loop of diagrams is a degenerate case of a distinguished rectangle of type (2).

an edge labeled by the identity of H_1. Then this is a distinguished rectangle of type (2), with two opposite edges being stabilizations, and the other two being α- or β-equivalences.

It might happen that at the two ends of a (k,l)-stabilization stratum, we need to use resolutions with different orientations of the curves α and β (recall that the resolution process depends on these orientations). Then we can interpolate between the different resolutions as follows.

LEMMA 7.8. *Suppose that the isotopy diagrams H and H' are related by a (k,l)-stabilization. Using the notation introduced at the beginning of Section 7.1, fix orientations of the curves α and β, and let*

$$H' = H_0, \dots, H_{k+l+1} = H$$

be the resolution where the diagrams H_i and H_{i+1} are related by an α-handleslide for $i \in \{0, \dots, l-1\}$, a β-handleslide for $i \in \{l, \dots, l+k-1\}$, and by a simple stabilization for $i = l+k$. If we reverse the orientation of α, we obtain a different resolution

$$H' = H_0, \dots, H_l, H'_{l+1}, \dots, H'_{k+l+1} = H$$

(the first $l+1$ diagrams are the same as we have kept the orientation of β). Then there is a regular CW decomposition $\mathcal{C}_k$ of D^2 and a labeling of its vertices with diagrams such that

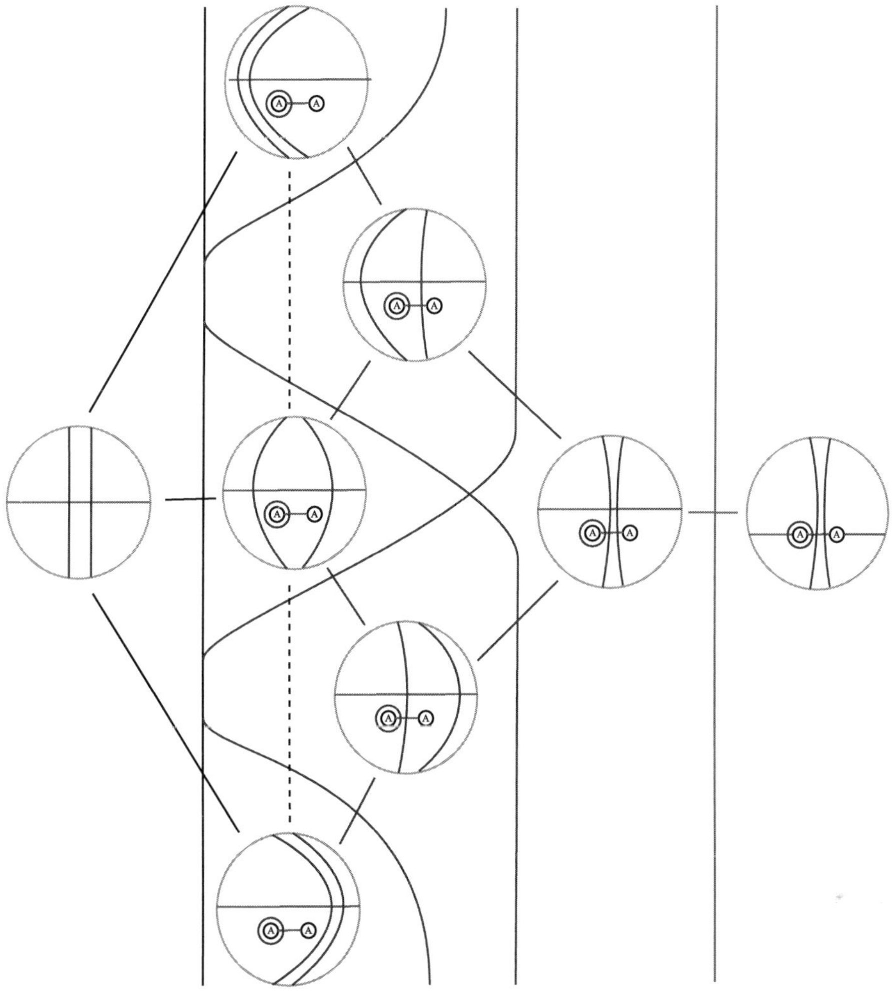

FIGURE 34. Switching the orientation involved in resolving a $(2,1)$-stabilization. On the top and bottom are two different ways of resolving a $(2,1)$-stabilization, with different choices for the orientation of α. The two different choices can be related by loops of β-equivalences and stabilization slides, as shown. Dashed edges are diagonals of rectangles whose other diagonal intersects the stabilization stratum.

(1) *the consecutive vertices along* S^1 *are labeled*

$$H, H_{k+l}, \dots, H_l, H'_{l+1}, \dots, H'_{k+l},$$

(2) *the boundary of every 2-cell is either labeled by a slide triangle, or all of its vertices are labeled by* β*-equivalent diagrams.*

Before proving Lemma 7.8, we consider the example of switching the orientation of α in a $(2,1)$-stabilization. Figure 34 illustrates how to modify the bordered polyhedral decomposition $\mathcal{R}'$ and the dual polyhedral decomposition $\mathcal{P}'$ along a

$(2,1)$-stabilization stratum of $\mathfrak{S}$ to obtain new decompositions $\mathcal{R}''$ and $\mathcal{P}_0''$ that interpolate between resolutions obtained using the two different orientations of α. Essentially, we cut $\mathcal{P}'$ and $\mathcal{R}'$ along an arc transverse to the simple stabilization stratum and all the β-handleslide strata, and glue in the CW decomposition $\mathcal{C}_2$ of the disk given by Lemma 7.8. We can also extend the stratification and its bordered polyhedral decomposition to the disk we glued in. Note that, in this figure, we have introduced four codimension-2 bifurcations of type (B3). The modified decomposition $\mathcal{R}''$ has some 1-cells corresponding to $(1,0)$-stabilizations. We obtain $\mathcal{P}_0''$ by taking the dual polyhedral decomposition $\mathcal{P}''$ of $\mathcal{R}''$, then, for each $(1,0)$-stabilization edge e of $\mathcal{R}''$, we delete the edge of $\mathcal{P}''$ passing through e, and replace it with the other diagonal of the quadrilateral in $\mathcal{P}''$ containing e. We indicated these new diagonals by dashed lines in the figure. Each such diagonal divides the corresponding quadrilateral in $\mathcal{P}_0''$ into a stabilization slide and a triangle all of whose vertices are β-equivalences. (A strong Heegaard invariant commutes along these by the Functoriality Axiom.)

PROOF OF LEMMA 7.8. Along S^1, the CW decomposition $\mathcal{C}_k$ and the labeling of the vertices by diagrams are given by condition (1). We denote each vertex on S^1 with the corresponding diagram. Assume that $H = (-1,0)$, $H_l = (1,0)$, while H_i has negative y-coordinate and H_i' has positive y-coordinate for every $i \in \{l+1,\dots,l+k\}$. We also suppose that the vertices are evenly spaced along S^1.

Before extending the CW decomposition to the interior of D^2, it is helpful to define a bordered stratification $\mathfrak{S}_k$ first: Connect the midpoint of the 1-cell between H and H_{k+l} with the midpoint of the 1-cell between H and H_{k+l}' by a straight arc a. Let $v_1,\dots,v_k$ be distinct consecutive points in the interior of a such that v_1 is closest to H_{k+l}. The straight arc b_i connects v_i with the midpoint of the 1-cell between $H_{k+l+1-i}$ and H_{k+l-i} for $i \in \{1,\dots,k\}$. Similarly, the straight arc b_i' connects v_i with the midpoint of the 1-cell between H_{l+i-1}' and H_{l+i}'. Then the set of vertices V of $\mathfrak{S}_k$ are the points $b_i \cap b_j'$ for $1 \le j \le i \le k$, and the 1-cells are the components of

$$(a \cup (b_1 \cup b_1') \cup \cdots \cup (b_k \cup b_k')) \setminus V;$$

see Figure 35 for an illustration. The bordered stratification $\mathfrak{S}_k$ can also be viewed as a bordered polyhedral decomposition. The vertices of $\mathcal{C}_k$ lying in the interior of D^2 are the centers of the 2-cells of $\mathfrak{S}_k$ disjoint from S^1. The open 1-cells of $\mathcal{C}_k$ lying in the interior of D^2 are obtained by connecting any two vertices that lie in 2-cells of $\mathfrak{S}_k$ that share an edge by a straight line. Finally, for every 2-cell of $\mathcal{C}_k$ that has H as a vertex (these are all quadrilaterals), we add the diagonal disjoint from H to the 1-skeleton of $\mathcal{C}_k$.

For $1 \le j \le i \le k$, we denote by $v_{i,j}$ the vertex of $\mathcal{C}_k$ that lies between b_i and b_{i+1} and between b_j' and b_{j+1}'. Every interior vertex of $\mathcal{C}_k$ is of this form. We label $v_{i,j}$ with the diagram $H_{i,j}$ that is obtained from H_l by handlesliding $\beta_k,\dots,\beta_{i+1}$, in this order, over β according to the positive orientation of α, and $\beta_1,\dots,\beta_j$ over β according to the negative orientation of α. The diagrams labeling the vertices of an edge of $\mathcal{C}_k$ that intersects

$$(b_1 \cup b_1') \cup \cdots \cup (b_k \cup b_k')$$

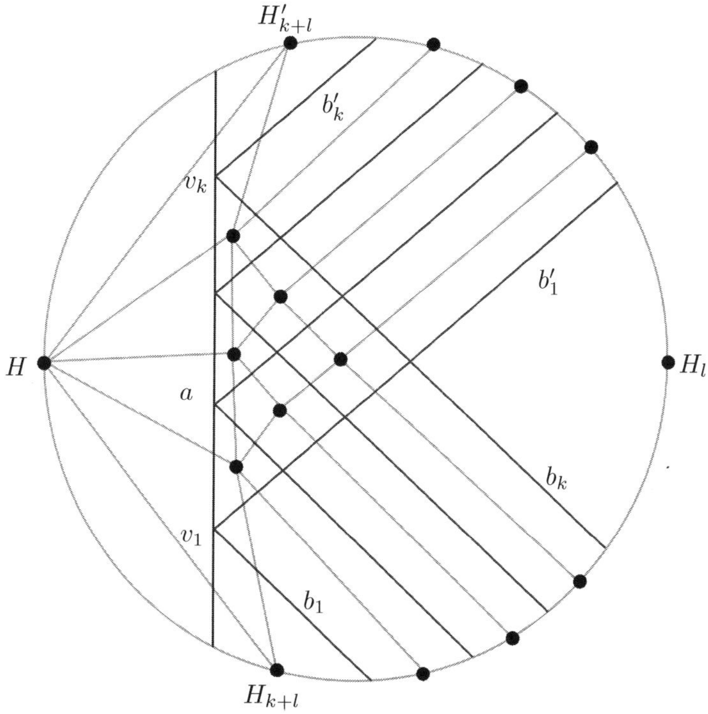

FIGURE 35. The CW decomposition $\mathcal{C}_k$ of D^2 from Lemma 7.8 is shown in grey, and the arcs $b_1, \dots, b_k$ and $b'_1, \dots, b'_k$ in blue.

in $m \in \{1,2\}$ points are related by m consecutive β-handleslides. Finally, the diagrams H, $H_{i,i}$, and $H_{i+1,i+1}$ form a slide triangle for $i \in \{0, \dots, k-1\}$, where $H_{0,0} = H_{k+l}$ and $H_{k,k} = H'_{k+l}$. □

In order to prove Theorem 2.38, it suffices to construct the modified resolution $\mathcal{P}''_0$ of $\mathcal{P}$ in a purely combinatorial manner, without actually showing the existence of a corresponding modification of the 2-parameter family of gradient-like vector fields. Thus, by cutting the parameter space along an arc transverse to the stabilization stratum and the k β-handleslide strata, and gluing in the CW decomposition $\mathcal{C}_k$ of D^2, together with the labeling of its vertices given in Lemma 7.8, we may assume that the orientations are picked conveniently along both ends of each (k,l)-stabilization stratum.

Now suppose that at one end of a (k,l)-stabilization stratum, we resolve by doing the β-handleslides first, while at the other end, we do the α-handleslides first. We can interpolate between these two choices by introducing a grid of distinguished rectangles of type (1) of Definition 2.29; see Figure 36.

If the diagram $\mathcal{H}'$ is obtained from $\mathcal{H}$ by a generalized α-handleslide of type (m,n), as in Definition 6.27, then $\mathcal{H}'$ can also be obtained from $\mathcal{H}$ by a simple (i.e., type $(0, m+n)$) handleslide, followed by an isotopy of the resulting α-curve. For an illustration, see Figure 37. Since we are passing to isotopy diagrams, we do not have to distinguish between simple and generalized handleslides.

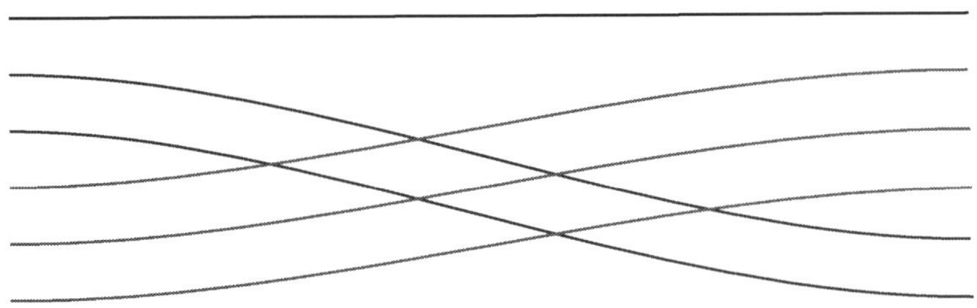

FIGURE 36. Interpolating in $\mathcal{R}'$ between the resolution of a $(2,3)$-stabilization stratum starting with the β-handleslides, and the one starting with the α-handleslides. For this, we introduce a grid of distinguished rectangles of type (1). As before, the handleslide stratum is black, the α-handleslide strata are red, and the β-handleslide strata are blue.

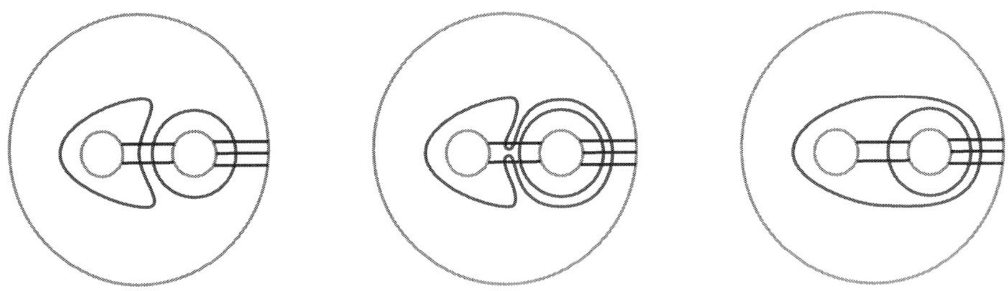

FIGURE 37. Writing a generalized handleslide of type $(2,3)$ as the composition of a simple handleslide and an isotopy of the resulting α-curve.

We now prove Proposition 2.36, which claims that, for any balanced sutured manifold (M,γ), in the graph $\mathcal{G}_{(M,\gamma)}$, any two vertices can be connected by an oriented path. Our argument refines the proofs of [**28**, Proposition 2.2] and [**16**, Proposition 2.15], in order to include diffeomorphisms isotopic to the identity among the moves connecting two diagrams.

PROOF OF PROPOSITION 2.36. Let H and H' be isotopy diagrams of (M,γ). Then pick representatives $\mathcal{H} = (\Sigma, \boldsymbol{\alpha}, \boldsymbol{\beta})$ and $\mathcal{H}' = (\Sigma', \boldsymbol{\alpha}', \boldsymbol{\beta}')$ such that $\boldsymbol{\alpha} \pitchfork \boldsymbol{\beta}$ and $\boldsymbol{\alpha}' \pitchfork \boldsymbol{\beta}'$.

By Proposition 6.17, there are simple Morse–Smale pairs (f,v) and $(f',v') \in \mathcal{FV}_0(M,\gamma)$ such that $H(f,v) = \mathcal{H}$ and $H(f',v') = \mathcal{H}'$. By Corollary 5.20, there exists a generic 1-parameter family

$$\{ (f_t, v_t) \in \mathcal{FV}_{\leq 1}(M,\gamma) : t \in I \}$$

of sutured functions and gradient-like vector fields such that $(f_0, v_0) = (f,v)$ and $(f_1,v_1) = (f',v')$. Let

$$0 < b_1 < \cdots < b_n < 1$$

be the set of parameter values such that $(f_t, v_t) \in \mathcal{FV}_1(M,\gamma)$ if and only if $t \in \{b_1, \ldots, b_n\}$.

Using Proposition 6.28 and the fact that, for a given splitting surface, any two attaching sets are α/β-equivalent, for every $i \in \{1, \ldots, n\}$, we can choose points $b_i^- < b_i < b_i^+$ close to b_i, separating surfaces

$$\Sigma_i^{\pm} \in \Sigma\left(f_{b_i^{\pm}}, v_{b_i^{\pm}}\right),$$

and spanning trees $T_i^{\pm}$ such that the diagrams $\mathcal{H}_i^-$ and $\mathcal{H}_i^+$ for

$$\mathcal{H}_i^{\pm} = H\left(f_{b_i^{\pm}}, v_{b_i^{\pm}}, \Sigma_i^{\pm}, T_i^{\pm}\right)$$

are related by an α- or β-equivalence, or a (k,l)-(de)stabilization. As explained at the beginning of Section 7.1, every (k,l)-stabilization can be written as a simple stabilization, followed by an α-equivalence and a β-equivalence.

Finally, by Lemma 6.21, $\mathcal{H}$ and $\mathcal{H}_1^-$, $\mathcal{H}_i^+$ and $\mathcal{H}_{i+1}^-$ for $i \in \{1,\dots,n-1\}$, and $\mathcal{H}_n^+$ and $\mathcal{H}'$ are related by a diffeomorphism isotopic to the identity in M, followed by an α-equivalence and a β-equivalence. □

7.2. Codimension-2

We consider the various types of singularities from Theorem 6.37 in Section 6.6, in an order that is more convenient for this section. For each type of singularity, we will construct a resolved bordered decomposition $\mathcal{R}'$, as described at the beginning of Chapter 7.

The links of singularities of type (A1a)–(A1d) and (A2) from Theorem 6.37 (involving pairs of handleslides) are easy, and we do not modify these during the resolution process. After choosing arbitrary spanning trees, we get a loop in $\mathcal{G}_{(M,\gamma)}$ where each edge is an α- or β-equivalence. Any strong Heegaard invariant F applied to this loop commutes. Indeed, such a loop can be subdivided into triangles where each edge is of the same color, and some rectangles with two opposite edges blue and two opposite edges red. The commutativity of F along a triangle is guaranteed by the Functoriality Axiom of Definition 2.32, whereas, for the rectangles – each of which is a distinguished rectangle of type (1) in the sense of Definition 2.29 – we can use the Commutativity Axiom.

Next, we consider a singularity of type (B3). This essentially is just changing the type of destabilization, and can be done with no singularities in the resolved bifurcation diagram $\mathcal{R}'$, as long as the choice of orientation for resolving the stabilization is consistent along the stabilization stratum. An example is shown in Figure 38.

For singularities of type (B1) (a birth-death singularity at p simultaneous with a handleslide of p_1 over p_2), recall that a crucial feature was the number $k = k_1 + k_2$ of flows from p_2 to p. If $k = 0$, the resolution can be done easily with several handleslide commutations (links of type (A1a)) and one stabilization-handleslide commutation. For $k > 0$, we need to introduce k slide pentagons (links of type (A2)), as $\alpha_1 = W^u(p_1) \cap \Sigma$ is sliding over $\alpha_2 = W^u(p_2) \cap \Sigma$, which in turn slides over the circle α introduced at the stabilization corresponding to p. Note that k_1 of these pentagons have exactly two vertices on the side of the α-handleslide stratum that existed before the resolution containing $C_3 \cup C_4$ (below the horizontal red line in Figure 39), while k_2 of them have exactly two vertices on the other side. See Figure 39 for an example with $k = 2$.

Before proceeding, we introduce handleswaps; loops of overcomplete diagrams that generalize the notion of simple handleswaps. As we shall see, these arise during the simplification procedure of links of type (E1), and are in fact quite close to them.

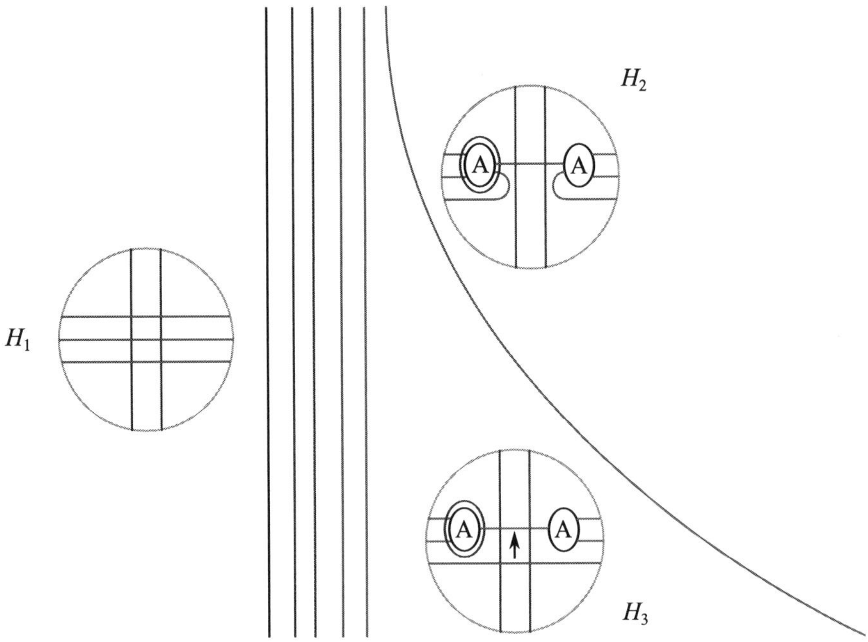

FIGURE 38. Resolving the singularity of type (B3) from Figure 25.

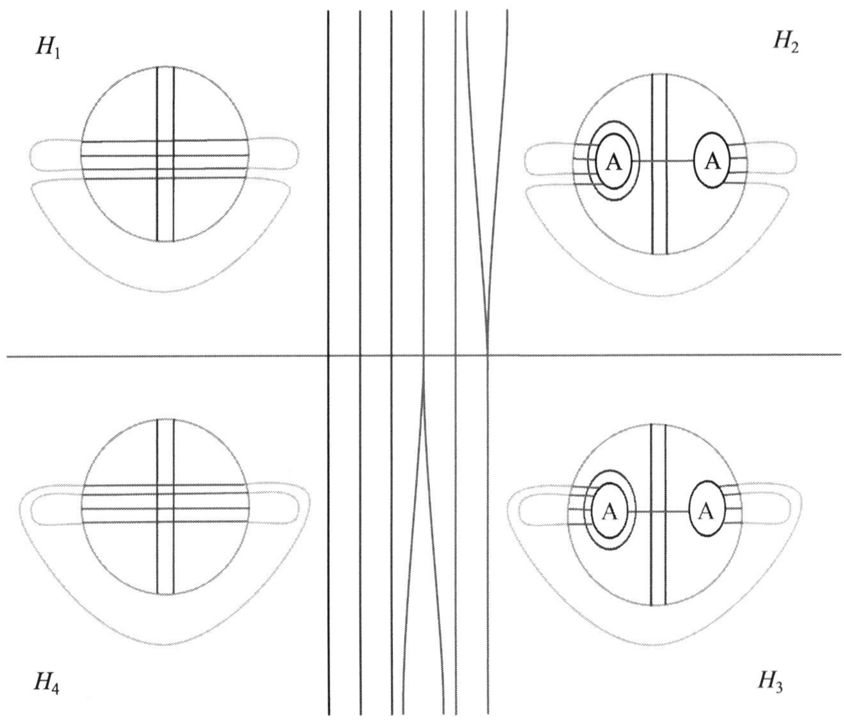

FIGURE 39. Resolving the singularity of type (B1) from Figure 23.

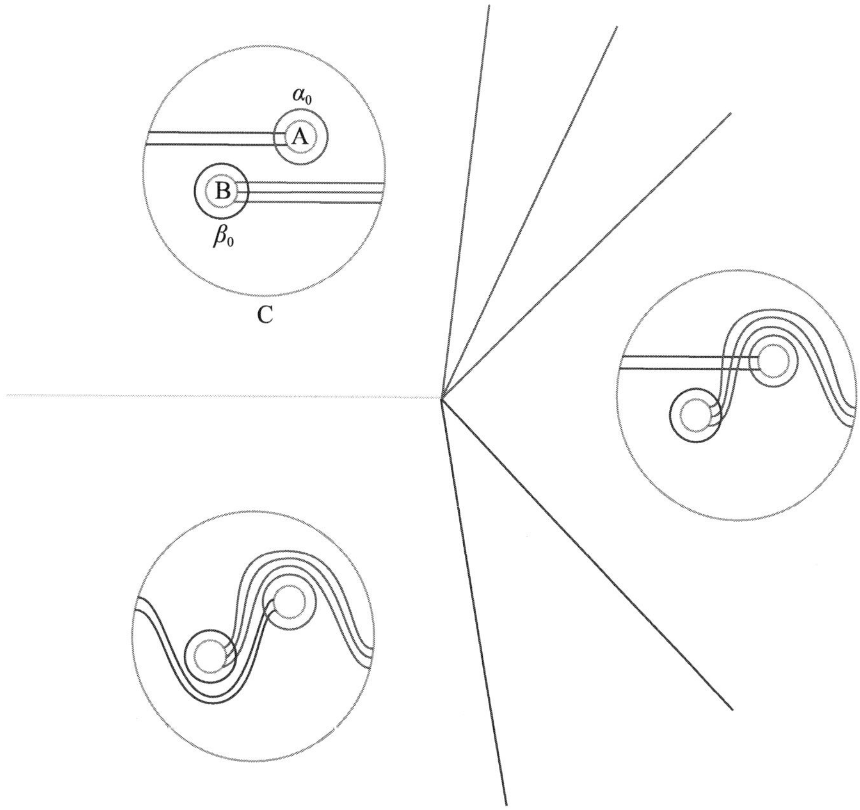

FIGURE 40. A (2; 3)-handleswap.

DEFINITION 7.9. A $(k;l)$-*handleswap* is a loop $\mathcal{H}_0, \dots, \mathcal{H}_{k+l}$ of overcomplete diagrams as follows. There is a surface Σ such that $\mathcal{H}_i = (\Sigma, \boldsymbol{\alpha}_i, \boldsymbol{\beta}_i)$ for every $i \in \{0, \dots, k+l\}$. Furthermore, there is a pair-of-pants $P \subset \Sigma$ such that

$$\boldsymbol{\alpha}_i \cap (\Sigma \setminus P) = \boldsymbol{\alpha}_j \cap (\Sigma \setminus P) \text{ and } \boldsymbol{\beta}_i \cap (\Sigma \setminus P) = \boldsymbol{\beta}_j \cap (\Sigma \setminus P)$$

for every i, j. Inside P, we have one full α-curve that we denote by α_0, and one full β-curve that we call β_0. The boundary ∂P consists of three curves, A being parallel to α_0, a curve B parallel to β_0, and the third we denote by C. The set $(\boldsymbol{\alpha}_0 \cap P) \setminus \alpha_0$ consists of l parallel arcs connecting B and C, while $(\boldsymbol{\beta}_0 \cap P) \setminus \beta_0$ consists of k parallel arcs connecting A and C. We also require that none of the α-arcs intersect the β-arcs in P. For $0 \leq i < l$, the diagram $\mathcal{H}_{i+1}$ is obtained from $\mathcal{H}_i$ by sliding one of the α-arcs over α_0, and for $l \leq i < k+l$, the diagram $\mathcal{H}_{i+1}$ is obtained from $\mathcal{H}_i$ by sliding one of the β-arcs over β_0. The diagram $\mathcal{H}_0$ is obtained from $\mathcal{H}_{k+l}$ by a diffeomorphism that is the composition of a left-handed Dehn twist about C and right-handed Dehn-twists about A and B. The case of a $(2;3)$-handleswap is depicted in Figure 40.

Similarly, we say that a loop $H_0, \dots, H_{k+l}$ of isotopy diagrams is a $(k;l)$-handleswap if every H_i has a representative $\mathcal{H}_i$ such that $\mathcal{H}_0, \dots, \mathcal{H}_{k+l}$ is a $(k;l)$-handleswap.

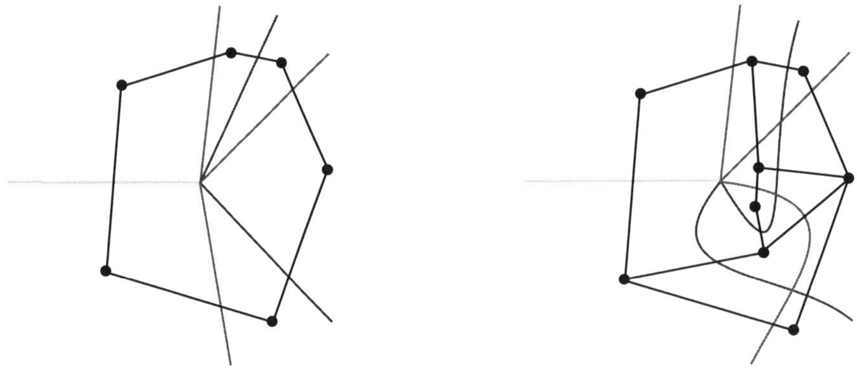

FIGURE 41. Rearranging the order of α- and β-handleslides in a handleswap using commutations. On the left, we see $\mathcal{P}$ and $\mathcal{R}$, on the right, the modified decompositions $\mathcal{P}'$ and $\mathcal{R}'$.

We will show in Chapter 8 that any $(k;l)$-handleswap can be resolved into a number of simple handleswaps, and thus that any strong Heegaard invariant applied to the $(k;l)$-handleswap commutes.

Suppose we have a loop of diagrams as in Definition 7.9, but where the α- and β-handleslides are not necessarily separated from each other. Using commutations, these can easily be rearranged in the standard form, so we will also refer to these as handleswaps. Figure 41 shows how to write a handleswap loop with mixed α- and β-handleslides as a product of a standard handleswap and some distinguished rectangles corresponding to commuting handleslides. The procedure is easier to understand on the level of the bordered polyhedral decomposition $\mathcal{R}$, where one spirals the blue strata and red strata in opposite directions to separate them.

As mentioned above, the link of a singularity of type (E1) (a flow from an index 2 critical point to an index 1 critical point) is quite close to a handleswap. We can make it exactly a handleswap by introducing some commutation moves between diffeomorphisms and handleslides. On the level of the bordered polyhedral decomposition $\mathcal{R}$, in a small neighborhood of the (E1) singularity, we spiral the yellow diffeomorphism strata corresponding to the diffeomorphisms $\varphi_1, \dots, \varphi_n$ to all lie next to each other, then compose the diffeomorphisms. For an illustration, see Figure 42. Recall that the composition

$$d = \varphi_n \circ \cdots \circ \varphi_1 \colon \Sigma_1 \to \Sigma_1$$

is the product of Dehn twists about the boundary components of the pair-of-pants T_1 (the three green circles in the figure). Finally, we rearrange the α- and β-handleswaps as above to first have the α-handleslides, followed by the β-handleslides. We denote this new surface enhanced polyhedral decomposition by $\mathcal{P}'$, and the dual bordered polyhedral decomposition by $\mathcal{R}'$.

The link of the (E1) singularity in $\mathcal{P}'$ appears slightly different from a standard handleswap, since, in the diagram H_1 right above the diffeomorphism stratum, the α- and β-arcs intersect each other. However, this is not an issue as we are dealing with isotopy diagrams. Indeed, consider the smaller pair-of-pants T_1' bounded by the dashed curve and the two small green circles in Figure 42. We choose T_1' so small that, inside it, all the α- and β-arcs are disjoint. If we now perform all handleslides

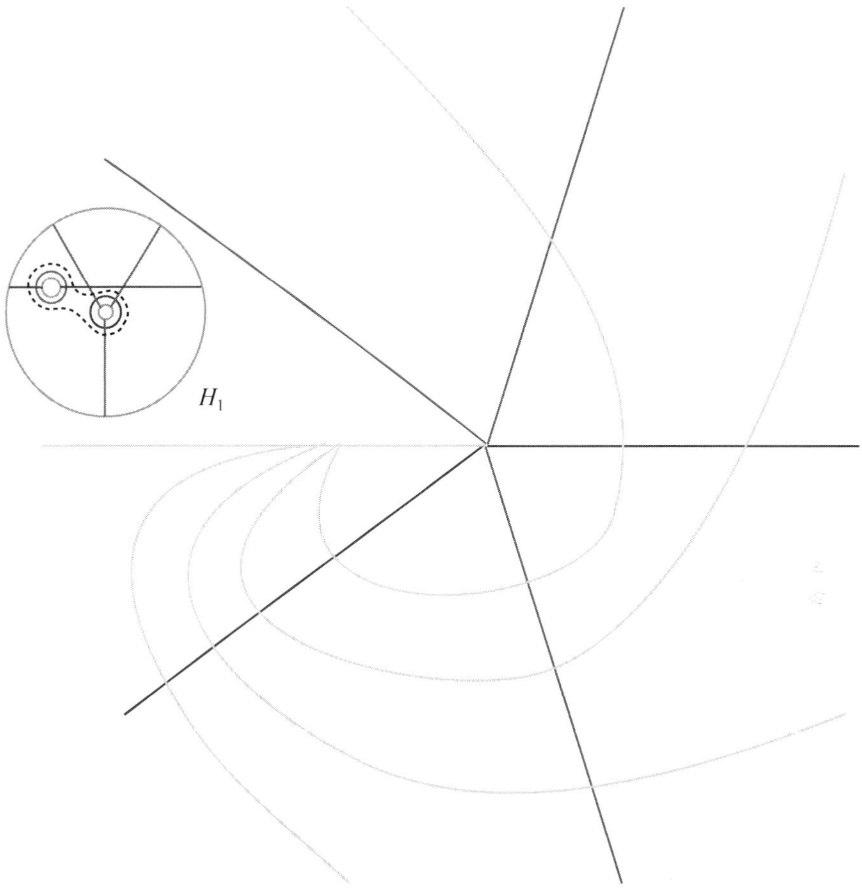

FIGURE 42. Simplifying a link of type (E1).

and the diffeomorphism within T_1', we get a standard handleswap loop, and each diagram is isotopic to the corresponding diagram in $\mathcal{P}'$. If we even replace the diffeomorphism d by the diffeomorphism d' that is a product of Dehn twists about the boundary components of T_1', then d and d' are isotopic, and by the Continuity Axiom of strong Heegaard invariants,

$$F(d) = F(d')\colon F(H_1) \to F(H_1).$$

Next, we consider singularities of type (B2), a flow from a birth-death singularity p to an index 1 critical point $\overline{p}$. As explained in the proof of Theorem 6.37 (also see Figure 24), the crucial features are the number k of flows from index 1 critical points to p, and the number $m = m_1 + m_2$ of flows from $\overline{p}$ to index 2 critical points. There are two codimension-1 stabilization strata of types $(l + m_1, k)$ and a $(l + m_2, k)$, respectively. We resolve both of these according to Section 7.1. This results in a simple stabilization stratum, followed by k consecutive α-handleslide strata, and $l + m_1$ consecutive β-handleslide strata in one of the resolutions, and $l+m_2$ consecutive β-handleslide strata in the other one. We connect the k consecutive α-handleslide strata, but the β-handleslide strata do not match up in number.

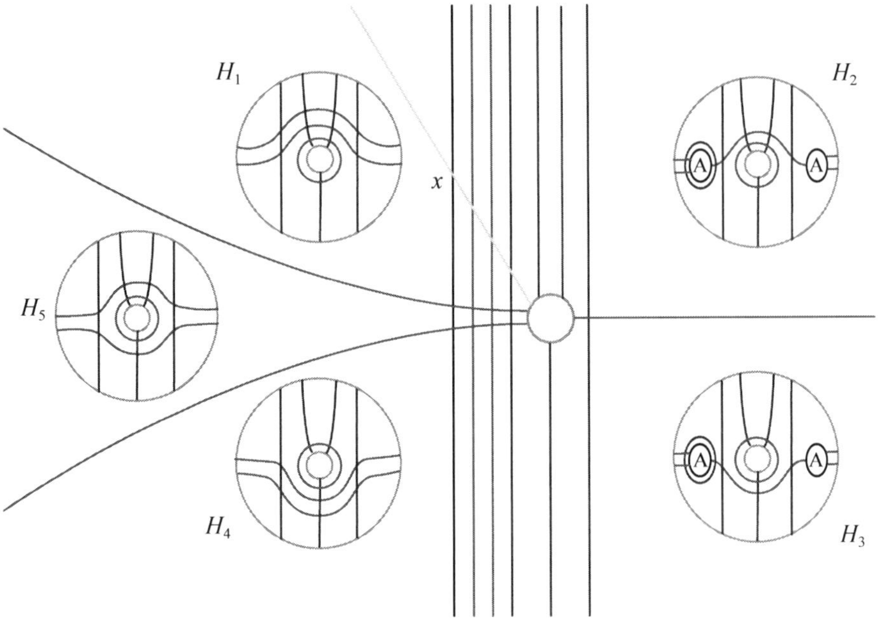

FIGURE 43. Resolving the singularity of type (B2) from Figure 24, which has $k = 2$ and $m = 3$. Here, for clarity, the handleswap has not yet been put in the standard form.

In order to obtain the resolution $\mathcal{R}'$, we remove a small disk D around the central bifurcation value in the parameter space D^2. See Figure 43 for an example, where we draw ∂D in green. In this step, we only construct $\mathcal{R}'$ and the dual polyhedral decomposition $\mathcal{P}'$ outside D. The simple stabilization stratum and the k consecutive α-handleslide strata are to one side of D. Let $\overline{\alpha}$ be the α-curve corresponding to $\overline{p}$. We connect the β-handleslide strata that correspond to β-curves disjoint from $\overline{\alpha}$ (there are l of these), and the $m_1 + m_2$ remaining β-handleslide strata end on ∂D. In a neighborhood of ∂D, the β-handleslide strata look the same as the corresponding β-curves in a neighborhood of $\overline{\alpha}$.

In the polyhedral decomposition $\mathcal{P}'$ dual to $\mathcal{R}'$, we have a number of commutations as usual, as well as an $(m; k+1)$-handleswap between the circles $\beta = W^s(p) \cap \Sigma$ and $\overline{\alpha} = W^u(\overline{p}) \cap \Sigma$. To obtain this handleswap, we add a single diffeomorphism stratum to $\mathcal{R}'$, drawn in yellow. The handleswap is the loop of diagrams in $\mathcal{P}'$ around the green circle; this loop is illustrated in Figure 44. We will explain in Chapter 8 how to extend $\mathcal{R}'$ to the interior of the green circle such that the $(m; k+1)$-handleswap is reduced to a simple handleswap. The edge e of $\mathcal{P}'$ dual to the yellow stratum on the destabilized side corresponds to a diffeomorphism that is isotopic to the identity of the Heegaard surface, hence the two vertices of e correspond to the same isotopy diagram. So we can terminate the yellow diffeomorphism stratum at a point x of the black stabilization stratum, giving rise to a triangle in the dual polyhedral decomposition $\mathcal{P}'$ containing x.

For singularities of type (D), a 2-1-2 birth-death-birth singularity, we can, as usual, replace the stabilization by a simple stabilization and a number of handleslides. This time, we can replace the cusp singularity by a slide triangle and

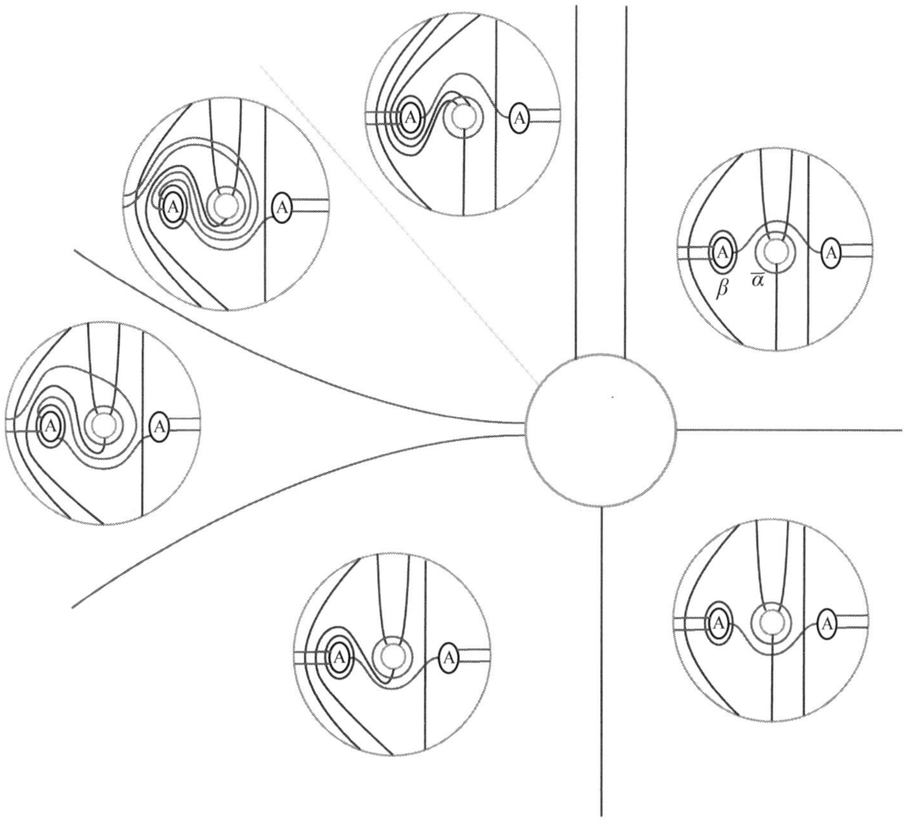

FIGURE 44. A closeup of the handleswap loop of diagrams in $\mathcal{P}'$ around the green circle from Figure 43.

a $(1; k+l)$-handleswap, as shown in Figure 45. As in case (B2), we add a diffeomorphism stratum passing through the stabilization stratum. The corresponding diffeomorphism is isotopic to the identity on the destabilized side. For a closeup of the handleslide loop, see Figure 46. The handleswap is between $\alpha_1 = W^u(p_1) \cap \Sigma$ and $\beta_2 = W^s(p_2) \cap \Sigma$.

Finally, we consider the case of a double stabilization, type (C). As shown in Figure 47, we can eliminate the diffeomorphism, and assume that we are dealing with the link in Figure 48. More precisely, we first remove from the parameter space D^2 a small disk D' around the codimension-2 bifurcation point. Let $D \subset D'$ be a smaller concentric disk whose boundary is shown in green in Figure 47. We extend the stratification $\mathcal{R}$ from $D^2 \setminus D'$ to $D^2 \setminus D$ by extending the stabilization strata straight to ∂D, and spiralling the yellow diffeomorphism stratum across the region (codimension-0 stratum) C_2 labelled by H_2, and into the region C_1 labeled by H_1, ending on ∂D. This creates two new regions. The first is split off C_2, which we label by the destabilization H_2' of H_3' along the tube B. The second one is split off C_1, and we label it by the diagram H_1' obtained from H_2' by destabilizing it along the tube A'. As explained in Remark 6.38, the diffeomorphism $d_3 \colon H_3 \to H_3'$ destabilizes to a diffeomorphism $d_3' \colon H_1 \to H_1'$ that is isotopic to Id_Σ, and so $H_1 = H_1'$ as isotopy diagrams. Consequently, we can remove the diffeomorphism

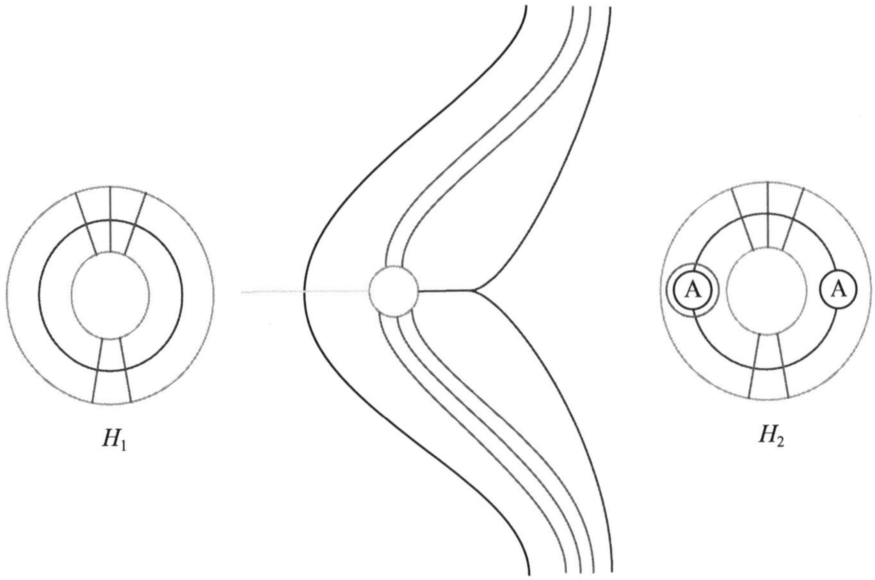

FIGURE 45. Resolving the singularity of type (D) from Figure 28.

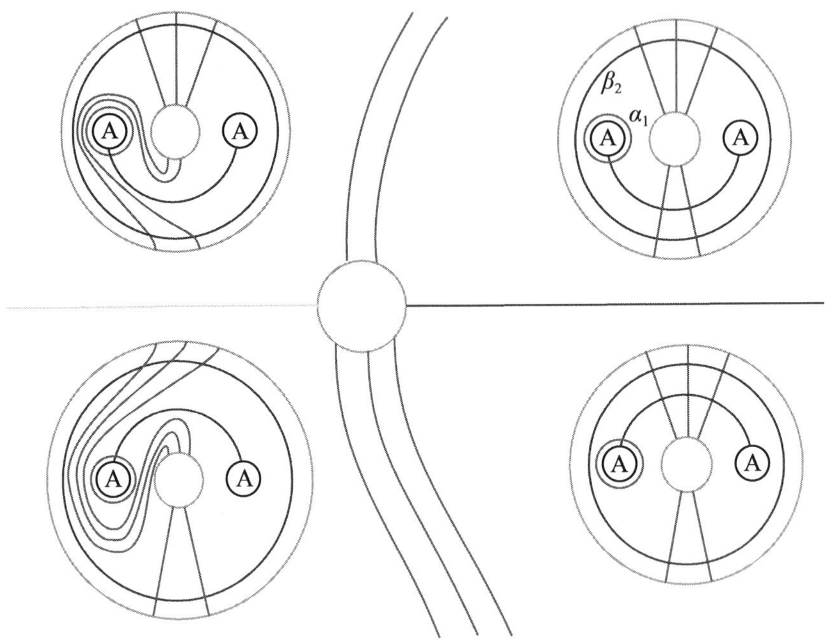

FIGURE 46. A closeup of the handleswap loop around the green circle from Figure 45. The handleswap is between α_1 and β_2.

stratum dividing the regions labeled by H_1 and H_1'. We can then extend the stabilization strata straight across the disk D, and their intersection point will have

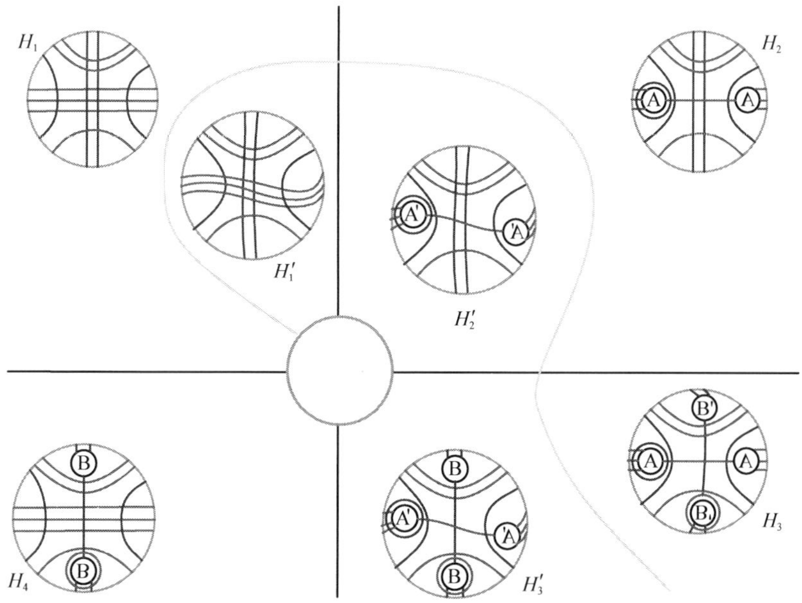

FIGURE 47. First step in resolving the singularity of type (C) from Figure 26. The two diagrams in the upper left quadrant are isotopic, so we manage to eliminate the diffeomorphism this way.

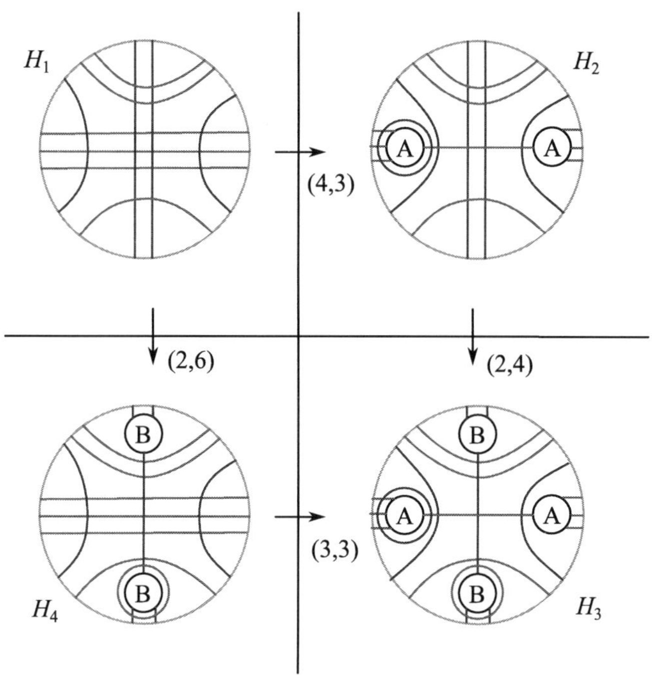

FIGURE 48. After the first reduction step in Figure 47, the link in Figure 26 can be replaced by this simpler link.

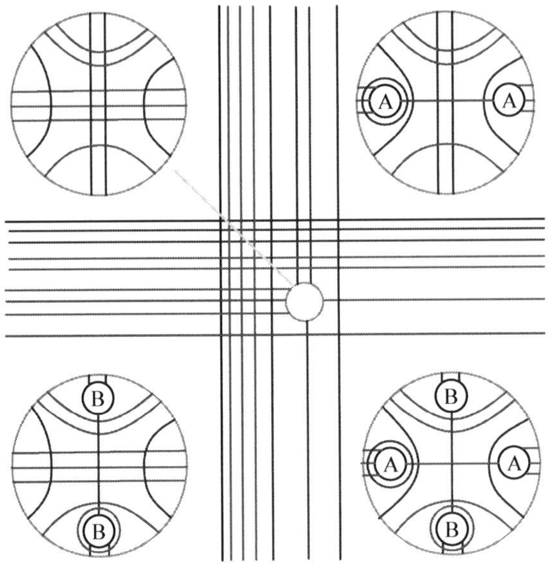

FIGURE 49. Resolving the singularity of type (C) with $t = 1$ from Figure 48.

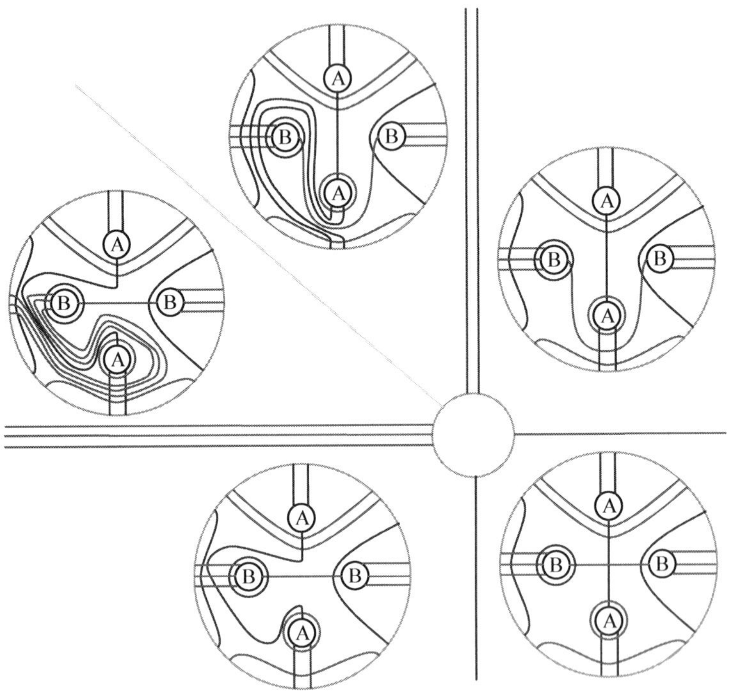

FIGURE 50. A closeup of the handleswap loop around the small green circle in Figure 49.

the link H_1, H_2', H_3', H_4 shown in Figure 48. The first step in the simplification of Figure 27 is shown in Figure 51.

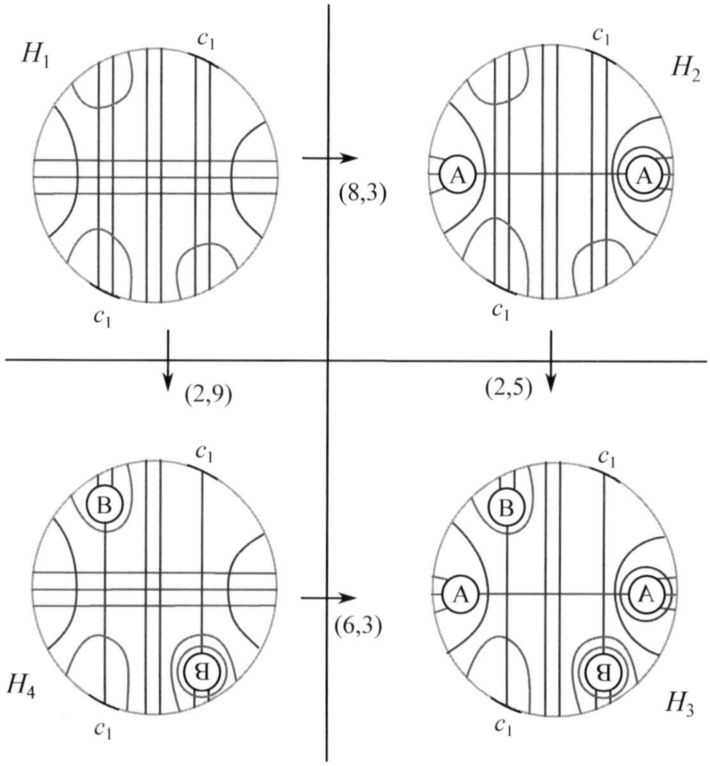

FIGURE 51. The first step in the simplification of the more complicated loop of Figure 27.

Recall that a key feature in case (C) was the number t of flows between the two stabilization points (from p_1 to p_2). If $t = 0$ and the resolutions of the generalized stabilization strata (the sequence of a stabilization and a number of handleslides) are compatible with each other, then we can fill in the link with a number of commuting squares. Otherwise, we will get a total of t different handleswaps, as shown by example in Figure 49 for $t = 1$. For a closeup of the handleswap loop, see Figure 50. An example for the $t = 2$ case is shown in Figure 52.

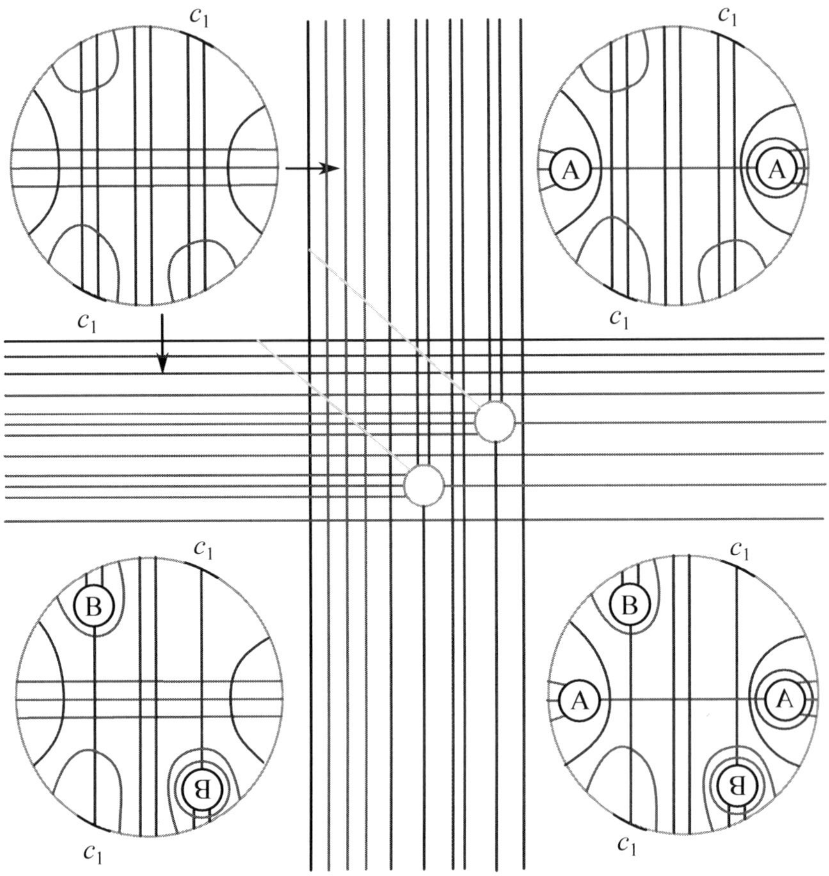

FIGURE 52. Resolving the singularity of type (C) with $t = 2$ from Figure 51.

CHAPTER 8

Simplifying Handleswaps

Note that, in Definition 7.9, a β-curve might intersect α_0 multiple times, hence several β-arcs in the pair-of-pants P might belong to the same β-curve.

DEFINITION 8.1. A $(k, 1; l)$-handleswap is a $(k + 1; l)$-handleswap between α_0 and β_0 such that there is a β-curve that intersects α_0 in a single point. Similarly, a $(k, 1; l, 1)$ handleswap is a $(k, 1; l + 1)$-handleswap such that there is an α-curve that intersects β_0 in a single point. A $(k; l, 1)$-handleswap is defined in an analogous manner.

The final ingredient in the proof of Theorem 2.38 is to replace an arbitrary $(k; l)$-handleswap by simple handleswaps. This proceeds in several stages:

- We first stabilize the diagram, to guarantee that, in each handleswap between α_0 and β_0, at least one of the β-circles meeting α_0 meets it exactly once, giving a $(k, 1; l)$-handleswap. Similarly, we do the same thing for the α-circles meeting β_0, giving a $(k, 1; l, 1)$-handleswap.
- Given a $(k, 1; l, 1)$-handleswap between α_0 and β_0 in which α_1 intersects β_0 once, we can perform handleslides of each of the α-circles intersecting β_0 over α_1 to get rid of these intersections, and reduce to the case of a $(k, 1; 1)$-handleswap. Similarly, we can perform handleslides on the $(k, 1; 1)$-handleswap to reduce it to the case of a $(1; 1)$-handleswap.
- Finally, in a $(1; 1)$-handleswap between α_0 and β_0 with β_1 intersecting α_0 and α_1 intersecting β_0, we can perform handleslides of each α-circle intersecting β_1 over α_0 to guarantee that β_1 has no intersections besides the one with α_0. We can similarly guarantee that α_1 has no intersections besides the one with β_0. This is now, by definition, a simple handleswap.

In this overview, we have talked rather loosely about "stabilizing" and "performing handleslides" on a codimension two singularity (the handleswap). In fact, we have to perform these operations consistently in 2-parameter families, and see that we reduce our original loop of Heegaard diagrams to the elementary loops of Section 2.4. We will carry this out in the following sections. Recall that, in the previous chapter, each time we encountered a handleswap in a figure we removed a disk – indicated by a green circle – from the parameter space and only drew $\mathcal{P}'$ and $\mathcal{R}'$ outside this disk. In $\mathcal{P}'$, the handleswap loop is parallel to this green circle. In each step, we extend $\mathcal{P}'$ and $\mathcal{R}'$ to an annulus in the interior of the disk removed, until we reduce to simple handleswaps.

Reducing to $(k, 1; l, 1)$-handleswaps. It is easiest to understand this reduction by using non-simple stabilizations. Specifically, we will reduce a $(k; l)$-handleswap to a $(k, 1; l)$-handleswap and a $(1; l)$-handleswap (which is, of course, a $(0, 1; l)$-handleswap). Start with a $(k; l)$-handleswap involving α_0 and β_0. Let the

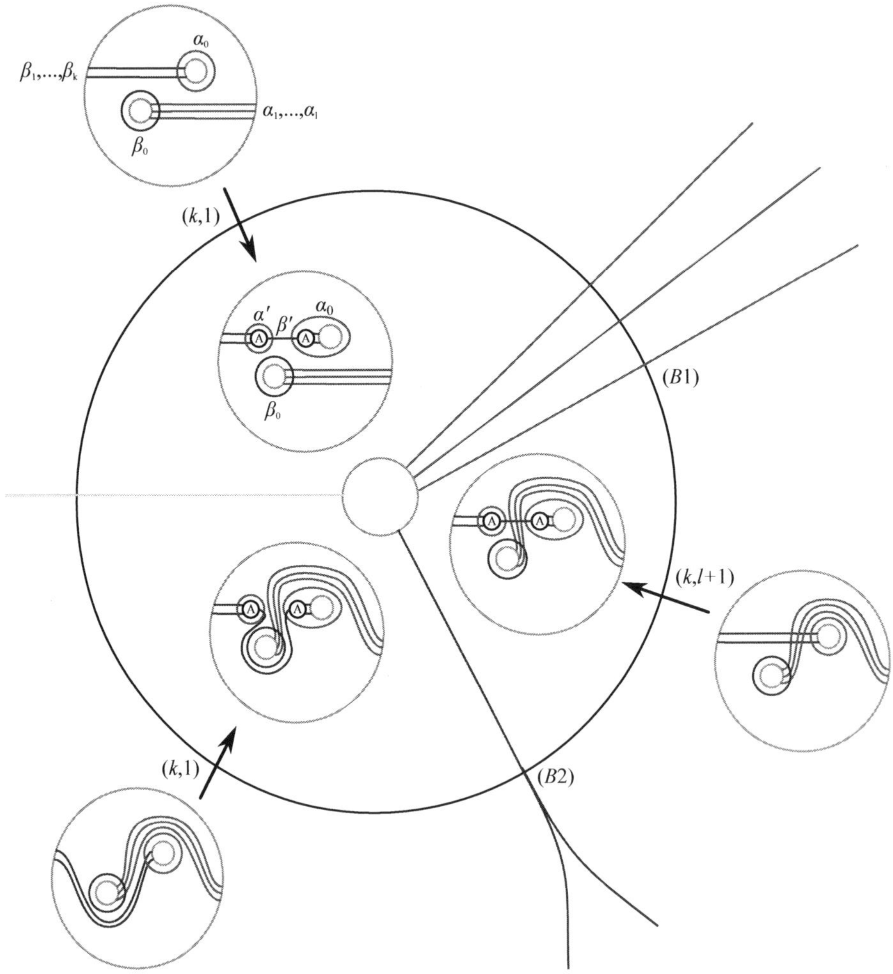

FIGURE 53. Reducing from a general $(k;l)$-handleswap to a $(k,1;l)$-handleswap, step 1. Here, we have introduced a circle of non-simple stabilizations to the $(2;3)$-handleswap from Figure 40.

β-strands crossing α_0 be $\beta_1,\ldots,\beta_k$, and let the α-strands crossing β_0 be $\alpha_1,\ldots,\alpha_l$ (both lists with multiplicities). In the diagrams involved in a handleswap without extra crossings (on the top and bottom in Figure 40), we can do a $(k,1)$-stabilization on α_0 and $\beta_1,\ldots,\beta_k$. Similarly, on the diagram with $\beta_1,\ldots,\beta_k$ crossing $\alpha_1,\ldots,\alpha_l$ (on the right in Figure 40), we can do a $(k,l+1)$-stabilization on $\beta_1,\ldots,\beta_k$ and α_0, $\alpha_1,\ldots,\alpha_l$. Let α' and β' be the new circles introduced in the stabilization. These two stabilizations in fact fit into a 2-parameter family (with the same boundary as the original $(k;l)$ handleswap): each α_i sliding over α_0 for $i\in\{1,\ldots,l\}$ introduces a singularity of type (B1), while β' sliding over β_0 introduces a singularity of type (B2). See Figure 53 for an example. Note that the original handleswap is now a $(1;l)$-handleswap.

We can resolve the non-simple stabilization introduced in this procedure, following the algorithm of Sections 7.1 and 7.2, to obtain a diagram involving only

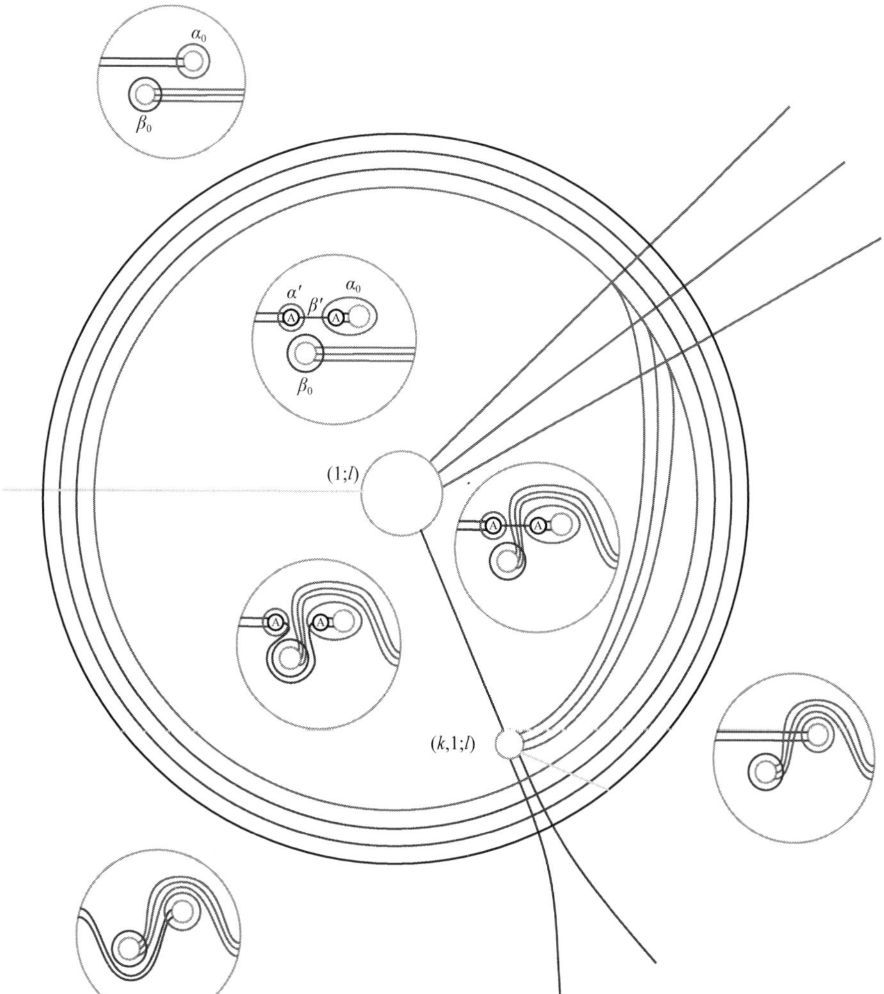

FIGURE 54. Reducing from a general $(k;l)$-handleswap to a $(1;l)$- and a $(k,1;l)$-handleswap, step 2. This is the resolution (following Chapter 7) of Figure 53.

simple stabilizations and handleslides. An example of the result is shown in Figure 54. Resolving the singularities of type (B1) introduces only slide pentagons, but resolving the singularity of type (B2) introduces another handleswap, of β_0 and α'. Since β' intersects α' in only one point, this is a $(k,1;l)$-handleswap, as desired.

Observe that the set of α-strands involved in these two handleswaps did not change. Thus, we can perform the same procedure again, but with the roles of α and β switched, to reduce to handleswaps of type $(k,1;l,1)$.

Reducing to $(1;1)$-handleswaps. Next, we reduce a $(k;l,1)$-handleswap between α_0 and β_0 to a $(k;1)$-handleswap. Again, let the β-strands intersecting α_0 be $\beta_1,\dots,\beta_k$, and let the α-strands intersecting β_0 be $\alpha_1,\dots,\alpha_{l+1}$. Assume that the circle containing α_1 intersects β_0 only once. Then, by sliding $\alpha_2,\dots,\alpha_{l+1}$ over α_1,

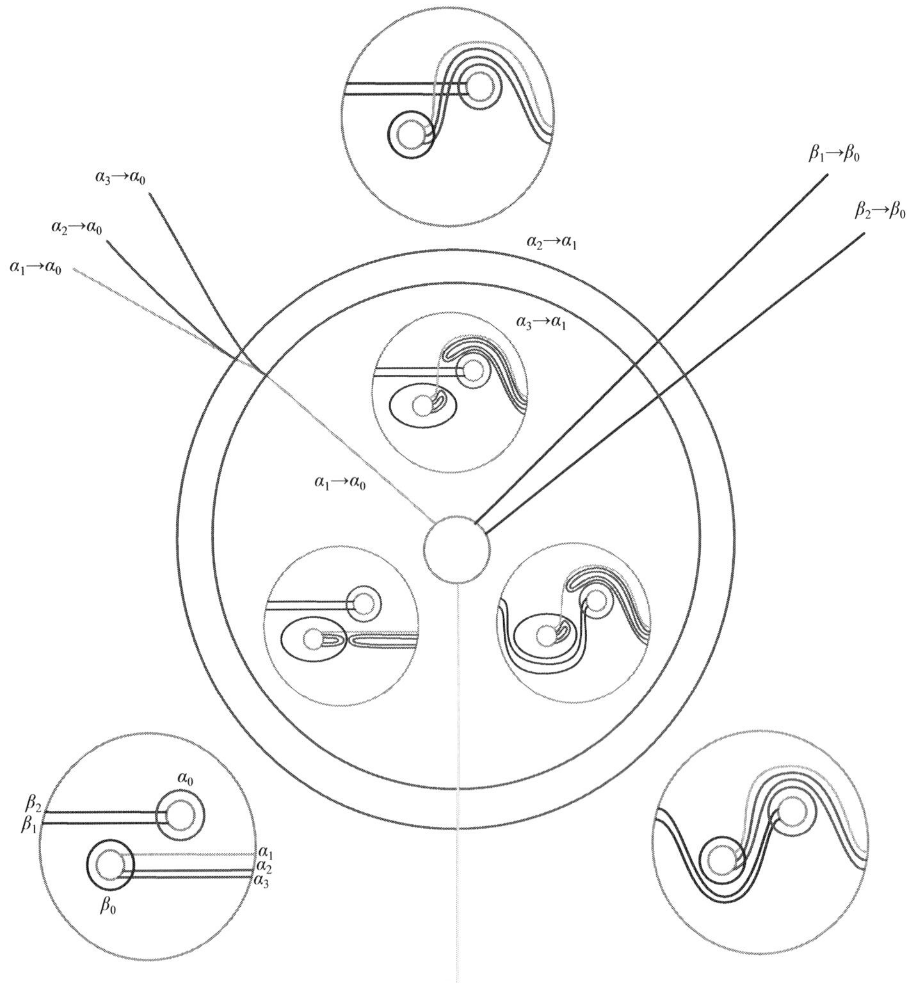

FIGURE 55. We illustrate reduction from a $(k; l, 1)$-handleswap to a $(k; 1)$-handleswap. In this example, $k = 2$ and $l = 2$. The curve α_1, which, by hypothesis, intersects β_0 only once, is shown in orange.

we can reduce all three stages of the handleswap to diagrams where only α_1 intersects β_0, which can in turn be related by a $(k; 1)$-handleswap. These handleslides can be done consistently in a family with the introduction of commuting squares and slide pentagons that arise when α_i for $i > 1$ slides over α_1, which in turn slides over α_0. See Figure 55 for an example.

This reduction did not affect the β-strands intersecting α_0. Thus, if we start with a $(k, 1; l, 1)$-handleswap, we can first reduce it to a $(k, 1; 1)$-handleswap as above, and then perform the same operation on the β-strands to reduce to a $(1; 1)$-handleswap.

Reducing to simple handleswaps. Finally, we reduce a $(1; 1)$-handleswap to a simple handleswap; this is illustrated in Figure 56. Suppose the $(1; 1)$-handleswap

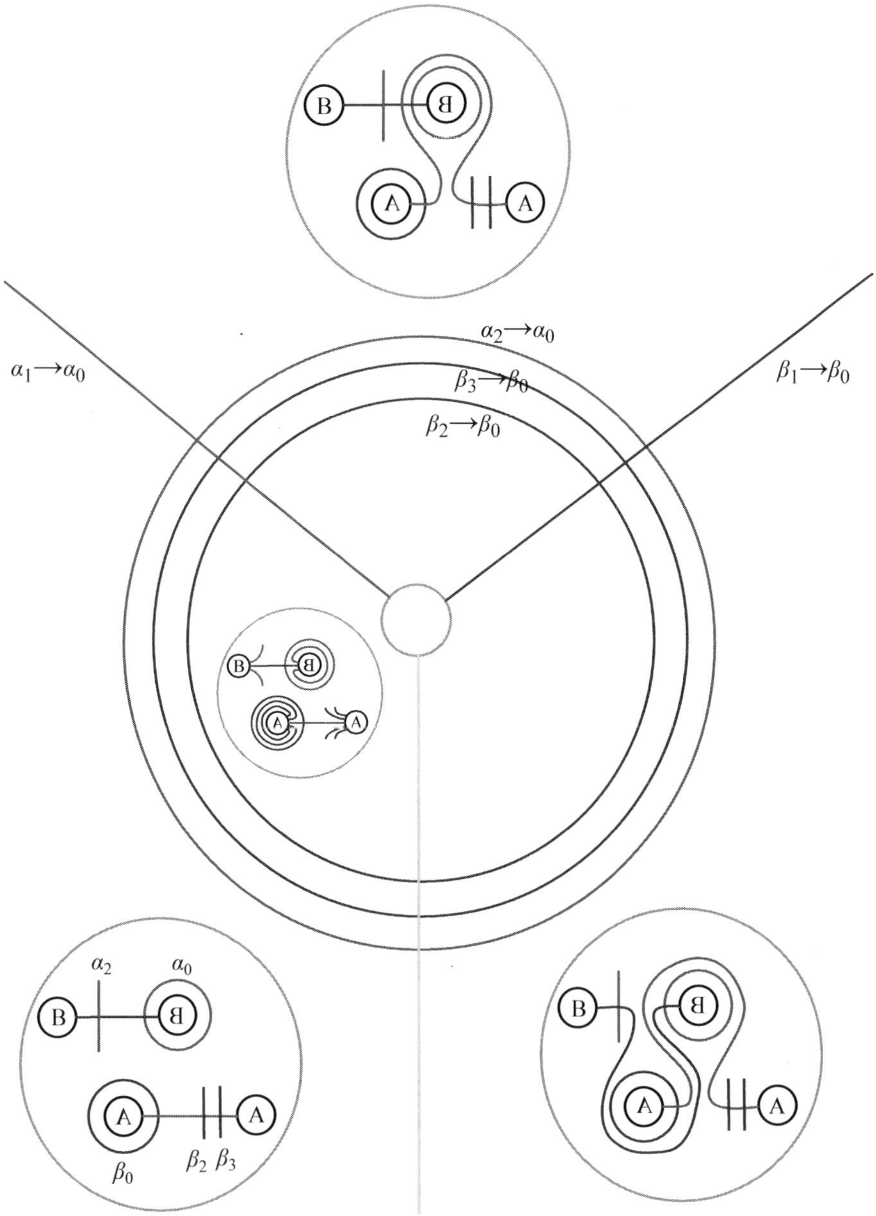

FIGURE 56. Reducing from a $(1;1)$-handleswap to a simple handleswap.

involves α_0 and β_0, a single curve β_1 intersecting α_0, and a single curve α_1 intersecting β_0. Let the other strands intersecting β_1 be $\alpha_2, \ldots, \alpha_{k+1}$, and let the other strands intersecting α_1 be $\beta_2, \ldots, \beta_{l+1}$, numbered such that, in the stage of the handleswap where α_1 and β_1 cross, the intersections along α_1 are β_0, β_1, $\beta_2, \ldots$ in that order, and similarly, the intersections along β_1 are α_0, α_1, $\alpha_2, \ldots$. We can now slide (in order) $\beta_{l+1}, \ldots, \beta_2$ over β_0, from the opposite side of the slide of β_1 over β_0 that appears in the handleswap. This commutes with all three moves in the handleswap. (It commutes with the slide of β_1 over β_0 because we are sliding $\beta_{l+1}, \ldots, \beta_2$ from the opposite side of β_0.) Similarly, slide $\alpha_{k+1}, \ldots, \alpha_2$ over α_0. Again, if we slide from the opposite side from the α_1 slide, this commutes with all three moves in the handleswap. But, after these slides, α_1 and β_1 do not intersect any other strands, and we have a simple handleswap, as in Figure 4.

CHAPTER 9

Strong Heegaard Invariants Have No Monodromy

We now have all the ingredients ready to prove Theorem 2.38. For the reader's convenience, we restate it here.

THEOREM. *Let $\mathcal{S}$ be a set of diffeomorphism types of sutured manifolds containing $[(M,\gamma)]$. Furthermore, let $F\colon \mathcal{G}(\mathcal{S}) \to \mathcal{C}$ be a strong Heegaard invariant. Given isotopy diagrams H, $H' \in |\mathcal{G}_{(M,\gamma)}|$ and any two oriented paths η and ν in $\mathcal{G}_{(M,\gamma)}$ connecting H to H', we have*

$$F(\eta) = F(\nu).$$

PROOF. Since F satisfies the Functoriality Axiom of Definition 2.32, it suffices to show that, for any loop η in $\mathcal{G}_{(M,\gamma)}$ of the form

$$H_0 \xrightarrow{e_1} H_1 \xrightarrow{e_2} \cdots \xrightarrow{e_{n-1}} H_{n-1} \xrightarrow{e_n} H_0,$$

we have $F(\eta) = \mathrm{Id}_{F(H_0)}$. By Lemma 2.11, every α- and β-equivalence between isotopy diagrams can be written as a product of handleslides. So, by the functoriality of F, we can assume that, for every $k \in \{1,\dots,n\}$, if e_k is an α- or β-equivalence, then it is actually a handleslide.

We are going to construct a generic 2-parameter family $\mathcal{F}\colon D^2 \to \mathcal{FV}(M,\gamma)$ of sutured functions and gradient-like vector fields, together with a surface enhanced polyhedral decomposition $\mathcal{P}$ of D^2 such that along S^1 we have the loop η. First, for every $k \in \{0,\dots,n-1\}$, pick a representative $\mathcal{H}_k = (\Sigma_k, \boldsymbol{\alpha}_k, \boldsymbol{\beta}_k)$ of the isotopy diagram H_k such that $\boldsymbol{\alpha}_k \pitchfork \boldsymbol{\beta}_k$. We can then apply Proposition 6.17 to obtain a simple Morse–Smale pair $(f_k, v_k) \in \mathcal{FV}_0(M,\gamma)$ such that $H(f_k,v_k) = \mathcal{H}_k$. Let $p_k = e^{2\pi i k/n}$ be a vertex of $\mathcal{P}$ for every $k \in \{0,\dots,n-1\}$. We define $\mathcal{F}(p_k) = (f_k, v_k)$, and the surface enhancement assigns $\Sigma_k \in \Sigma(f_k,v_k)$ to p_k. In fact, the vertices of $\mathcal{P}$ along S^1 are precisely $p_0,\dots,p_{n-1}$, and the edges are the arcs in between them. We extend $\mathcal{F}$ to the edge

$$\overline{p_k p_{k+1}} = \{\, e^{2\pi i t/n} : t \in [k,k+1] \,\}$$

between p_k and p_{k+1} using Proposition 6.35. By construction, each edge $\overline{p_k p_{k+1}}$ contains at most one bifurcation point of $\mathcal{F}$. Furthermore, if $\overline{p_k p_{k+1}}$ does contain a bifurcation point p, then at least one of Σ_k and Σ_{k+1} is in $\Sigma(\mathcal{F}(p))$ and is transverse to v_μ for every $\mu \in \overline{p_k p_{k+1}}$, so Proposition 6.28 applies to the whole edge for this separating surface. Hence $\mathcal{P}$ and $\mathcal{F}$ satisfy the boundary conditions of Lemma 7.4.

For $\mu \in S^1$, let $\mathcal{F}(\mu) = (f_\mu, v_\mu)$. By Proposition 5.19, the space $G(f_\mu, v_\mu)$ of Riemannian metrics g on M for which $v_\mu = \mathrm{grad}_g(f_\mu)$ is non-empty and contractible. So we can choose a generic family of metrics

$$\{\, g_\mu \in G(f_\mu, v_\mu) : \mu \in S^1 \,\}.$$

Choose a generic extension of $\{ f_\mu : \mu \in S^1 \}$ to a family of sutured functions $\{ f_\mu : \mu \in D^2 \}$, and similarly, extend $\{ g_\mu : \mu \in S^1 \}$ to a generic family of metrics $\{ g_\mu : \mu \in D^2 \}$. For $\mu \in D^2$, let

$$v_\mu = \text{grad}_{g_\mu}(f_\mu),$$

modified near γ such that it becomes a gradient-like vector field; see condition (3) of Definition 5.13. Then, away from a neighborhood of γ, the family $\{ v_\mu : \mu \in D^2 \}$ is a generic 2-parameter family of gradients, as in Definition 5.9. The possible bifurcations of generic 2-parameter families of gradients were all listed in Section 5.2.2. Even though the boundary behavior of v_μ on γ is not generic, this will not cause any problems since γ is an invariant subset of v_μ containing no singular points. Finally, let

$$\mathcal{F}(\mu) = (f_\mu, v_\mu) \in \mathcal{FV}(M, \gamma)$$

for every $\mu \in D^2$. By Lemma 7.4, we can extend $\mathcal{P}$ to a polyhedral decomposition of S^1 adapted to $\mathcal{F}$. The surface enhancement assigning Σ_k to the boundary vertices $p_k \in \text{sk}_0(\mathcal{P}) \cap S^1$ can be extended to a choice of Heegaard surfaces

$$\{ \Sigma_\mu \in \Sigma(\mathcal{F}(\mu)) : \mu \in \text{sk}_0(\mathcal{P}) \}$$

coherent with $\mathcal{P}$ according to Lemma 7.6.

As in Chapter 7, let

$$\mathfrak{S} = \mathfrak{S}(\mathcal{F}) = V_0 \sqcup V_1 \sqcup V_2$$

be the bordered stratification given by the bifurcation strata of the family $\mathcal{F}$. Furthermore, pick a bordered polyhedral decomposition $\mathcal{R}$ of D^2 refining $\mathfrak{S}$ that is dual to $\mathcal{P}$. After applying the resolution process of Chapter 7, we obtain a new surface enhanced polyhedral decomposition $\mathcal{P}'$ of D^2, with dual bordered polyhedral decomposition $\mathcal{R}'$. Since we only have simple stabilizations along S^1, and because we can assume that none of the 2-cells of $\mathcal{P}$ that intersect S^1 contain codimension-2 bifurcations of $\mathcal{F}$, after the resolution, $\mathcal{P} \cap S^1$ and $\mathcal{P}' \cap S^1$ coincide, and have the same surface enhancement. Note that we no longer claim that $\mathcal{P}'$ is adapted to some family of gradient-like vector fields, but along the boundary of each 2-cell of $\mathcal{P}'$, we have a loop of overcomplete diagrams that appears in Definition 2.32 (or a stabilization slide, which is a degenerate distinguished rectangle; see Definition 7.7). So it is either a loop of α-equivalences, a loop of β-equivalences, a loop of diffeomorphisms, a distinguished rectangle, a simple handleswap, or a stabilization slide. In addition, if we have a loop of diffeomorphisms, their composition is isotopic to the identity. Indeed, the composition d of the diffeomorphisms around a 2-cell σ is the same as the one induced by $\mathcal{F}|_{\partial\sigma} \colon \partial\sigma \to \mathcal{FV}_0(M, \gamma)$. Since $\mathcal{F}$ has no bifurcations inside σ, the loop $\mathcal{F}|_{\partial\sigma}$ is null-homotopic in $\mathcal{FV}_0(M, \gamma)$, so d is isotopic to the identity by Lemma 6.24.

The diagrams assigned to the vertices of $\mathcal{P}'$ might be overcomplete (except along the boundary). We now explain how to pass to a polyhedral decomposition $\mathcal{P}''$ that is decorated by actual (non-overcomplete) isotopy diagrams without altering anything along S^1. We obtain $\mathcal{P}''$ as follows. Let v be a vertex of $\mathcal{P}'$ lying in the interior of D^2 that is the endpoint of k_v 1-cells. Then pick a k_v-gon σ_v centered at v such that it has one vertex in each component of $D_\epsilon(v) \setminus \text{sk}_1(\mathcal{P}')$ for some ϵ very small. For every such v, the polygon σ_v is a 2-cell of $\mathcal{P}''$. Then, for each edge e of $\mathcal{P}'$ in the interior of D^2 with $\partial e = v - w$, connect the sides of σ_v and σ_w that intersect e by two arcs parallel to e; these will be edges of $\mathcal{P}''$. If e is an

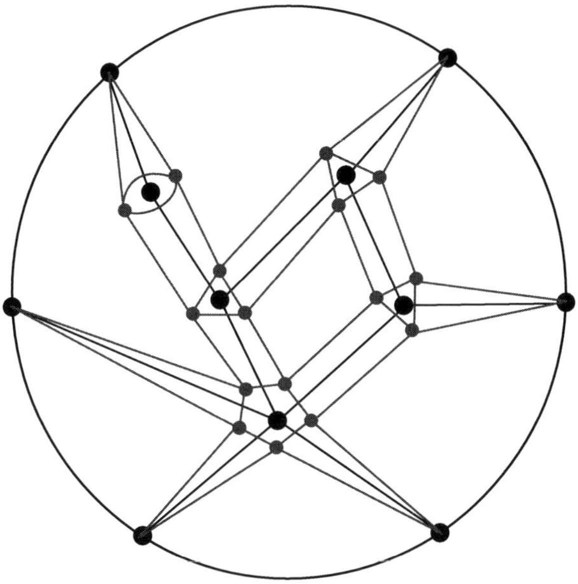

FIGURE 57. The polyhedral decomposition $\mathcal{P}'$ of D^2 is shown in black, and the "blown-up" decomposition $\mathcal{P}''$ in blue (along the boundary S^1 the two coincide).

edge with one endpoint w in S^1 and the other endpoint v in the interior of D^2, then we connect the side of σ_v intersecting e with w, forming a 2-cell of $\mathcal{P}''$ that is a triangle. So each 2-cell of $\mathcal{P}'$ is replaced by a smaller 2-cell in $\mathcal{P}''$, each interior vertex of $\mathcal{P}'$ is "blown up" to a 2-cell, and each edge to a rectangle or triangle. For an illustration, see Figure 57.

We are going to decorate the vertices of $\mathcal{P}''$ with (non-overcomplete) isotopy diagrams by choosing spanning trees for the overcomplete diagram at the "nearest" vertex of $\mathcal{P}'$. If σ is a 2-cell of $\mathcal{P}''$ with r vertices, then we will write $K_1, \dots, K_r$ for the loop of overcomplete diagrams along $\partial\sigma$.

Recall that, to a Morse–Smale gradient $(f, v) \in \mathcal{FV}_0(M, \gamma)$, we assigned the graphs $\Gamma_\pm(f, v)$, and any separating surface $\Sigma \in \Sigma(f, v)$ gives rise to an overcomplete diagram

$$H(f, v, \Sigma) = (\Sigma, \boldsymbol{\alpha}, \boldsymbol{\beta}).$$

However, we can obtain graphs $\Gamma_\pm(\Sigma, \boldsymbol{\alpha}, \boldsymbol{\beta})$ directly from the overcomplete diagram $(\Sigma, \boldsymbol{\alpha}, \boldsymbol{\beta})$ as follows. First, consider the graph whose vertices correspond to the components of $\Sigma \setminus \boldsymbol{\alpha}$, and for each component α of $\boldsymbol{\alpha}$, connect the vertices corresponding to the components on the two sides of α by an edge (possibly introducing a loop). Then $\Gamma_-(\Sigma, \boldsymbol{\alpha}, \boldsymbol{\beta})$ is obtained by identifying all the vertices that correspond to a component of $\Sigma \setminus \boldsymbol{\alpha}$ that intersects $\partial\Sigma$ non-trivially. We define $\Gamma_+(\Sigma, \boldsymbol{\alpha}, \boldsymbol{\beta})$ in an analogous manner. In case $(\Sigma, \boldsymbol{\alpha}, \boldsymbol{\beta}) = H(f, v)$, we have

$$\Gamma_\pm(\Sigma, \boldsymbol{\alpha}, \boldsymbol{\beta}) = \Gamma_\pm(f, v).$$

If D is an overcomplete diagram and $T_\pm$ is a spanning tree of $\Gamma_\pm(D)$, then we denote by $H(D, T_\pm)$ the diagram obtained from D by removing the α- and β-curves corresponding to edges in $T_\pm$. A diffeomorphism of isotopy diagrams $d\colon D_1 \to D_2$ induces a map $d_*\colon \Gamma_\pm(D_1) \to \Gamma_\pm(D_2)$.

Note that each vertex of $\mathcal{P}''$ in the interior of D^2 lies in a unique 2-cell σ that corresponds to a 2-cell of $\mathcal{P}'$. Hence, we can pick spanning trees for each such 2-cell separately to make F commute along their boundaries. Then we need to check that F also commutes along 2-cells of $\mathcal{P}''$ corresponding to 0-cells and 1-cells of $\mathcal{P}'$.

DEFINITION 9.1. The isotopy diagrams (Σ_1, A_1, B_1) and (Σ_2, A_2, B_2) are α/β-equivalent if $\Sigma_1 = \Sigma_2$, $A_1 \sim A_2$, and $B_1 \sim B_2$.

Clearly, an α-equivalence or a β-equivalence is a special case of an α/β-equivalence. From $\mathcal{G}(\mathcal{S})$, we obtain a graph $\mathcal{G}'(\mathcal{S})$ by adding an edge for every α/β-equivalence that is not an α-equivalence or a β-equivalence, and similarly, from $\mathcal{G}_{(M,\gamma)}$ we obtain the graph $\mathcal{G}'_{(M,\gamma)}$. The strong Heegaard invariant $F\colon \mathcal{G}(\mathcal{S}) \to \mathcal{C}$ extends to $\mathcal{G}'(\mathcal{S})$ as follows. Given an edge e from (Σ, A_1, B_1) to (Σ, A_2, B_2), there is an α-equivalence h from (Σ, A_1, B_1) to (Σ, A_2, B_1) and a β-equivalence g from (Σ, A_2, B_1) to (Σ, A_2, B_2). We let

$$F(e) = F(g) \circ F(h).$$

Note that we could have taken the intermediate diagram to be (Σ, A_1, B_2), but that gives the same map by the Commutativity Axiom of strong Heegaard invariants applied to a distinguished rectangle of type (1).

LEMMA 9.2. *Suppose that*

$$D_1 \xrightarrow{a_1} D_2 \xrightarrow{a_2} \cdots \xrightarrow{a_{r-1}} D_r \xrightarrow{a_r} D_1$$

is a loop of isotopy diagrams in $\mathcal{G}'(\mathcal{S})$ such that each edge a_i is an α/β-equivalence. Furthermore, let $F\colon \mathcal{G}(\mathcal{S}) \to \mathcal{C}$ be a strong Heegaard invariant. Then

$$F(a_r) \circ \cdots \circ F(a_1) = Id_{F(D_1)}.$$

PROOF. As above, we can write every α/β-equivalence as a product of an α-equivalence and a β-equivalence. By the Commutativity Axiom, it suffices to prove the lemma when $a_1, \dots, a_{i-1}$ are α-equivalences and $a_i, \dots, a_r$ are β-equivalences for some i. However, in this case $D_1 = D_i$, so we only have to prove the lemma when $a_1, \dots, a_r$ are all α-equivalences, or when they are all β-equivalences. This is a simple consequence of the Functoriality Axiom of strong Heegaard invariants. □

If σ is a 2-cell of $\mathcal{P}''$ corresponding to a vertex v of $\mathcal{P}'$ and v is decorated by the overcomplete diagram K, then choosing arbitrary spanning trees $T_\pm^1, \dots, T_\pm^r$ for $\Gamma_\pm(K)$ gives diagrams $D_i = H(K, T_\pm^i)$ for $i \in \{1, \dots, r\}$ such that any two of them are α/β-equivalent. Hence F applied to the loop of diagrams $D_1, \dots, D_r$ along $\partial\sigma$ commutes by Lemma 9.2.

Next, suppose that σ is a 2-cell of $\mathcal{P}''$ that corresponds to a 2-cell σ_0 of $\mathcal{P}'$. We distinguish several cases. In all the cases, we make sure that if the edge between K_i and K_{i+1} is a diffeomorphism, then we choose spanning trees $T_\pm^i$ and $T_\pm^{i+1}$ such that $T_\pm^{i+1} = d_*(T_\pm^i)$. Furthermore, if this edge is an index 1-2 stabilization, then $T_\pm^{i+1}$ is the same as $T_\pm^i$ (in particular, it does not contain the edges corresponding to the new α- and β-curve).

If all the edges of $\partial\sigma_0$ are diffeomorphisms $d_1, \dots, d_r$, then we showed above that their composition is isotopic to the identity. Choose a spanning tree $T_\pm^1$ for $\Gamma_\pm(K_1)$. Given $T_\pm^i$, we define $T_\pm^{i+1} = d_{i*}(T_\pm^i)$ for $i \in \{1, \dots, r-1\}$. Note that $T_\pm^1 = d_{r*}(T_\pm^r)$, since $d_r \circ \cdots \circ d_1$ is isotopic to the identity and hence it cannot permute the α-curves or the β-curves, which are both linearly independent in $H_1(\Sigma_1)$. By taking $D_i = H(K_i, T^i_\pm)$ at the vertices of $\partial\sigma$, we obtain the loop of diffeomorphisms

$$D_1 \xrightarrow{d_1} D_2 \xrightarrow{d_2} \cdots \xrightarrow{d_{r-1}} D_r \xrightarrow{d_r} D_1$$

in $\mathcal{G}_{(M,\gamma)}$. The invariant F commutes along this loop, since

$$F(d_r) \circ \cdots \circ F(d_1) = F(d_r \circ \cdots \circ d_1) = \mathrm{Id}_{F(D_1)}$$

by the Functoriality and Continuity Axioms.

If $\partial\sigma_0$ is a loop of α- or β-equivalences (e.g., a link of a singularity of type (A)), or a commutative rectangle of type (1), then, after choosing arbitrary spanning trees, we get a loop of α/β-equivalences along $\partial\sigma$. Then the strong Heegaard invariant F commutes by Lemma 9.2.

Suppose that, along σ_0, we have the distinguished rectangle

$$\begin{array}{ccc} K_1 & \xrightarrow{e} & K_2 \\ \downarrow{\scriptstyle f} & & \downarrow{\scriptstyle g} \\ K_3 & \xrightarrow{h} & K_4. \end{array}$$

If this is of type (2), with e, h being α-equivalences and f, g being stabilizations, then we choose a spanning tree $T^1_\pm$ of $\Gamma_\pm(K_1)$, followed by a spanning tree $T^2_\pm$ of $\Gamma_\pm(K_2)$ such that $T^2_+ = T^1_+$. We can view $T^1_\pm$ as a spanning tree $T^3_\pm$ of $\Gamma_\pm(K_3)$, and we can view $T^2_\pm$ as a spanning tree $T^4_\pm$ of $\Gamma_\pm(K_4)$. Then the vertices of σ are decorated by the diagrams $D_i = H(K_i, T^i_\pm)$ for $i \in \{1, \dots, 4\}$, which also form a distinguished rectangle of type (2). A distinguished rectangle of overcomplete diagrams of type (3), where f and g are diffeomorphisms, can be reduced to a distinguished rectangle of non-overcomplete diagrams of the same type in an analogous manner. In case of a rectangle of type (4) including only stabilizations, we start with a spanning tree $T^1_\pm$ for $\Gamma_\pm(K_1)$, which then gives rise to $T^2_\pm$ and $T^3_\pm$ in a natural manner. Both $T^2_\pm$ and $T^3_\pm$ give the same spanning tree $T^4_\pm$ of $\Gamma_\pm(K_4)$, as this is also the image of $T^1_\pm$ under the embedding of $\Gamma_\pm(K_1)$ into $\Gamma_\pm(K_4)$. Finally, for a rectangle of type (5), where e and h are stabilizations and f, g are diffeomorphisms, we first choose $T^1_\pm$, then let $T^3_\pm = f_*(T^1_\pm)$. We let $T^2_\pm$ be the image of $T^1_\pm$ under the embedding of $\Gamma_\pm(K_1)$ into $\Gamma_\pm(K_2)$, and $T^4_\pm$ is the image of $T^3_\pm$ under the embedding of $\Gamma_\pm(K_3)$ into $\Gamma_\pm(K_4)$. By construction, $T^4_\pm = g_*(T^2_\pm)$, hence reducing to a loop of non-overcomplete diagrams of type (5) along $\partial\sigma$.

The last possible type of loop along $\partial\sigma_0$ is a simple handleswap, a triangle in $\mathcal{G}_{(M,\gamma)}$ with vertices decorated by isotopy diagrams K_1, K_2, and K_3 on the common Heegaard surface Σ. Let the α- and β-curves involved in the handleswap be α_1, α_2, and β_1, β_2. Recall that the other α- and β-curves coincide in K_1, K_2, and K_3, so the graphs $\Gamma_\pm(K_i)$ only differ in the 4 edges corresponding to α_1, α_2, β_1, β_2. Since $\Sigma \setminus (\alpha_1 \cup \alpha_2)$ has the same number of components as Σ, there exists a common spanning tree T_- of $\Gamma_-(K_i)$ for $i \in \{1,2,3\}$ not containing the edges corresponding to α_1 and α_2. Similarly, $\beta_1 \cup \beta_2$ is non-separating, so there is a common spanning tree T_+ of $\Gamma_+(K_i)$ for $i \in \{1,2,3\}$. If we take the non-overcomplete sutured diagrams $D_i = H(K_i, T_\pm)$ for $i \in \{1,2,3\}$, then D_1, D_2, and D_3 also form a simple handleswap. Indeed, they all contain α_1, α_2, β_1, and β_2, and all other curves coincide.

Finally, let σ be a 2-cell of $\mathcal{P}''$ that corresponds to an edge e of $\mathcal{P}'$ not lying entirely in S^1. Then σ is a rectangle if e lies in the interior of D^2, and is a triangle if $e \cap S^1 \neq \emptyset$. In the latter case, we view σ as a rectangle in $\mathcal{G}_{(M,\gamma)}$ with one edge being the identity. Let σ_0 and σ_1 be the 2-cells of $\mathcal{P}'$ lying on the two sides of e, and the edges corresponding to e in $\mathcal{P}''$ are $g_0 \subset \sigma_0$ and $g_1 \subset \sigma_1$. We denote the other two edges of σ by h_0 and h_1. The vertices of e are decorated by the overcomplete diagrams K_0 and K_1. We distinguish three cases, depending on the type of e. If e is an α- or β-equivalence, then no matter how we choose trees for K_0 and K_1 in σ_0 and σ_1, along σ we get a loop of α/β-equivalences for which F commutes by Lemma 9.2.

If e is a stabilization, then, in both σ_0 and σ_1, we chose trees such that g_0 and g_1 are decorated by stabilizations. Furthermore, the edges h_0 and h_1 are decorated by α/β-equivalences, coming from the fact that we chose spanning trees for the same overcomplete diagram to decorate the endpoints of h_i. If h_1 is on the stabilized side, then this α/β-equivalence leaves the α- and β-curve involved in the stabilizations unchanged. To see that applying F to σ we get a commutative square, bisect both h_0 and h_1, and write them as a product of an α-equivalence and a β-equivalence. Connect the midpoints of h_0 and h_1 by a stabilization edge, hence decomposing σ into two distinguished rectangles of type (2). Then F commutes when applied to each of these distinguished rectangles. If e is a diffeomorphism, then we proceed in a way analogous to the previous case; we can decompose σ into two distinguished rectangles.

So we now have a polyhedral decomposition $\mathcal{P}''$ of D^2, together with a morphism of graphs

$$H\colon \mathrm{sk}_0(\mathcal{P}'') \to \mathcal{G}_{(M,\gamma)},$$

such that $F \circ H$ commutes along the boundary of each 2-cell of $\mathcal{P}''$. What remains to show is that this implies that F commutes along the boundary of D^2; i.e.,

$$F(\eta) = F(e_n) \circ \cdots \circ F(e_1) = \mathrm{Id}_{F(H_0)}.$$

The proof of this requires some care as the composition of morphisms is not commutative. For this, we show that there is a "combinatorial 0-homotopy" from S^1 to the boundary of a 2-cell of $\mathcal{P}''$. By this, we mean that there is a sequence of curves $\eta_0, \dots, \eta_k$ in D^2 such that

(1) $\eta_0 = \eta$ and $\eta_k = \partial\sigma_0$ for some two-cell σ_0 of $\mathcal{P}''$,
(2) every η_i is a properly embedded curve in $\mathrm{sk}_1(\mathcal{P}'')$, and
(3) the 1-chain $\eta_i - \eta_{i+1}$ is the boundary of a single 2-cell σ_i of $\mathcal{P}''$.

This clearly implies that $F(\eta) = \mathrm{Id}_{F(H_0)}$, since

$$F(\eta_i) \circ F(\eta_{i+1})^{-1} = F(\partial\sigma_i) = \mathrm{Id}$$

for every $i \in \{1, \dots, k-1\}$, and $F(\eta_k) = F(\partial\sigma_0) = \mathrm{Id}$.

To construct the combinatorial 0-homotopy, we proceed recursively. Suppose we have already obtained η_i. Then η_i bounds a disk D_i^2 in D^2, and $\mathcal{P}''$ restricts to a polyhedral decomposition of D_i^2. It suffices to show that if D_i^2 has more than one 2-cells, then there exists a 2-cell σ_i in D_i^2 that intersects η_i in a single arc. Indeed, we then take $\eta_{i+1} = \eta_i - \partial\sigma_i$, which is a simple closed curve. The existence of such a σ_i follows from the following lemma.

LEMMA 9.3. *For any polyhedral decomposition of D^2 with more than one 2-cells, there exists a 2-cell that intersects S^1 in a single arc.*

PROOF. We proceed by induction on the number t of 2-cells. If $t = 2$, then let the 2-cells be σ_1 and σ_2. Since the attaching map of each 2-cell is an embedding, $\sigma_1 \cap \sigma_2$ consists of some disjoint arcs, and to obtain D^2, it has to be a single arc a. Hence

$$\sigma_i \cap S^1 = \partial\sigma_i \setminus \mathrm{Int}(a)$$

is a single arc for $i \in \{1, 2\}$.

Now suppose that the statement holds for polyhedral decompositions for which the number of 2-cells is less than t for some $t > 2$, and consider a decomposition where the number of 2-cells is t. There is a 2-cell σ_1 such that $\mathrm{Int}(\sigma_1 \cap S^1) \neq \emptyset$. If $\sigma_1 \cap S^1$ has a single component, it has to be an arc, and we are done. Otherwise, $D^2 \setminus \sigma_1$ consists of at least two components, each of whose closure is homeomorphic to a disk; let D_1 be one of these. Observe that $s_1 = D_1 \cap \sigma_1$ is an arc. If there are at least two 2-cells in D_1, then, by induction, there is a 2-cell σ_2 in D_1 for which $\sigma_2 \cap \partial D_1$ is a single arc a_1. Since

$$\sigma_2 \cap S^1 = a_1 \setminus \mathrm{Int}(s_1),$$

we are done if this is a single arc. Otherwise, either $a_1 \subset s_1$ or $a_1 \supset s_1$. In both cases, we merge σ_1 and σ_2 by removing all the vertices and edges in $\mathrm{Int}(a_1 \cap s_1)$. We obtain a polyhedral decomposition of D^2 where the number of 2-cells is $t - 1 \geq 2$, so, by the induction hypothesis, there is a 2-cell σ_3 that intersects S^1 in a single arc. Since $\sigma_1 \cup \sigma_2$ intersects S^1 in the same number of components as σ_1, which is more than one, $\sigma_3 \neq \sigma_1 \cup \sigma_2$, and so σ_3 is also a 2-cell of the original decomposition. Finally, if D_1 consists of a single 2-cell σ_2, then $\sigma_2 \cap S^1 = \partial\sigma_2 \setminus \mathrm{Int}(s_1)$, which is a single arc. □

Since the polyhedral decomposition $\mathcal{P}''|_{D^2_{i+1}}$ contains one less 2-cell than $\mathcal{P}''|_{D^2_i}$, the process ends when $\mathcal{P}''|_{D^2_k}$ consists of a single 2-cell, and we obtain the combinatorial 0-homotopy. This concludes the proof of Theorem 2.38. □

CHAPTER 10

Heegaard Floer Homology

In this chapter, we prove Theorem 2.33. We first explain how the different versions of Heegaard Floer homology fit into the framework of weak Heegaard invariants in the sense of Definition 2.24. We then show they are strong Heegaard invariants in the sense of Definition 2.32. Most of our constructions build on the work of Ozsváth and Szabó [**28**, Section 2.5]. To show Theorem 2.33, we perform the following additional checks:

(1) We keep track of the embedding of the Heegaard surface in the 3-manifold, and include diffeomorphisms isotopic to the identity among the Heegaard moves, which allows us to define functorial diffeomorphism maps on Heegaard Floer homology.
(2) We verify the Continuity Axiom (3) of Definition 2.32.
(3) Most importantly, we verify simple handleswap invariance.

For concreteness, we explain the case of sutured Floer homology in detail as it includes $\widehat{HF}$ and $\widehat{HFL}$ as special cases, and only remark on the differences for the other versions. In particular, we show that SFH is a strong Heegaard invariant of the class $\mathcal{S}_{\text{bal}}$. However, to emphasize that all the arguments are essentially the same for the other versions of Heegaard Floer homology, we will write HF° instead of SFH. All Heegaard diagrams appearing in this chapter are assumed to be balanced.

10.1. Heegaard Floer homology as a weak Heegaard invariant

We start by explaining how Heegaard Floer homology fits into the framework of weak Heegaard invariants. This was proved, in a different form, by Ozsváth and Szabó (with the exception of SFH); we remind the reader of the proof in order to fill in details and because we will later extend the arguments to prove that Heegaard Floer homology is a strong Heegaard invariant. One complication arises from the fact that Heegaard Floer homology is, in fact, not an invariant associated to arbitrary Heegaard diagrams; rather, these Heegaard diagrams must satisfy the additional property of *admissibility*. There are several forms of admissibility. We will focus presently on the case of *weak admissibility* in the sense of Ozsváth and Szabó [**27**, Definition 4.10] and Juhász [**16**, Definition 3.11], which is sufficient for defining $\widehat{HF}$, SFH, and HF^+. The stronger variant, used in the construction of HF^- and HF^∞, is defined in reference to an auxiliary Spin^c structure.

We briefly discuss Spin^c structures. Let $\mathcal{H} = (\Sigma, \boldsymbol{\alpha}, \boldsymbol{\beta})$ be an abstract (i.e., non-embedded) Heegaard diagram. Our aim is to explain what we mean by a Spin^c-structure for $\mathcal{H}$. If $\mathcal{H}$ is a diagram of the sutured manifolds (M, γ) and (M', γ'), then there is a diffeomorphism $\phi\colon (M, \gamma) \to (M', \gamma')$ that is well-defined up to isotopy fixing Σ. So ϕ induces a bijection

$$b_{(M,\gamma),(M',\gamma')}\colon \operatorname{Spin}^c(M,\gamma) \to \operatorname{Spin}^c(M',\gamma')$$

that intertwines the $H_1(M)$-action on the set $\operatorname{Spin}^c(M,\gamma)$ and the $H_1(M')$-action on the set $\operatorname{Spin}^c(M',\gamma')$. For $\mathfrak{s} \in \operatorname{Spin}^c(M,\gamma)$ and $\mathfrak{s}' \in \operatorname{Spin}^c(M',\gamma')$, we write $\mathfrak{s} \sim \mathfrak{s}'$ if and only if $b_{(M,\gamma),(M',\gamma')}(\mathfrak{s}) = \mathfrak{s}'$. Then "$\sim$" defines an equivalence relation on the class of elements of $\operatorname{Spin}^c(M,\gamma)$ for all (M,γ) such that $\mathcal{H}$ is a diagram of (M,γ). We define $\operatorname{Spin}^c(\mathcal{H})$ to be the collection of these equivalence classes. Given $\mathfrak{s} \in \operatorname{Spin}^c(M,\gamma)$, we denote its equivalence class by $[\mathfrak{s}]$; this is an element of $\operatorname{Spin}^c(\Sigma,\boldsymbol{\alpha},\boldsymbol{\beta})$.

Let $\mathbf{x} \in \mathbb{T}_\alpha \cap \mathbb{T}_\beta$ be a Heegaard Floer generator. By Proposition 6.17, there exists a simple pair $(f,v) \in \mathcal{FV}_0(M,\gamma)$ such that $H(f,v) = \mathcal{H}$. We can associate to $\mathbf{x}$ a nowhere vanishing vector field $v_{\mathbf{x}}$ on M as follows. Let $\gamma_{\mathbf{x}}$ be the union of the flow lines of v passing through the points of $\mathbf{x}$. Then we delete v on a thin regular neighborhood $N_{\mathbf{x}}$ of $\gamma_{\mathbf{x}}$, and extend it to $N_{\mathbf{x}}$ as a nowhere vanishing vector field. This is possible since each component of $N_{\mathbf{x}}$ contains two critical points of v of opposite sign, and hence the degree of v is zero along each component of $\partial N_{\mathbf{x}}$. If we take a different simple pair $(\overline{f},\overline{v}) \in \mathcal{FV}_0(M,\gamma)$ with $H(\overline{f},\overline{v}) = \mathcal{H}$, then Proposition 6.18 implies that (f,v) and $(\overline{f},\overline{v})$ can be connected by a path within $\mathcal{FV}_0(M,\gamma)$. This show that the vector fields $v_{\mathbf{x}}$ and $\overline{v}_{\mathbf{x}}$ are homologous relative to ∂M. (Recall that two vector fields are *homologous* relative to ∂M if they are homotopic in the complement of a ball, where the homotopy is the identity on ∂M.) In particular, the Spin^c structures defined by $v_{\mathbf{x}}$ and $\overline{v}_{\mathbf{x}}$ coincide; we denote it by $\mathfrak{s}_{(M,\gamma)}(\mathbf{x})$. As above, we can also assign to $\mathbf{x}$ an element $\mathfrak{s}_{(M',\gamma')}(\mathbf{x}) \in \operatorname{Spin}^c(M',\gamma')$. By construction, $\mathfrak{s}_{(M,\gamma)}(\mathbf{x}) \sim \mathfrak{s}_{(M',\gamma')}(\mathbf{x})$, so we can define $\mathfrak{s}(\mathbf{x}) \in \operatorname{Spin}^c(\mathcal{H})$ to be $[\mathfrak{s}_{(M,\gamma)}(\mathbf{x})]$ for any (M,γ) such that $\mathcal{H}$ is a diagram of (M,γ).

As explained by Ozsváth and Szabó [**27**, Section 4 and Theorem 6.1], the Floer homology groups depend on a choice of complex structure $\mathfrak{j}$ on Σ and a generic path $J_s \subset \mathcal{U}$ of perturbations of the induced complex structure over $\operatorname{Sym}^g(\Sigma)$, where $\mathcal{U}$ is a certain contractible set of almost complex structures. The following result is due to Ozsváth and Szabó [**28**, Lemma 2.11].

LEMMA 10.1. *Let $(\Sigma,\boldsymbol{\alpha},\boldsymbol{\beta})$ be admissible. Fix two different choices $(\mathfrak{j},J_s)$ and $(\mathfrak{j}',J_s')$ of complex structures and perturbations. Then there is an isomorphism*

$$\Phi_{J_s\to J_s'}\colon HF^\circ_{J_s}(\Sigma,\boldsymbol{\alpha},\boldsymbol{\beta},\mathfrak{s}) \to HF^\circ_{J_s'}(\Sigma,\boldsymbol{\alpha},\boldsymbol{\beta},\mathfrak{s}).$$

These isomorphisms are natural in the sense that

$$\Phi_{J_s'\to J_s''} \circ \Phi_{J_s\to J_s'} = \Phi_{J_s\to J_s''},$$

and $\Phi_{J_s\to J_s}$ is the identity.

Hence, we can define

$$HF^\circ(\Sigma,\boldsymbol{\alpha},\boldsymbol{\beta},\mathfrak{s}) = \coprod_{(\mathfrak{j},J_s)} HF^\circ_{J_s}(\Sigma,\boldsymbol{\alpha},\boldsymbol{\beta},\mathfrak{s})\Big/\sim,$$

where $x \sim y$ if and only if $y = \Phi_{J_s\to J_s'}(x)$ for some $(\mathfrak{j},J_s)$ and $(\mathfrak{j}',J_s')$.

Let $(\Sigma,\boldsymbol{\alpha},\boldsymbol{\beta},\boldsymbol{\gamma})$ be an admissible triple diagram. Then, by counting pseudo-holomorphic triangles, Ozsváth and Szabó [**27**, Theorem 8.12] in the case of ordinary

Heegaard triple-diagrams, and Grigsby and Wehrli [**12**, Section 3.3] for sutured triple-diagrams defined maps

$$F_{\alpha,\beta,\gamma}\colon HF^\circ(\Sigma,\boldsymbol{\alpha},\boldsymbol{\beta})\otimes HF^\circ(\Sigma,\boldsymbol{\beta},\boldsymbol{\gamma})\to HF^\circ(\Sigma,\boldsymbol{\alpha},\boldsymbol{\gamma}).$$

LEMMA 10.2. *Let $(\Sigma,\boldsymbol{\beta},\boldsymbol{\gamma})$ be an admissible sutured diagram such that $\boldsymbol{\beta}\sim\boldsymbol{\gamma}$. Then there is a unique* Spin^c *structure $\mathfrak{s}_0\in \mathrm{Spin}^c(\Sigma,\boldsymbol{\beta},\boldsymbol{\gamma})$ such that $c_1(\mathfrak{s}_0)=0$. Furthermore, in the highest non-zero relative homological grading $HF^\circ(\Sigma,\boldsymbol{\beta},\boldsymbol{\gamma},\mathfrak{s}_0)$ is isomorphic to $\mathbb{F}_2$; we denote its generator by $\Theta_{\beta,\gamma}$.*

PROOF. Suppose we have a diagram $(\Sigma,\boldsymbol{\beta},\boldsymbol{\gamma})$ such that $\boldsymbol{\beta}\sim\boldsymbol{\gamma}$, and let $k=|\boldsymbol{\beta}|=|\boldsymbol{\gamma}|$. Then $(\Sigma,\boldsymbol{\beta},\boldsymbol{\gamma})$ defines a sutured manifold diffeomorphic to

$$M(R_+,k)=(R_+\times I,\partial R_+\times I)\,\#\,\left(\#^k(S^1\times S^2)\right)$$

for some compact oriented surface R_+. There is a unique Spin^c structure $\mathfrak{s}_0$ on $M(R_+,k)$ such that

$$c_1(\mathfrak{s}_0)=0\in H^2(M(R_+,k)),$$

and which can be represented by a vector field that is vertical on the summand $(R_+\times I,\partial R_+\times I)$. By the connected sum formula for sutured manifolds of Juhász [**16**, Proposition 9.15],

$$HF^\circ(M(R_+,k),\mathfrak{s}_0)\cong\Lambda^*H_1(S^1\times S^2;\mathbb{F}_2)$$

as relatively $\mathbb{Z}$-graded groups. Here, we do not use naturality, only that Heegaard Floer homology is well-defined up to isomorphism in each Spin^c structure and relative homological grading, as shown by Ozsváth and Szabó [**27**]. Hence, in the "top" non-zero homological grading, the group

$$HF^\circ(\Sigma,\boldsymbol{\beta},\boldsymbol{\gamma},[\mathfrak{s}_0])\cong HF^\circ(M(R_+,k),\mathfrak{s}_0)$$

is isomorphic to $\mathbb{F}_2$; we denote its generator by $\Theta_{\beta,\gamma}$. Since $[\mathfrak{s}_0]$ is independent of the concrete manifold representing $M(R_+,k)$, we see that $\Theta_{\beta,\gamma}$ is a well-defined element of $HF^\circ(\Sigma,\boldsymbol{\beta},\boldsymbol{\gamma})$. □

DEFINITION 10.3. Let $(\Sigma,\boldsymbol{\alpha},\boldsymbol{\beta},\boldsymbol{\gamma})$ be an admissible triple diagram. If $\boldsymbol{\beta}\sim\boldsymbol{\gamma}$, then we write $\Psi^{\boldsymbol{\alpha}}_{\boldsymbol{\beta}\to\boldsymbol{\gamma}}$ for the map

$$F_{\alpha,\beta,\gamma}(-\otimes\Theta_{\beta,\gamma})\colon HF^\circ(\Sigma,\boldsymbol{\alpha},\boldsymbol{\beta})\to HF^\circ(\Sigma,\boldsymbol{\alpha},\boldsymbol{\gamma}).$$

Similarly, if $\boldsymbol{\alpha}\sim\boldsymbol{\beta}$, then let

$$\Psi^{\boldsymbol{\alpha}\to\boldsymbol{\beta}}_{\boldsymbol{\gamma}}(-)=F_{\beta,\alpha,\gamma}(\Theta_{\beta,\alpha}\otimes-)\colon HF^\circ(\Sigma,\boldsymbol{\alpha},\boldsymbol{\gamma})\to HF^\circ(\Sigma,\boldsymbol{\beta},\boldsymbol{\gamma}).$$

LEMMA 10.4. *Let $(\Sigma,\boldsymbol{\alpha},\boldsymbol{\beta},\boldsymbol{\gamma})$ be an admissible triple diagram such that $\boldsymbol{\beta}\sim\boldsymbol{\gamma}$. Then there is an identification*

$$\mathrm{Spin}^c(\Sigma,\boldsymbol{\alpha},\boldsymbol{\beta})\cong\mathrm{Spin}^c(\Sigma,\boldsymbol{\alpha},\boldsymbol{\gamma})$$

such that, for every $\mathfrak{s}\in\mathrm{Spin}^c(\Sigma,\boldsymbol{\alpha},\boldsymbol{\beta})$, the map $\Psi^{\boldsymbol{\alpha}}_{\boldsymbol{\beta}\to\boldsymbol{\gamma}}$ maps $HF^\circ(\Sigma,\boldsymbol{\alpha},\boldsymbol{\beta},\mathfrak{s})$ to $HF^\circ(\Sigma,\boldsymbol{\alpha},\boldsymbol{\gamma},\mathfrak{s})$, and preserves the relative $\mathbb{Z}_{\mathfrak{d}(\mathfrak{s})}$-grading, where $\mathfrak{d}(\mathfrak{s})$ is the divisibility of $c_1(\mathfrak{s})\in H^2(M)$.

PROOF. If $(\Sigma,\boldsymbol{\alpha},\boldsymbol{\beta})$ is a diagram of (M,γ), then $\boldsymbol{\beta}\sim\boldsymbol{\gamma}$ implies that $(\Sigma,\boldsymbol{\alpha},\boldsymbol{\gamma})$ is also a diagram of (M,γ). Hence, for $\mathfrak{s}\in\mathrm{Spin}^c(M,\gamma)$, we identify its equivalence class $[\mathfrak{s}]\in\mathrm{Spin}^c(\Sigma,\boldsymbol{\alpha},\boldsymbol{\beta})$ with $[\mathfrak{s}]\in\mathrm{Spin}^c(\Sigma,\boldsymbol{\alpha},\boldsymbol{\gamma})$.

Suppose that the index 0 homotopy class $\psi\in\pi_2(\mathbf{x},\theta,\mathbf{y})$ contributes to $\Psi^{\boldsymbol{\alpha}}_{\boldsymbol{\beta}\to\boldsymbol{\gamma}}$, where $\mathbf{x}\in\mathbb{T}_\alpha\cap\mathbb{T}_\beta$ is such that $\mathfrak{s}(\mathbf{x})=\mathfrak{s}$, while $\mathbf{y}\in\mathbb{T}_\alpha\cap\mathbb{T}_\gamma$, and $\theta\in\mathbb{T}_\beta\cap$

$\mathbb{T}_\gamma$ represents $\Theta_{\beta,\gamma}$. By definition, $\mathfrak{s}(\theta) = \mathfrak{s}_0$. There is a cobordism of sutured manifolds $\mathcal{W} = \mathcal{W}_{\alpha,\beta,\gamma}$ from $M_{\alpha,\beta} \sqcup M_{\beta,\gamma}$ to $M_{\alpha,\gamma}$ associated to the triple diagram $(\Sigma, \boldsymbol{\alpha}, \boldsymbol{\beta}, \boldsymbol{\gamma})$ (where $M_{i,j}$ is the sutured manifold defined by (Σ, i, j) for every i, $j \in \{\boldsymbol{\alpha}, \boldsymbol{\beta}, \boldsymbol{\gamma}\}$), together with a set of Spin^c structures $\underline{\mathrm{Spin}^c}(\mathcal{W})$; see [**17**, p. 961]. By [**17**, Proposition 5.5], there is a map

$$\underline{\mathfrak{s}}\colon \pi_2(\mathbf{x}, \theta, \mathbf{y}) \to \underline{\mathrm{Spin}^c}(\mathcal{W})$$

such that $\underline{\mathfrak{s}}(\psi)|_{M_{\alpha,\beta}} = \mathfrak{s}$, while $\underline{\mathfrak{s}}(\psi)|_{M_{\beta,\gamma}} = \mathfrak{s}_0$, and $\underline{\mathfrak{s}}(\psi)|_{M_{\alpha,\gamma}} = \mathfrak{s}(\mathbf{y})$. Since $\boldsymbol{\beta} \sim \boldsymbol{\gamma}$, we can glue a cobordism from $\emptyset$ to $M_{\beta,\gamma}$ to $\mathcal{W}$ and obtain a cobordism $\mathcal{W}'$ from $M_{\alpha,\beta}$ to $M_{\alpha,\gamma}$ diffeomorphic to the product $M_{\alpha,\beta} \times I$. Since $\underline{\mathfrak{s}}(\psi)|_{M_{\beta,\gamma}} = \mathfrak{s}_0$, it follows that $\underline{\mathfrak{s}}(\psi)|_{M_{\beta,\gamma}}$ extends to a Spin^c structure $\mathfrak{t}'$ on $\mathcal{W}'$. As $\mathfrak{t}'|_{M_{\alpha,\beta}} = \mathfrak{s}$, we have $\mathfrak{t}'|_{M_{\alpha,\gamma}} = \mathfrak{s}$. Consequently, $\mathfrak{s}(\mathbf{y}) = \underline{\mathfrak{s}}(\psi)|_{M_{\alpha,\gamma}} = \mathfrak{s}$, hence $\Psi^{\boldsymbol{\alpha}}_{\boldsymbol{\beta}\to\boldsymbol{\gamma}}$ maps $HF^\circ(\Sigma, \boldsymbol{\alpha}, \boldsymbol{\beta}, \mathfrak{s})$ to $HF^\circ(\Sigma, \boldsymbol{\alpha}, \boldsymbol{\gamma}, \mathfrak{s})$.

It also follows from the above argument that there is a unique Spin^c structure $\mathfrak{t}$ on $\mathcal{W}$ that satisfies $\mathfrak{t}|_{M_{\alpha,\beta}} = \mathfrak{s}$ and $\mathfrak{t}|_{M_{\beta,\gamma}} = \mathfrak{s}_0$. Let $\psi' \in \pi_2(\mathbf{x}', \theta', \mathbf{y}')$ be another index zero homotopy class that contributes to $\Psi^{\boldsymbol{\alpha}}_{\boldsymbol{\beta}\to\boldsymbol{\gamma}}$, where $\mathbf{x}' \in \mathbb{T}_\alpha \cap \mathbb{T}_\beta$ is such that $\mathfrak{s}(\mathbf{x}') = \mathfrak{s}$, while $\mathbf{y}' \in \mathbb{T}_\alpha \cap \mathbb{T}_\gamma$, and $\theta' \in \mathbb{T}_\beta \cap \mathbb{T}_\gamma$ represents $\Theta_{\beta,\gamma}$. Choose a domain D_x from $\mathbf{x}$ to $\mathbf{x}'$, a domain D_y from $\mathbf{y}$ to $\mathbf{y}'$, and a domain D_θ from θ to θ'. We write D for the domain of ψ and D' for the domain of ψ'. Then

$$D_1 := D + D_x + D_\theta - D_y \in \pi_2(\mathbf{x}', \theta', \mathbf{y}'),$$

and $\underline{\mathfrak{s}}(D_1) = \underline{\mathfrak{s}}(D')$. Hence, by [**17**, Proposition 5.9], we can write the domain $D_1 - D'$ as a sum $P_{\alpha,\beta} + P_{\beta,\gamma} + P_{\alpha,\gamma}$ for a periodic domain $P_{\alpha,\beta}$ in $(\Sigma, \boldsymbol{\alpha}, \boldsymbol{\beta})$, a periodic domain $P_{\beta,\gamma}$ in $(\Sigma, \boldsymbol{\beta}, \boldsymbol{\gamma})$, and a periodic domain $P_{\alpha,\gamma}$ in $(\Sigma, \boldsymbol{\alpha}, \boldsymbol{\gamma})$. Then, on one hand,

$$\mu(D_1 - D') = \mu(D) + \mu(D_x) + \mu(D_\theta) - \mu(D_y) - \mu(D') = \mu(D_x) - \mu(D_y)$$

since $\mu(D) = \mu(D') = 0$ as they contribute to the map $\Psi^{\boldsymbol{\alpha}}_{\boldsymbol{\beta}\to\boldsymbol{\gamma}}$, and $\mu(D_\theta) = 0$ since $\mathcal{D}_\theta$ connects θ and θ' that have the same homological grading as they are homologous. On the other hand,

$$\mu(D_1 - D') = \mu(P_{\alpha,\beta}) + \mu(P_{\beta,\gamma}) + \mu(P_{\alpha,\gamma}).$$

As $\mathfrak{s}(\theta') = \mathfrak{s}_0$ and $c_1(\mathfrak{s}_0) = 0$, we have

$$\mu(P_{\beta,\gamma}) = \langle\, c_1(\mathfrak{s}(\theta')), H(P_{\beta,\gamma})\,\rangle = 0.$$

Similarly, $\mu(P_{\alpha,\beta}) = \langle\, c_1(\mathfrak{s}), H(P_{\alpha,\beta})\,\rangle$ and $\mu(P_{\alpha,\gamma}) = \langle\, c_1(\mathfrak{s}), H(P_{\alpha,\gamma})\,\rangle$ are divisible by $\mathfrak{d}(\mathfrak{s})$. It follows that $\mathfrak{d}(\mathfrak{s})$ divides

$$\mu(D_x) - \mu(D_y) = \mathrm{gr}(\mathbf{x}, \mathbf{x}') - \mathrm{gr}(\mathbf{y}, \mathbf{y}'),$$

hence $\Psi^{\boldsymbol{\alpha}}_{\boldsymbol{\beta}\to\boldsymbol{\gamma}}$ preserves the relative $\mathbb{Z}_{\mathfrak{d}}(\mathfrak{s})$-grading. □

Before proceeding, we state two key lemmas that will be used multiple times. Recall that a *Hamiltonian isotopy* of a symplectic manifold (Σ, ω) is a 1-parameter family of diffeomorphisms of Σ obtained by integrating a time-dependent Hamiltonian vector field; see Ozsváth and Szabó [**27**, Section 7.3].

LEMMA 10.5. *For every $i \in \{1, \dots, k\}$, let*

$$(\Sigma, \boldsymbol{\eta}_0^i, \dots, \boldsymbol{\eta}_{n-1}^i, \boldsymbol{\eta}_n)$$

be a sutured multi-diagram such that $(\Sigma, \boldsymbol{\eta}_0^i, \dots, \boldsymbol{\eta}_{n-1}^i)$ *is admissible. Then there is a Hamiltonian isotopic translate* $\boldsymbol{\eta}_n'$ *of* $\boldsymbol{\eta}_n$ *such that* $(\Sigma, \boldsymbol{\eta}_0^i, \dots, \boldsymbol{\eta}_{n-1}^i, \boldsymbol{\eta}_n')$ *is admissible for every* $i \in \{1, \dots, k\}$.

PROOF. The case $i = 1$ was shown by Grigsby and Wehrli [**12**, proof of Lemma 3.13]. We proceed the same way, and isotope $\boldsymbol{\eta}_n$ to $\boldsymbol{\eta}_n'$ using finger moves along oriented arcs representing a basis of $H_1(\Sigma, \partial\Sigma)$ and their parallel opposites. Since this isotopy, and hence $\boldsymbol{\eta}_n'$ is independent of i, the diagram $(\Sigma, \boldsymbol{\eta}_0^i, \dots, \boldsymbol{\eta}_{n-1}^i, \boldsymbol{\eta}_n')$ is admissible for every $i \in \{1, \dots, k\}$. Note that the finger moves of $\boldsymbol{\eta}_n$ can be achieved by a Hamiltonian isotopy. □

LEMMA 10.6. *Suppose that the quadruple diagram* $(\Sigma, \boldsymbol{\alpha}, \boldsymbol{\beta}_1, \boldsymbol{\beta}_2, \boldsymbol{\beta}_3)$ *is admissible,* $\boldsymbol{\beta}_1 \sim \boldsymbol{\beta}_2 \sim \boldsymbol{\beta}_3$, *and* $\Psi^{\boldsymbol{\beta}_1}_{\boldsymbol{\beta}_2 \to \boldsymbol{\beta}_3}$ *is an isomorphism. Then*

$$\Psi^{\boldsymbol{\alpha}}_{\boldsymbol{\beta}_1 \to \boldsymbol{\beta}_3} = \Psi^{\boldsymbol{\alpha}}_{\boldsymbol{\beta}_2 \to \boldsymbol{\beta}_3} \circ \Psi^{\boldsymbol{\alpha}}_{\boldsymbol{\beta}_1 \to \boldsymbol{\beta}_2}.$$

PROOF. Pick an element $x \in HF^\circ(\Sigma, \boldsymbol{\alpha}, \boldsymbol{\beta}_1)$. Since $(\Sigma, \boldsymbol{\alpha}, \boldsymbol{\beta}_1, \boldsymbol{\beta}_2, \boldsymbol{\beta}_3)$ is admissible, we can use the associativity of the triangle maps, which was proved by Ozsváth and Szabó [**27**, Theorem 8.16], to conclude that

$$\begin{aligned}\Psi^{\boldsymbol{\alpha}}_{\boldsymbol{\beta}_2 \to \boldsymbol{\beta}_3} \circ \Psi^{\boldsymbol{\alpha}}_{\boldsymbol{\beta}_1 \to \boldsymbol{\beta}_2}(x) &= F_{\alpha,\beta_2,\beta_3}\left(F_{\alpha,\beta_1,\beta_2}(x \otimes \Theta_{\beta_1,\beta_2}) \otimes \Theta_{\beta_2,\beta_3}\right) \\ &= F_{\alpha,\beta_1,\beta_3}\left(x \otimes F_{\beta_1,\beta_2,\beta_3}(\Theta_{\beta_1,\beta_2} \otimes \Theta_{\beta_2,\beta_3})\right) \\ &= F_{\alpha,\beta_1,\beta_3}\left(x \otimes \Psi^{\boldsymbol{\beta}_1}_{\boldsymbol{\beta}_2 \to \boldsymbol{\beta}_3}(\Theta_{\beta_1,\beta_2})\right).\end{aligned}$$

So it suffices to show that $\Psi^{\boldsymbol{\beta}_1}_{\boldsymbol{\beta}_2 \to \boldsymbol{\beta}_3}(\Theta_{\beta_1,\beta_2}) = \Theta_{\beta_1,\beta_3}$. We assumed that $\Psi^{\boldsymbol{\beta}_1}_{\boldsymbol{\beta}_2 \to \boldsymbol{\beta}_3}$ is an isomorphism. Hence, by Lemma 10.4, it induces an isomorphism between the top groups

$$HF^\circ_{\text{top}}(\Sigma, \boldsymbol{\beta}_1, \boldsymbol{\beta}_2, \mathfrak{s}_0) = \mathbb{F}_2\langle\Theta_{\beta_1,\beta_2}\rangle \text{ and } HF^\circ_{\text{top}}(\Sigma, \boldsymbol{\beta}_1, \boldsymbol{\beta}_3, \mathfrak{s}_0) = \mathbb{F}_2\langle\Theta_{\beta_1,\beta_3}\rangle,$$

where $\mathfrak{s}_0$ is the torsion Spin^c structure, and has to map the generator Θ_{β_1,β_2} to the generator Θ_{β_1,β_3}. □

LEMMA 10.7. *Let* $(\Sigma, \boldsymbol{\alpha}, \boldsymbol{\beta})$ *and* $(\Sigma, \boldsymbol{\alpha}, \boldsymbol{\beta}')$ *be two admissible diagrams, let* ω *be a symplectic form on* Σ, *and suppose we are given a Hamiltonian isotopy* I *from* $\boldsymbol{\beta}$ *to* $\boldsymbol{\beta}'$. *Then the isotopy* I *induces an isomorphism*

$$\Gamma^{\boldsymbol{\alpha}}_{\boldsymbol{\beta} \to \boldsymbol{\beta}'} \colon HF^\circ(\Sigma, \boldsymbol{\alpha}, \boldsymbol{\beta}) \to HF^\circ(\Sigma, \boldsymbol{\alpha}, \boldsymbol{\beta}').$$

These isomorphisms compose under juxtaposition of isotopies, and we have $\Gamma^{\boldsymbol{\alpha}}_{\boldsymbol{\beta} \to \boldsymbol{\beta}} = Id_{HF^\circ(\Sigma,\boldsymbol{\alpha},\boldsymbol{\beta})}$. *If, moreover, the triple* $(\Sigma, \boldsymbol{\alpha}, \boldsymbol{\beta}, \boldsymbol{\beta}')$ *is admissible, then*

$$\Gamma^{\boldsymbol{\alpha}}_{\boldsymbol{\beta} \to \boldsymbol{\beta}'} = \Psi^{\boldsymbol{\alpha}}_{\boldsymbol{\beta} \to \boldsymbol{\beta}'}, \tag{10.8}$$

and, in particular, it is independent of the isotopy I *(i.e., it depends only on the endpoints). Analogous results hold for admissible diagrams* $(\Sigma, \boldsymbol{\alpha}, \boldsymbol{\beta})$ *and* $(\Sigma, \boldsymbol{\alpha}', \boldsymbol{\beta})$, *and isomorphisms* $\Gamma^{\boldsymbol{\alpha} \to \boldsymbol{\alpha}'}_{\boldsymbol{\beta}} \colon HF^\circ(\Sigma, \boldsymbol{\alpha}, \boldsymbol{\beta}) \to HF^\circ(\Sigma, \boldsymbol{\alpha}', \boldsymbol{\beta})$.

PROOF. The continuation maps $\Gamma^{\boldsymbol{\alpha}}_{\boldsymbol{\beta} \to \boldsymbol{\beta}'}$ and $\Gamma^{\boldsymbol{\alpha} \to \boldsymbol{\alpha}'}_{\boldsymbol{\beta}}$ were defined by Ozsváth and Szabó [**27**, Section 7.3], and they showed they are natural under juxtaposition in [**28**, Lemma 2.12]. The result $\Gamma^{\boldsymbol{\alpha}}_{\boldsymbol{\beta} \to \boldsymbol{\beta}} = \text{Id}_{HF^\circ(\Sigma,\boldsymbol{\alpha},\boldsymbol{\beta})}$ follows immediately from the definition of Γ on [**27**, p. 1087], as we count Maslov index zero disks in $\text{Sym}^g(\Sigma)$, with half of the boundary on $\mathbb{T}_\alpha$, and the other half on the now constant torus $\mathbb{T}_\beta$, which are exactly the constant disks whith image being some $\mathbf{x} \in \mathbb{T}_\alpha \cap \mathbb{T}_\beta$.

Equation (10.8) follows from commutativity of the continuation and triangle maps. Indeed, if $(\Sigma, \boldsymbol{\alpha}, \boldsymbol{\beta}, \boldsymbol{\gamma})$ is an admissible triple, and $\boldsymbol{\beta}'$ is a Hamiltonian translate of $\boldsymbol{\beta}$ such that $(\Sigma, \boldsymbol{\alpha}, \boldsymbol{\beta}', \boldsymbol{\gamma})$ is also admissible, then the last commutative diagram in [**28**, Theorem 2.3] is

$$\begin{CD} HF^\circ(\Sigma, \boldsymbol{\alpha}, \boldsymbol{\beta}) \otimes HF^\circ(\Sigma, \boldsymbol{\beta}, \boldsymbol{\gamma}) @>{F_{\alpha,\beta,\gamma}}>> HF^\circ(\Sigma, \boldsymbol{\alpha}, \boldsymbol{\gamma}) \\ @V{\Gamma^{\boldsymbol{\alpha}}_{\boldsymbol{\beta}\to\boldsymbol{\beta}'} \otimes \Gamma^{\boldsymbol{\beta}\to\boldsymbol{\beta}'}_{\boldsymbol{\gamma}}}VV @VV{\mathrm{Id}}V \\ HF^\circ(\Sigma, \boldsymbol{\alpha}, \boldsymbol{\beta}') \otimes HF^\circ(\Sigma, \boldsymbol{\beta}', \boldsymbol{\gamma}) @>{F_{\alpha,\beta',\gamma}}>> HF^\circ(\Sigma, \boldsymbol{\alpha}, \boldsymbol{\gamma}); \end{CD}$$

see also [**27**, Proposition 8.14]. If $\boldsymbol{\beta} \sim \boldsymbol{\gamma}$, then

$$\Gamma^{\boldsymbol{\beta}\to\boldsymbol{\beta}'}_{\boldsymbol{\gamma}} : HF^\circ(\Sigma, \boldsymbol{\beta}, \boldsymbol{\gamma}, \mathfrak{s}_0) \to HF^\circ(\Sigma, \boldsymbol{\beta}', \boldsymbol{\gamma}, \mathfrak{s}_0)$$

is an isomorphism preserving the relative $\mathbb{Z}$-gradings, hence $\Gamma^{\boldsymbol{\beta}\to\boldsymbol{\beta}'}_{\boldsymbol{\gamma}}(\Theta_{\beta,\gamma}) = \Theta_{\beta',\gamma}$. Together with Definition 10.3, we obtain that the diagram

$$\begin{CD} HF^\circ(\Sigma, \boldsymbol{\alpha}, \boldsymbol{\beta}) @>{\Psi^{\boldsymbol{\alpha}}_{\boldsymbol{\beta}\to\boldsymbol{\gamma}}}>> HF^\circ(\Sigma, \boldsymbol{\alpha}, \boldsymbol{\gamma}) \\ @V{\Gamma^{\boldsymbol{\alpha}}_{\boldsymbol{\beta}\to\boldsymbol{\beta}'}}VV @VV{\mathrm{Id}}V \\ HF^\circ(\Sigma, \boldsymbol{\alpha}, \boldsymbol{\beta}') @>{\Psi^{\boldsymbol{\alpha}}_{\boldsymbol{\beta}'\to\boldsymbol{\gamma}}}>> HF^\circ(\Sigma, \boldsymbol{\alpha}, \boldsymbol{\gamma}) \end{CD} \tag{10.9}$$

is commutative if $\boldsymbol{\beta} \sim \boldsymbol{\gamma}$.

A priori, the maps $\Psi^{\boldsymbol{\alpha}}_{\boldsymbol{\beta}\to\boldsymbol{\gamma}}$ and $\Psi^{\boldsymbol{\alpha}}_{\boldsymbol{\beta}'\to\boldsymbol{\gamma}}$ might not be isomorphisms; we choose $\boldsymbol{\gamma}$ such that they are, as follows: Let $\boldsymbol{\gamma}$ be a sufficiently small Hamiltonian translate of $\boldsymbol{\beta}'$ such that each component of $\boldsymbol{\gamma}$ intersects the corresponding component of $\boldsymbol{\beta}'$ transversely in two points. Since the triple $(\Sigma, \boldsymbol{\alpha}, \boldsymbol{\beta}, \boldsymbol{\beta}')$ is admissible, we can choose $\boldsymbol{\gamma}$ such that the quadruple $(\Sigma, \boldsymbol{\alpha}, \boldsymbol{\beta}, \boldsymbol{\beta}', \boldsymbol{\gamma})$ is also admissible. Indeed, if $\mathcal{P}$ is a quadruply-periodic domain in $(\Sigma, \boldsymbol{\alpha}, \boldsymbol{\beta}, \boldsymbol{\beta}', \boldsymbol{\gamma})$ and $\mathcal{D}_i$ is the doubly-periodic domain traced by the isotopy from $\beta_i' \in \boldsymbol{\beta}'$ to $\gamma_i \in \boldsymbol{\gamma}$ for $i \in \{1, \dots, d\}$ (in particular, $\partial \mathcal{D}_i = \beta_i' - \gamma_i$), then

$$\mathcal{P}' = \mathcal{P} + a_1 \mathcal{D}_1 + \dots + a_d \mathcal{D}_d$$

is a triply-periodic domain in $(\Sigma, \boldsymbol{\alpha}, \boldsymbol{\beta}, \boldsymbol{\beta}')$, where a_i is the multiplicity of γ_i in $\partial\mathcal{P}$. Since $(\Sigma, \boldsymbol{\alpha}, \boldsymbol{\beta}, \boldsymbol{\beta}')$ is admissible, $\mathcal{P}'$ has both positive and negative multiplicities, and hence so does $\mathcal{P}$ if $\boldsymbol{\gamma}$ is so close to $\boldsymbol{\beta}'$ that none of the components of $\Sigma \setminus (\boldsymbol{\alpha} \cup \boldsymbol{\beta} \cup \boldsymbol{\beta}')$ is contained in a bigon between $\boldsymbol{\beta}'$ and $\boldsymbol{\gamma}$. In particular, both $(\Sigma, \boldsymbol{\alpha}, \boldsymbol{\beta}, \boldsymbol{\gamma})$ and $(\Sigma, \boldsymbol{\alpha}, \boldsymbol{\beta}', \boldsymbol{\gamma})$ are admissible, satisfying the conditions for the above diagram to be commutative. Since $\boldsymbol{\gamma}$ is close to $\boldsymbol{\beta}'$, a result of Ozsváth and Szabó [**27**, Proposition 9.8] implies that the maps

$$\Psi^{\boldsymbol{\alpha}}_{\boldsymbol{\beta}'\to\boldsymbol{\gamma}} : HF^\circ(\Sigma, \boldsymbol{\alpha}, \boldsymbol{\beta}) \to HF^\circ(\Sigma, \boldsymbol{\alpha}, \boldsymbol{\gamma}) \text{ and}$$
$$\Psi^{\boldsymbol{\beta}}_{\boldsymbol{\beta}'\to\boldsymbol{\gamma}} : HF^\circ(\Sigma, \boldsymbol{\beta}, \boldsymbol{\beta}') \to HF^\circ(\Sigma, \boldsymbol{\beta}, \boldsymbol{\gamma})$$

are isomorphisms. Then the commutativity of diagram (10.9) gives that

$$\Gamma^{\boldsymbol{\alpha}}_{\boldsymbol{\beta}\to\boldsymbol{\beta}'} = \left(\Psi^{\boldsymbol{\alpha}}_{\boldsymbol{\beta}'\to\boldsymbol{\gamma}}\right)^{-1} \circ \Psi^{\boldsymbol{\alpha}}_{\boldsymbol{\beta}\to\boldsymbol{\gamma}}.$$

Since the quadruple $(\Sigma, \boldsymbol{\alpha}, \boldsymbol{\beta}, \boldsymbol{\beta}', \boldsymbol{\gamma})$ is admissible and $\Psi^{\boldsymbol{\beta}}_{\boldsymbol{\beta}'\to\boldsymbol{\gamma}}$ is an isomorphism, we can apply Lemma 10.6 to conclude that

$$\left(\Psi^{\boldsymbol{\alpha}}_{\boldsymbol{\beta}'\to\boldsymbol{\gamma}}\right)^{-1} \circ \Psi^{\boldsymbol{\alpha}}_{\boldsymbol{\beta}\to\boldsymbol{\gamma}} = \Psi^{\boldsymbol{\alpha}}_{\boldsymbol{\beta}\to\boldsymbol{\beta}'}.$$

An alternate elegant argument can be given using monogons; see the work of Lipshitz [**19**, Proposition 11.4]. □

REMARK 10.10. Continuation maps in general symplectic manifolds do depend on the homotopy class of the isotopy, and hence cannot be written in terms of triangle maps; see Seidel [**36**]. The above lemma is specific to Heegaard Floer homology for two reasons: First, in Lemma 10.6, we used that, if $\boldsymbol{\beta} \sim \boldsymbol{\gamma}$, then $HF^\circ(\Sigma, \boldsymbol{\beta}, \boldsymbol{\gamma})$ is relatively $\mathbb{Z}$-graded as opposed to just being $\mathbb{Z}_2$-graded, and hence the top-graded generator $\Theta_{\beta,\gamma}$ is well-defined up to grading-preserving automorphism (cf. Lemma 10.4). Secondly, the admissibility conditions allow us to work over a polynomial ring instead of Novikov ring coefficients (cf. [**27**, Section 10.0.1]).

Another way to view Lemma 10.7 is that the triangle map $\Psi^{\boldsymbol{\alpha}}_{\boldsymbol{\beta}\to\boldsymbol{\beta}'}$ is an isomorphism whenever the triple $(\Sigma, \boldsymbol{\alpha}, \boldsymbol{\beta}, \boldsymbol{\beta}')$ is admissible and $\boldsymbol{\beta}$ and $\boldsymbol{\beta}'$ are Hamiltonian isotopic. Our next goal is to relax the second condition and show that $\Psi^{\boldsymbol{\alpha}}_{\boldsymbol{\beta}\to\boldsymbol{\beta}'}$ is also an isomorphism whenever $\boldsymbol{\beta} \sim \boldsymbol{\beta}'$.

PROPOSITION 10.11. (1) *Suppose that $(\Sigma, \boldsymbol{\alpha}, \boldsymbol{\beta}, \boldsymbol{\beta}')$ is an admissible triple and we have $\boldsymbol{\beta} \sim \boldsymbol{\beta}'$. Then the map*

$$\Psi^{\boldsymbol{\alpha}}_{\boldsymbol{\beta}\to\boldsymbol{\beta}'} \colon HF^\circ(\Sigma, \boldsymbol{\alpha}, \boldsymbol{\beta}) \to HF^\circ(\Sigma, \boldsymbol{\alpha}, \boldsymbol{\beta}')$$

is an isomorphism.

(2) *These isomorphisms are compatible in the sense that, if the triple diagrams $(\Sigma, \boldsymbol{\alpha}, \boldsymbol{\beta}, \boldsymbol{\beta}')$, $(\Sigma, \boldsymbol{\alpha}, \boldsymbol{\beta}', \boldsymbol{\beta}'')$, and $(\Sigma, \boldsymbol{\alpha}, \boldsymbol{\beta}, \boldsymbol{\beta}'')$ are admissible, then*

$$\Psi^{\boldsymbol{\alpha}}_{\boldsymbol{\beta}'\to\boldsymbol{\beta}''} \circ \Psi^{\boldsymbol{\alpha}}_{\boldsymbol{\beta}\to\boldsymbol{\beta}'} = \Psi^{\boldsymbol{\alpha}}_{\boldsymbol{\beta}\to\boldsymbol{\beta}''}.$$

(3) *Similarly, if $(\Sigma, \boldsymbol{\alpha}', \boldsymbol{\alpha}, \boldsymbol{\beta})$ is admissible and $\boldsymbol{\alpha} \sim \boldsymbol{\alpha}'$, then the map*

$$\Psi^{\boldsymbol{\alpha}\to\boldsymbol{\alpha}'}_{\boldsymbol{\beta}} \colon HF^\circ(\Sigma, \boldsymbol{\alpha}, \boldsymbol{\beta}) \to HF^\circ(\Sigma, \boldsymbol{\alpha}', \boldsymbol{\beta})$$

is an isomorphism, and satisfies the analogue of (2). *Finally, we have*

$$\Psi^{\boldsymbol{\alpha}\to\boldsymbol{\alpha}'}_{\boldsymbol{\beta}'} \circ \Psi^{\boldsymbol{\alpha}}_{\boldsymbol{\beta}\to\boldsymbol{\beta}'} = \Psi^{\boldsymbol{\alpha}'}_{\boldsymbol{\beta}\to\boldsymbol{\beta}'} \circ \Psi^{\boldsymbol{\alpha}\to\boldsymbol{\alpha}'}_{\boldsymbol{\beta}},$$

assuming all four triple diagrams involved are admissible.

PROOF. We first show (1). By Lemma 2.11, we can get from $\boldsymbol{\beta}$ to $\boldsymbol{\beta}'$ by a sequence of isotopies and handleslides; let $h(\boldsymbol{\beta}, \boldsymbol{\beta}')$ be the minimal number of handleslides required in such a sequence. We prove the claim by induction on $h(\boldsymbol{\beta}, \boldsymbol{\beta}')$.

Suppose that $h(\boldsymbol{\beta}, \boldsymbol{\beta}') = 0$. Since the triple $(\Sigma, \boldsymbol{\alpha}, \boldsymbol{\beta}, \boldsymbol{\beta}')$ is admissible, the pair $(\Sigma, \boldsymbol{\beta}, \boldsymbol{\beta}')$ is also admissible. According to Ozsváth and Szabó [**27**, Lemma 4.12], there exists a volume form ω on Σ for which every periodic domain has signed area equal to zero. Let $\boldsymbol{\beta} = \{\beta_1, \dots, \beta_k\}$ and $\boldsymbol{\beta}' = \{\beta_1', \dots, \beta_k'\}$, labeled such that β_i and β_i' are isotopic for every $i \in \{1, \dots, k\}$. Then the cycle $\beta_i - \beta_i'$ is the boundary of a 2-chain $\mathcal{P}_i$, which can be viewed as a periodic domain. Since $\mathcal{P}_i$ has signed area zero with respect to ω, it follows that β_i and β_i' are Hamiltonian isotopic for every $i \in \{1, \dots, k\}$. Let d_1 be a diffeomorphism Hamiltonian isotopic to the identity such that $d_1(\beta_1) = \beta_1'$. Then $d_1(\beta_2)$ and β_2' are also Hamiltonian isotopic, and since they are both disjoint and homologically linearly independent from $d_1(\beta_1) = \beta_1'$, they can be connected by a Hamiltonian isotopy that is the identity in a neighborhood of β_1'. Let d_2 be a diffeomorphism Hamiltonian isotopic to the identity that fixes a neighborhood of β_1', and such that $d_2 \circ d_1(\beta_2) = \beta_2'$. Continuing this process, we recursively obtain diffeomorphisms $d_1, \dots, d_k$ such that d_i is Hamiltonian isotopic to the identity, and fixes a neighborhood of $\beta_1', \dots, \beta_{i-1}'$ for every $i \in \{1, \dots, k\}$. Hence

$$d_k \circ \dots \circ d_1(\boldsymbol{\beta}) = \boldsymbol{\beta}',$$

and $d_k \circ \dots \circ d_1$ is Hamiltonian isotopic to the identity. By Lemma 10.7, the triangle map $\Psi^{\boldsymbol{\alpha}}_{\boldsymbol{\beta} \to \boldsymbol{\beta}'}$ is an isomorphism for any complex structure compatible with ω. However, the triangle maps commute with the maps $\Phi_{J_s \to J_s'}$, hence it is an isomorphism for any complex structure and perturbation.

Suppose we know the statement for $h(\boldsymbol{\beta}, \boldsymbol{\beta}') < n$ for some $n > 0$. If $h(\boldsymbol{\beta}, \boldsymbol{\beta}') = n$, then we can choose an attaching set $\boldsymbol{\gamma}$ such that $h(\boldsymbol{\beta}, \boldsymbol{\gamma}) = 1$ and $h(\boldsymbol{\gamma}, \boldsymbol{\beta}') = n-1$; furthermore, $\boldsymbol{\gamma}$ is obtained from $\boldsymbol{\beta}$ by a model handleslide as described by Ozsváth and Szabó [**27**, Section 9]. Then, according to Ozsváth and Szabó [**27**, Theorem 9.5], the triple $(\Sigma, \boldsymbol{\alpha}, \boldsymbol{\beta}, \boldsymbol{\gamma})$ is admissible and the map $\Psi^{\boldsymbol{\alpha}}_{\boldsymbol{\beta} \to \boldsymbol{\gamma}}$ is an isomorphism. The triple diagram $(\Sigma, \boldsymbol{\alpha}, \boldsymbol{\gamma}, \boldsymbol{\beta}')$ might not be admissible, but, by Lemma 10.5, there is a Hamiltonian translate $\boldsymbol{\gamma}'$ of $\boldsymbol{\gamma}$ for which both $(\Sigma, \boldsymbol{\alpha}, \boldsymbol{\beta}, \boldsymbol{\beta}', \boldsymbol{\gamma}')$ and $(\Sigma, \boldsymbol{\alpha}, \boldsymbol{\beta}, \boldsymbol{\gamma}, \boldsymbol{\gamma}')$ are admissible. Then consider the following diagram:

$$\begin{array}{ccc}
HF^\circ(\Sigma, \boldsymbol{\alpha}, \boldsymbol{\beta}) & \xrightarrow{\Psi^{\boldsymbol{\alpha}}_{\boldsymbol{\beta} \to \boldsymbol{\beta}'}} & HF^\circ(\Sigma, \boldsymbol{\alpha}, \boldsymbol{\beta}') \\
{\scriptstyle \Psi^{\boldsymbol{\alpha}}_{\boldsymbol{\beta} \to \boldsymbol{\gamma}}} \big\downarrow & \searrow^{\Psi^{\boldsymbol{\alpha}}_{\boldsymbol{\beta} \to \boldsymbol{\gamma}'}} & \big\uparrow {\scriptstyle \Psi^{\boldsymbol{\alpha}}_{\boldsymbol{\gamma}' \to \boldsymbol{\beta}'}} \\
HF^\circ(\Sigma, \boldsymbol{\alpha}, \boldsymbol{\gamma}) & \xrightarrow[\Psi^{\boldsymbol{\alpha}}_{\boldsymbol{\gamma} \to \boldsymbol{\gamma}'}]{} & HF^\circ(\Sigma, \boldsymbol{\alpha}, \boldsymbol{\gamma}').
\end{array}$$

We will prove that it is commutative. Since $(\Sigma, \boldsymbol{\alpha}, \boldsymbol{\gamma}, \boldsymbol{\gamma}')$ is admissible and $\boldsymbol{\gamma}'$ is a Hamiltonian translate of $\boldsymbol{\gamma}$, Lemma 10.7 implies that the map $\Psi^{\boldsymbol{\alpha}}_{\boldsymbol{\gamma} \to \boldsymbol{\gamma}'} = \Gamma^{\boldsymbol{\alpha}}_{\boldsymbol{\gamma} \to \boldsymbol{\gamma}'}$ is an isomorphism. Similarly, $\Psi^{\boldsymbol{\beta}}_{\boldsymbol{\gamma} \to \boldsymbol{\gamma}'}$ is also an isomorphism, and as the tuple $(\Sigma, \boldsymbol{\alpha}, \boldsymbol{\beta}, \boldsymbol{\gamma}, \boldsymbol{\gamma}')$ is admissible, we can apply Lemma 10.6 to conclude that

$$\Psi^{\boldsymbol{\alpha}}_{\boldsymbol{\gamma} \to \boldsymbol{\gamma}'} \circ \Psi^{\boldsymbol{\alpha}}_{\boldsymbol{\beta} \to \boldsymbol{\gamma}} = \Psi^{\boldsymbol{\alpha}}_{\boldsymbol{\beta} \to \boldsymbol{\gamma}'}.$$

We have seen that both $\Psi^{\boldsymbol{\alpha}}_{\boldsymbol{\gamma} \to \boldsymbol{\gamma}'}$ and $\Psi^{\boldsymbol{\alpha}}_{\boldsymbol{\beta} \to \boldsymbol{\gamma}}$ are isomorphisms, so $\Psi^{\boldsymbol{\alpha}}_{\boldsymbol{\beta} \to \boldsymbol{\gamma}'}$ is an isomorphism. Since $h(\boldsymbol{\gamma}', \boldsymbol{\beta}') = n - 1$, the map $\Psi^{\boldsymbol{\alpha}}_{\boldsymbol{\gamma}' \to \boldsymbol{\beta}'}$ is an isomorphism by the induction hypothesis. So we are done if we show that

$$\Psi^{\boldsymbol{\alpha}}_{\boldsymbol{\beta} \to \boldsymbol{\beta}'} = \Psi^{\boldsymbol{\alpha}}_{\boldsymbol{\gamma}' \to \boldsymbol{\beta}'} \circ \Psi^{\boldsymbol{\alpha}}_{\boldsymbol{\beta} \to \boldsymbol{\gamma}'}.$$

This also follows from Lemma 10.6. Indeed, the quadruple diagram $(\Sigma, \boldsymbol{\alpha}, \boldsymbol{\beta}, \boldsymbol{\beta}', \boldsymbol{\gamma}')$ is admissible; furthermore, the map $\Psi^{\boldsymbol{\beta}}_{\boldsymbol{\gamma}' \to \boldsymbol{\beta}'}$ is an isomorphism by the induction hypothesis (the diagram $(\Sigma, \boldsymbol{\beta}, \boldsymbol{\gamma}', \boldsymbol{\beta}')$ is admissible and $h(\boldsymbol{\gamma}', \boldsymbol{\beta}') = n - 1$). It follows that $\Psi^{\boldsymbol{\alpha}}_{\boldsymbol{\beta} \to \boldsymbol{\beta}'}$ is an isomorphism, concluding the proof of (1).

A useful consequence of (1) is that, in Lemma 10.6, the condition that $\Psi^{\boldsymbol{\beta}_1}_{\boldsymbol{\beta}_2 \to \boldsymbol{\beta}_3}$ is an isomorphism automatically follows from the others (we only need that the triple $(\Sigma, \boldsymbol{\beta}_1, \boldsymbol{\beta}_2, \boldsymbol{\beta}_3)$ is admissible and $\boldsymbol{\beta}_2 \sim \boldsymbol{\beta}_3$). Armed with this fact, we proceed to the proof of (2). By Lemma 10.5, there is a Hamiltonian translate $\boldsymbol{\beta}_1'$ of $\boldsymbol{\beta}'$ such that the quadruple diagrams $(\Sigma, \boldsymbol{\alpha}, \boldsymbol{\beta}, \boldsymbol{\beta}', \boldsymbol{\beta}_1')$, $(\Sigma, \boldsymbol{\alpha}, \boldsymbol{\beta}', \boldsymbol{\beta}'', \boldsymbol{\beta}_1')$, and $(\Sigma, \boldsymbol{\alpha}, \boldsymbol{\beta}, \boldsymbol{\beta}'', \boldsymbol{\beta}_1')$ are all admissible. Then consider the following diagram:

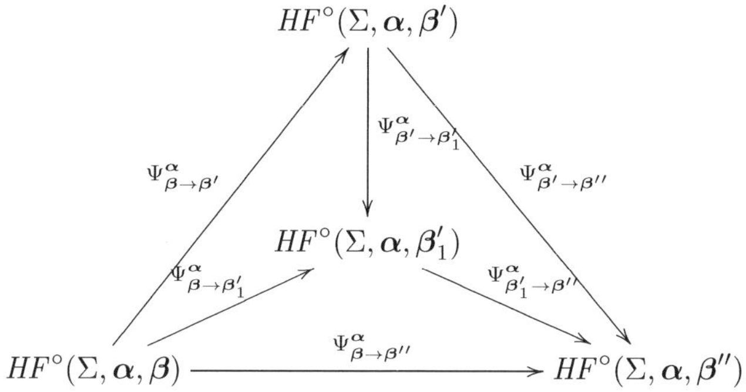

Commutativity of the three small triangles follows from the above improved version of Lemma 10.6. Hence the large triangle is also commutative; i.e.,

$$\Psi^{\boldsymbol{\alpha}}_{\boldsymbol{\beta} \to \boldsymbol{\beta}''} = \Psi^{\boldsymbol{\alpha}}_{\boldsymbol{\beta}' \to \boldsymbol{\beta}''} \circ \Psi^{\boldsymbol{\alpha}}_{\boldsymbol{\beta} \to \boldsymbol{\beta}'},$$

which concludes the proof of (2).

We finally prove (3). First, we verify this when $(\Sigma, \boldsymbol{\alpha}, \boldsymbol{\alpha}', \boldsymbol{\beta}, \boldsymbol{\beta}')$ is admissible. Pick an element $x \in HF^\circ(\Sigma, \boldsymbol{\alpha}, \boldsymbol{\beta})$. Then, using the associativity of the triangle maps [**27**, second part of Theorem 8.16],

$$\begin{aligned} \Psi^{\boldsymbol{\alpha} \to \boldsymbol{\alpha}'}_{\boldsymbol{\beta}'} \circ \Psi^{\boldsymbol{\alpha}}_{\boldsymbol{\beta} \to \boldsymbol{\beta}'}(x) &= F_{\alpha', \alpha, \beta'}(\Theta_{\alpha', \alpha} \otimes F_{\alpha, \beta, \beta'}(x \otimes \Theta_{\beta, \beta'})) \\ &= F_{\alpha', \beta, \beta'}(F_{\alpha', \alpha, \beta}(\Theta_{\alpha', \alpha} \otimes x) \otimes \Theta_{\beta, \beta'}) \\ &= \Psi^{\boldsymbol{\alpha}'}_{\boldsymbol{\beta} \to \boldsymbol{\beta}'} \circ \Psi^{\boldsymbol{\alpha} \to \boldsymbol{\alpha}'}_{\boldsymbol{\beta}}(x). \end{aligned}$$

We now consider the general case. By Lemma 10.5, there is an isotopic copy $\overline{\boldsymbol{\beta}}$ of $\boldsymbol{\beta}$ for which both $(\Sigma, \boldsymbol{\alpha}, \boldsymbol{\alpha}', \boldsymbol{\beta}, \overline{\boldsymbol{\beta}})$ and $(\Sigma, \boldsymbol{\alpha}, \boldsymbol{\alpha}', \boldsymbol{\beta}', \overline{\boldsymbol{\beta}})$ are admissible. Then

$$\begin{aligned} \Psi^{\boldsymbol{\alpha} \to \boldsymbol{\alpha}'}_{\boldsymbol{\beta}'} \circ \Psi^{\boldsymbol{\alpha}}_{\boldsymbol{\beta} \to \boldsymbol{\beta}'} &= \Psi^{\boldsymbol{\alpha} \to \boldsymbol{\alpha}'}_{\boldsymbol{\beta}'} \circ \Psi^{\boldsymbol{\alpha}}_{\overline{\boldsymbol{\beta}} \to \boldsymbol{\beta}'} \circ \Psi^{\boldsymbol{\alpha}}_{\boldsymbol{\beta} \to \overline{\boldsymbol{\beta}}} \\ &= \Psi^{\boldsymbol{\alpha}'}_{\overline{\boldsymbol{\beta}} \to \boldsymbol{\beta}'} \circ \Psi^{\boldsymbol{\alpha} \to \boldsymbol{\alpha}'}_{\overline{\boldsymbol{\beta}}} \circ \Psi^{\boldsymbol{\alpha}}_{\boldsymbol{\beta} \to \overline{\boldsymbol{\beta}}} \\ &= \Psi^{\boldsymbol{\alpha}'}_{\overline{\boldsymbol{\beta}} \to \boldsymbol{\beta}'} \circ \Psi^{\boldsymbol{\alpha}'}_{\boldsymbol{\beta} \to \overline{\boldsymbol{\beta}}} \circ \Psi^{\boldsymbol{\alpha} \to \boldsymbol{\alpha}'}_{\boldsymbol{\beta}} \\ &= \Psi^{\boldsymbol{\alpha}'}_{\boldsymbol{\beta} \to \boldsymbol{\beta}'} \circ \Psi^{\boldsymbol{\alpha} \to \boldsymbol{\alpha}'}_{\boldsymbol{\beta}}. \end{aligned}$$

Here, the first and fourth equalities follow from (2), while the second and third follow from the previous special case, assuming the admissibility conditions. This concludes the proof of (3). □

DEFINITION 10.12. Suppose that the quadruple diagram $(\Sigma, \boldsymbol{\alpha}, \boldsymbol{\alpha}', \boldsymbol{\beta}, \boldsymbol{\beta}')$ is admissible, $\boldsymbol{\alpha} \sim \boldsymbol{\alpha}'$, and $\boldsymbol{\beta} \sim \boldsymbol{\beta}'$. Then let

$$\Psi^{\boldsymbol{\alpha} \to \boldsymbol{\alpha}'}_{\boldsymbol{\beta} \to \boldsymbol{\beta}'} = \Psi^{\boldsymbol{\alpha} \to \boldsymbol{\alpha}'}_{\boldsymbol{\beta}'} \circ \Psi^{\boldsymbol{\alpha}}_{\boldsymbol{\beta} \to \boldsymbol{\beta}'} = \Psi^{\boldsymbol{\alpha}'}_{\boldsymbol{\beta} \to \boldsymbol{\beta}'} \circ \Psi^{\boldsymbol{\alpha} \to \boldsymbol{\alpha}'}_{\boldsymbol{\beta}}.$$

Note that the second equality holds by part (3) of Proposition 10.11.

LEMMA 10.13. *Suppose that the six-tuple* $(\Sigma, \boldsymbol{\alpha}, \boldsymbol{\alpha}', \boldsymbol{\alpha}'', \boldsymbol{\beta}, \boldsymbol{\beta}', \boldsymbol{\beta}'')$ *is admissible. Then*

$$\Psi^{\boldsymbol{\alpha}' \to \boldsymbol{\alpha}''}_{\boldsymbol{\beta}' \to \boldsymbol{\beta}''} \circ \Psi^{\boldsymbol{\alpha} \to \boldsymbol{\alpha}'}_{\boldsymbol{\beta} \to \boldsymbol{\beta}'} = \Psi^{\boldsymbol{\alpha} \to \boldsymbol{\alpha}''}_{\boldsymbol{\beta} \to \boldsymbol{\beta}''}.$$

PROOF. By parts (2) and (3) of Proposition 10.11,

$$\begin{aligned}\Psi^{\boldsymbol{\alpha}'\to\boldsymbol{\alpha}''}_{\boldsymbol{\beta}'\to\boldsymbol{\beta}''}\circ\Psi^{\boldsymbol{\alpha}\to\boldsymbol{\alpha}'}_{\boldsymbol{\beta}\to\boldsymbol{\beta}'} &= \Psi^{\boldsymbol{\alpha}''}_{\boldsymbol{\beta}'\to\boldsymbol{\beta}''}\circ\Psi^{\boldsymbol{\alpha}'\to\boldsymbol{\alpha}''}_{\boldsymbol{\beta}'}\circ\Psi^{\boldsymbol{\alpha}\to\boldsymbol{\alpha}'}_{\boldsymbol{\beta}'}\circ\Psi^{\boldsymbol{\alpha}}_{\boldsymbol{\beta}\to\boldsymbol{\beta}'}\\ &= \Psi^{\boldsymbol{\alpha}''}_{\boldsymbol{\beta}'\to\boldsymbol{\beta}''}\circ\Psi^{\boldsymbol{\alpha}\to\boldsymbol{\alpha}''}_{\boldsymbol{\beta}'}\circ\Psi^{\boldsymbol{\alpha}}_{\boldsymbol{\beta}\to\boldsymbol{\beta}'}\\ &= \Psi^{\boldsymbol{\alpha}''}_{\boldsymbol{\beta}'\to\boldsymbol{\beta}''}\circ\Psi^{\boldsymbol{\alpha}''}_{\boldsymbol{\beta}\to\boldsymbol{\beta}'}\circ\Psi^{\boldsymbol{\alpha}\to\boldsymbol{\alpha}''}_{\boldsymbol{\beta}}\\ &= \Psi^{\boldsymbol{\alpha}''}_{\boldsymbol{\beta}\to\boldsymbol{\beta}''}\circ\Psi^{\boldsymbol{\alpha}\to\boldsymbol{\alpha}''}_{\boldsymbol{\beta}} = \Psi^{\boldsymbol{\alpha}\to\boldsymbol{\alpha}''}_{\boldsymbol{\beta}\to\boldsymbol{\beta}''},\end{aligned}$$

as claimed. □

So triangle maps give canonical isomorphisms $\Psi^{\boldsymbol{\alpha}\to\boldsymbol{\alpha}'}_{\boldsymbol{\beta}\to\boldsymbol{\beta}'}$ between $HF^\circ(\Sigma,\boldsymbol{\alpha},\boldsymbol{\beta})$ and $HF^\circ(\Sigma,\boldsymbol{\alpha}',\boldsymbol{\beta}')$ (in the sense of Definition 1.1) whenever $(\Sigma,\boldsymbol{\alpha},\boldsymbol{\alpha}',\boldsymbol{\beta},\boldsymbol{\beta}')$ is admissible, $\boldsymbol{\alpha}\sim\boldsymbol{\alpha}'$, and $\boldsymbol{\beta}\sim\boldsymbol{\beta}'$. But what do we do when the admissibility condition fails? If the triple $(\Sigma,\boldsymbol{\alpha},\boldsymbol{\beta},\boldsymbol{\beta}')$ is not admissible, then the triangle count in $\Psi^{\boldsymbol{\alpha}}_{\boldsymbol{\beta}\to\boldsymbol{\beta}'}$ might not be finite, and even if it is, there are simple examples where it does not give a natural isomorphism, even though $\boldsymbol{\beta}$ and $\boldsymbol{\beta}'$ are isotopic. To overcome this obstacle, we first apply a Hamiltonian isotopy to $\boldsymbol{\alpha}$ and $\boldsymbol{\beta}$ so that the quadruple $(\boldsymbol{\alpha},\boldsymbol{\alpha}',\boldsymbol{\beta},\boldsymbol{\beta}')$ becomes admissible. According to Lemma 10.5, this is always possible.

PROPOSITION 10.14. *Suppose that the diagrams $(\Sigma,\boldsymbol{\alpha},\boldsymbol{\beta})$ and $(\Sigma,\boldsymbol{\alpha}',\boldsymbol{\beta}')$ are both admissible, $\boldsymbol{\alpha}\sim\boldsymbol{\alpha}'$, and $\boldsymbol{\beta}\sim\boldsymbol{\beta}'$. According to Lemma 10.5, there exist attaching sets $\overline{\boldsymbol{\alpha}}$ and $\overline{\boldsymbol{\beta}}$ isotopic to $\boldsymbol{\alpha}$ and $\boldsymbol{\beta}$, respectively, and such that the quadruples $(\Sigma,\boldsymbol{\alpha},\overline{\boldsymbol{\alpha}},\boldsymbol{\beta},\overline{\boldsymbol{\beta}})$ and $(\Sigma,\overline{\boldsymbol{\alpha}},\boldsymbol{\alpha}',\overline{\boldsymbol{\beta}},\boldsymbol{\beta}')$ are both admissible. Then the map*

$$\Psi^{\overline{\boldsymbol{\alpha}}\to\boldsymbol{\alpha}'}_{\overline{\boldsymbol{\beta}}\to\boldsymbol{\beta}'}\circ\Psi^{\boldsymbol{\alpha}\to\overline{\boldsymbol{\alpha}}}_{\boldsymbol{\beta}\to\overline{\boldsymbol{\beta}}}\colon HF^\circ(\Sigma,\boldsymbol{\alpha},\boldsymbol{\beta})\to HF^\circ(\Sigma,\boldsymbol{\alpha}',\boldsymbol{\beta}')$$

is an isomorphism. Furthermore, it is independent of the choice of $\overline{\boldsymbol{\alpha}}$ an $\overline{\boldsymbol{\beta}}$; we denote it by $\Phi^{\boldsymbol{\alpha}\to\boldsymbol{\alpha}'}_{\boldsymbol{\beta}\to\boldsymbol{\beta}'}$. Finally, if $(\Sigma,\boldsymbol{\alpha}'',\boldsymbol{\beta}'')$ is also admissible, $\boldsymbol{\alpha}''\sim\boldsymbol{\alpha}$, and $\boldsymbol{\beta}''\sim\boldsymbol{\beta}$, then

$$\Phi^{\boldsymbol{\alpha}'\to\boldsymbol{\alpha}''}_{\boldsymbol{\beta}'\to\boldsymbol{\beta}''}\circ\Phi^{\boldsymbol{\alpha}\to\boldsymbol{\alpha}'}_{\boldsymbol{\beta}\to\boldsymbol{\beta}'} = \Phi^{\boldsymbol{\alpha}\to\boldsymbol{\alpha}''}_{\boldsymbol{\beta}\to\boldsymbol{\beta}''}. \tag{10.15}$$

PROOF. The map $\Psi^{\overline{\boldsymbol{\alpha}}\to\boldsymbol{\alpha}'}_{\overline{\boldsymbol{\beta}}\to\boldsymbol{\beta}'}\circ\Psi^{\boldsymbol{\alpha}\to\overline{\boldsymbol{\alpha}}}_{\boldsymbol{\beta}\to\overline{\boldsymbol{\beta}}}$ is an isomorphism by part (1) of Proposition 10.11. We now show that it is independent of the choice of $\overline{\boldsymbol{\alpha}}$ and $\overline{\boldsymbol{\beta}}$. Let $\overline{\boldsymbol{\alpha}}_1$, $\overline{\boldsymbol{\beta}}_1$ and $\overline{\boldsymbol{\alpha}}_2$, $\overline{\boldsymbol{\beta}}_2$ be two different choices. Using Lemma 10.5, we isotope $\boldsymbol{\alpha}$ and $\boldsymbol{\beta}$ until we get attaching sets $\overline{\boldsymbol{\alpha}}$ and $\overline{\boldsymbol{\beta}}$ such that the six-tuples obtained by adding them to the quadruples $(\Sigma,\boldsymbol{\alpha},\overline{\boldsymbol{\alpha}}_1,\boldsymbol{\beta},\overline{\boldsymbol{\beta}}_1)$, $(\Sigma,\boldsymbol{\alpha},\overline{\boldsymbol{\alpha}}_2,\boldsymbol{\beta},\overline{\boldsymbol{\beta}}_2)$, $(\Sigma,\overline{\boldsymbol{\alpha}}_1,\boldsymbol{\alpha}',\overline{\boldsymbol{\beta}}_1,\boldsymbol{\beta}')$, and $(\Sigma,\overline{\boldsymbol{\alpha}}_2,\boldsymbol{\alpha}',\overline{\boldsymbol{\beta}}_2,\boldsymbol{\beta}')$ are all admissible. Then we can consider the following diagram:

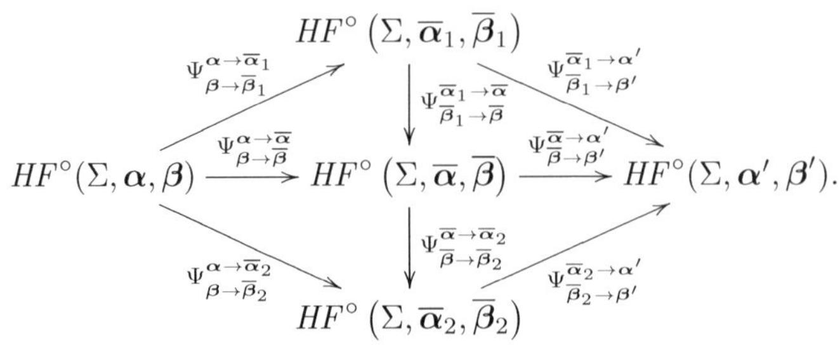

Each of the four small triangles is commutative by part (2) of Proposition 10.11. Hence, the outer square also commutes; i.e.,

$$\Psi^{\overline{\boldsymbol{\alpha}}_1\to\boldsymbol{\alpha}'}_{\overline{\boldsymbol{\beta}}_1\to\boldsymbol{\beta}'}\circ\Psi^{\boldsymbol{\alpha}\to\overline{\boldsymbol{\alpha}}_1}_{\boldsymbol{\beta}\to\overline{\boldsymbol{\beta}}_1} = \Psi^{\overline{\boldsymbol{\alpha}_2}\to\boldsymbol{\alpha}'}_{\overline{\boldsymbol{\beta}}_2\to\boldsymbol{\beta}'}\circ\Psi^{\boldsymbol{\alpha}\to\overline{\boldsymbol{\alpha}}_2}_{\boldsymbol{\beta}\to\overline{\boldsymbol{\beta}}_2},$$

so the map for $\overline{\boldsymbol{\alpha}}_1$, $\overline{\boldsymbol{\beta}}_1$ is the same as the map for $\overline{\boldsymbol{\alpha}}_2$, $\overline{\boldsymbol{\beta}}_2$.

Finally, we show equation (10.15). Using Lemma 10.5, pick isotopic copies $\overline{\boldsymbol{\alpha}}$, $\overline{\boldsymbol{\beta}}$, $\overline{\boldsymbol{\alpha}}'$, $\overline{\boldsymbol{\beta}}'$ of $\boldsymbol{\alpha}$, $\boldsymbol{\beta}$, $\boldsymbol{\alpha}'$, and $\boldsymbol{\beta}'$, respectively, such that the six-tuples obtained by adding these four attaching sets to the diagrams $(\Sigma, \boldsymbol{\alpha}, \boldsymbol{\beta})$, $(\Sigma, \boldsymbol{\alpha}', \boldsymbol{\beta}')$, and $(\Sigma, \boldsymbol{\alpha}'', \boldsymbol{\beta}'')$ are all admissible. Applying part (2) of Proposition 10.11 to the left-hand side of equation (10.15),

$$\begin{aligned}\Psi^{\overline{\boldsymbol{\alpha}}'\to\boldsymbol{\alpha}''}_{\overline{\boldsymbol{\beta}}'\to\boldsymbol{\beta}''} \circ \Psi^{\boldsymbol{\alpha}'\to\overline{\boldsymbol{\alpha}}'}_{\boldsymbol{\beta}'\to\overline{\boldsymbol{\beta}}'} \circ \Psi^{\overline{\boldsymbol{\alpha}}\to\boldsymbol{\alpha}'}_{\overline{\boldsymbol{\beta}}\to\boldsymbol{\beta}'} \circ \Psi^{\boldsymbol{\alpha}\to\overline{\boldsymbol{\alpha}}}_{\boldsymbol{\beta}\to\overline{\boldsymbol{\beta}}} &= \Psi^{\overline{\boldsymbol{\beta}}'\to\boldsymbol{\alpha}''}_{\overline{\boldsymbol{\beta}}'\to\boldsymbol{\beta}''} \circ \Psi^{\overline{\boldsymbol{\alpha}}\to\overline{\boldsymbol{\alpha}}'}_{\overline{\boldsymbol{\beta}}\to\overline{\boldsymbol{\beta}}'} \circ \Psi^{\boldsymbol{\alpha}\to\overline{\boldsymbol{\alpha}}}_{\boldsymbol{\beta}\to\overline{\boldsymbol{\beta}}} \\ &= \Psi^{\overline{\boldsymbol{\alpha}}\to\boldsymbol{\alpha}''}_{\overline{\boldsymbol{\beta}}\to\boldsymbol{\beta}''} \circ \Psi^{\boldsymbol{\alpha}\to\overline{\boldsymbol{\alpha}}}_{\boldsymbol{\beta}\to\overline{\boldsymbol{\beta}}} = \Phi^{\boldsymbol{\alpha}\to\boldsymbol{\alpha}''}_{\boldsymbol{\beta}\to\boldsymbol{\beta}''},\end{aligned}$$

as required. □

DEFINITION 10.16. Suppose that the diagrams $(\Sigma, \boldsymbol{\alpha}, \boldsymbol{\beta})$ and $(\Sigma, \boldsymbol{\alpha}, \boldsymbol{\beta}')$ are both admissible and $\boldsymbol{\beta} \sim \boldsymbol{\beta}'$. Then let

$$\Phi^{\boldsymbol{\alpha}}_{\boldsymbol{\beta}\to\boldsymbol{\beta}'} = \Phi^{\boldsymbol{\alpha}\to\boldsymbol{\alpha}}_{\boldsymbol{\beta}\to\boldsymbol{\beta}'}.$$

Similarly, when we have admissible diagrams $(\Sigma, \boldsymbol{\alpha}, \boldsymbol{\beta})$ and $(\Sigma, \boldsymbol{\alpha}', \boldsymbol{\beta})$ such that $\boldsymbol{\alpha} \sim \boldsymbol{\alpha}'$, we write

$$\Phi^{\boldsymbol{\alpha}\to\boldsymbol{\alpha}'}_{\boldsymbol{\beta}} = \Phi^{\boldsymbol{\alpha}\to\boldsymbol{\alpha}'}_{\boldsymbol{\beta}\to\boldsymbol{\beta}}.$$

LEMMA 10.17. *Suppose that the diagrams $(\Sigma, \boldsymbol{\alpha}, \boldsymbol{\beta})$ and $(\Sigma, \boldsymbol{\alpha}, \boldsymbol{\beta}')$ are both admissible and $\boldsymbol{\beta} \sim \boldsymbol{\beta}'$. Let $\overline{\boldsymbol{\beta}}$ be an isotopic copy of $\boldsymbol{\beta}$ such that the triples $(\Sigma, \boldsymbol{\alpha}, \boldsymbol{\beta}, \overline{\boldsymbol{\beta}})$ and $(\Sigma, \boldsymbol{\alpha}, \boldsymbol{\beta}', \overline{\boldsymbol{\beta}})$ are admissible. Then*

$$\Phi^{\boldsymbol{\alpha}}_{\boldsymbol{\beta}\to\boldsymbol{\beta}'} = \Psi^{\boldsymbol{\alpha}}_{\overline{\boldsymbol{\beta}}\to\boldsymbol{\beta}'} \circ \Psi^{\boldsymbol{\alpha}}_{\boldsymbol{\beta}\to\overline{\boldsymbol{\beta}}}.$$

An analogous statement holds for $\Phi^{\boldsymbol{\alpha}\to\boldsymbol{\alpha}'}_{\boldsymbol{\beta}}$. Finally,

$$\Phi^{\boldsymbol{\alpha}\to\boldsymbol{\alpha}}_{\boldsymbol{\beta}\to\boldsymbol{\beta}} = \Phi^{\boldsymbol{\alpha}}_{\boldsymbol{\beta}\to\boldsymbol{\beta}} = \Phi^{\boldsymbol{\alpha}\to\boldsymbol{\alpha}}_{\boldsymbol{\beta}} = Id_{HF^\circ(\Sigma,\boldsymbol{\alpha},\boldsymbol{\beta})}. \tag{10.18}$$

PROOF. Let $\overline{\boldsymbol{\alpha}}$ be a Hamiltonian translate of $\boldsymbol{\alpha}$ such that the quadruples $(\Sigma, \boldsymbol{\alpha}, \overline{\boldsymbol{\alpha}}, \boldsymbol{\beta}, \overline{\boldsymbol{\beta}})$ and $(\Sigma, \boldsymbol{\alpha}, \overline{\boldsymbol{\alpha}}, \boldsymbol{\beta}', \overline{\boldsymbol{\beta}})$ are admissible. By Lemma 10.7 and the naturality of the continuation maps under juxtaposition,

$$\Psi^{\overline{\boldsymbol{\alpha}}\to\boldsymbol{\alpha}}_{\overline{\boldsymbol{\beta}}} \circ \Psi^{\boldsymbol{\alpha}\to\overline{\boldsymbol{\alpha}}}_{\overline{\boldsymbol{\beta}}} = \Gamma^{\overline{\boldsymbol{\alpha}}\to\boldsymbol{\alpha}}_{\overline{\boldsymbol{\beta}}} \circ \Gamma^{\boldsymbol{\alpha}\to\overline{\boldsymbol{\alpha}}}_{\overline{\boldsymbol{\beta}}} = \mathrm{Id}_{HF^\circ(\Sigma,\boldsymbol{\alpha},\overline{\boldsymbol{\beta}})}.$$

It follows that

$$\begin{aligned}\Phi^{\boldsymbol{\alpha}}_{\boldsymbol{\beta}\to\boldsymbol{\beta}'} = \Phi^{\boldsymbol{\alpha}\to\boldsymbol{\alpha}}_{\boldsymbol{\beta}\to\boldsymbol{\beta}'} = \Psi^{\overline{\boldsymbol{\alpha}}\to\boldsymbol{\alpha}}_{\overline{\boldsymbol{\beta}}\to\boldsymbol{\beta}'} \circ \Psi^{\boldsymbol{\alpha}\to\overline{\boldsymbol{\alpha}}}_{\boldsymbol{\beta}\to\overline{\boldsymbol{\beta}}} = \Psi^{\boldsymbol{\alpha}}_{\overline{\boldsymbol{\beta}}\to\boldsymbol{\beta}'} \circ \Psi^{\overline{\boldsymbol{\alpha}}\to\boldsymbol{\alpha}}_{\overline{\boldsymbol{\beta}}} \circ \Psi^{\boldsymbol{\alpha}\to\overline{\boldsymbol{\alpha}}}_{\overline{\boldsymbol{\beta}}} \circ \Psi^{\boldsymbol{\alpha}}_{\boldsymbol{\beta}\to\overline{\boldsymbol{\beta}}} = \\ = \Psi^{\boldsymbol{\alpha}}_{\overline{\boldsymbol{\beta}}\to\boldsymbol{\beta}'} \circ \Psi^{\boldsymbol{\alpha}}_{\boldsymbol{\beta}\to\overline{\boldsymbol{\beta}}},\end{aligned}$$

as claimed. The statement for $\Phi^{\boldsymbol{\alpha}\to\boldsymbol{\alpha}'}_{\boldsymbol{\beta}}$ follows similarly.

We now prove the last statement regarding $\Phi^{\boldsymbol{\alpha}\to\boldsymbol{\alpha}}_{\boldsymbol{\beta}\to\boldsymbol{\beta}}$. Let $\overline{\boldsymbol{\beta}}$ be a Hamiltonian translate of $\boldsymbol{\beta}$ such that $(\Sigma, \boldsymbol{\alpha}, \boldsymbol{\beta}, \overline{\boldsymbol{\beta}})$ is admissible. If we apply the first part with $\boldsymbol{\beta} = \boldsymbol{\beta}'$, we get that

$$\Phi^{\boldsymbol{\alpha}}_{\boldsymbol{\beta}\to\boldsymbol{\beta}} = \Psi^{\boldsymbol{\alpha}}_{\overline{\boldsymbol{\beta}}\to\boldsymbol{\beta}} \circ \Psi^{\boldsymbol{\alpha}}_{\boldsymbol{\beta}\to\overline{\boldsymbol{\beta}}}.$$

Using Lemma 10.7, the right-hand side is $\Gamma^{\boldsymbol{\alpha}}_{\overline{\boldsymbol{\beta}}\to\boldsymbol{\beta}} \circ \Gamma^{\boldsymbol{\alpha}}_{\boldsymbol{\beta}\to\overline{\boldsymbol{\beta}}}$. By Lemma 10.7, the naturality of the continuation maps under juxtaposition, this is $\Gamma^{\boldsymbol{\alpha}}_{\boldsymbol{\beta}\to\boldsymbol{\beta}} = \mathrm{Id}_{HF^\circ(\Sigma,\boldsymbol{\alpha},\boldsymbol{\beta})}$. □

Corollary 10.19. *Let* $(\Sigma, \boldsymbol{\alpha}, \boldsymbol{\beta}, \boldsymbol{\beta}')$ *be an admissible triple such that* $\boldsymbol{\beta} \sim \boldsymbol{\beta}'$. *Then*

$$\left(\Psi^{\boldsymbol{\alpha}}_{\boldsymbol{\beta}\to\boldsymbol{\beta}'}\right)^{-1} = \Psi^{\boldsymbol{\alpha}}_{\boldsymbol{\beta}'\to\boldsymbol{\beta}}.$$

An analogous result holds for the maps $\Psi^{\boldsymbol{\alpha}\to\boldsymbol{\alpha}'}_{\boldsymbol{\beta}}$.

Proof. By Lemma 10.17,

$$\Psi^{\boldsymbol{\alpha}}_{\boldsymbol{\beta}'\to\boldsymbol{\beta}} \circ \Psi^{\boldsymbol{\alpha}}_{\boldsymbol{\beta}\to\boldsymbol{\beta}'} = \Phi^{\boldsymbol{\alpha}}_{\boldsymbol{\beta}\to\boldsymbol{\beta}} = \mathrm{Id}_{HF^\circ(\Sigma,\boldsymbol{\alpha},\boldsymbol{\beta})}. \qquad \square$$

Let (Σ, A, B) be an isotopy diagram. Then we denote by $M_{(\Sigma,A,B)}$ the set of admissible diagrams $(\Sigma, \boldsymbol{\alpha}, \boldsymbol{\beta})$ such that $[\boldsymbol{\alpha}] = A$ and $[\boldsymbol{\beta}] = B$. This is non-empty by Lemma 10.5. It follows from equations (10.15) and (10.18) that the groups $HF^\circ(\Sigma, \boldsymbol{\alpha}, \boldsymbol{\beta})$ for $(\Sigma, \boldsymbol{\alpha}, \boldsymbol{\beta}) \in M_{(\Sigma,A,B)}$, together with the isomorphisms $\Phi^{\boldsymbol{\alpha}\to\boldsymbol{\alpha}'}_{\boldsymbol{\beta}\to\boldsymbol{\beta}'}$ form a transitive system of groups, as in Definition 1.1.

Definition 10.20. Given an isotopy diagram H, let $HF^\circ(H)$ be the colimit of the transitive system of groups $HF^\circ(\Sigma, \boldsymbol{\alpha}, \boldsymbol{\beta})$ for $(\Sigma, \boldsymbol{\alpha}, \boldsymbol{\beta}) \in M_H$ and $\Phi^{\boldsymbol{\alpha}\to\boldsymbol{\alpha}'}_{\boldsymbol{\beta}\to\boldsymbol{\beta}'}$. In other words,

$$HF^\circ(H) = \coprod_{(\Sigma,\boldsymbol{\alpha},\boldsymbol{\beta})\in M_H} HF^\circ(\Sigma, \boldsymbol{\alpha}, \boldsymbol{\beta})\Big/\sim,$$

where $x \in HF^\circ(\Sigma, \boldsymbol{\alpha}, \boldsymbol{\beta})$ and $x' \in HF^\circ(\Sigma, \boldsymbol{\alpha}', \boldsymbol{\beta}')$ are equivalent if and only if $x' = \Phi^{\boldsymbol{\alpha}\to\boldsymbol{\alpha}'}_{\boldsymbol{\beta}\to\boldsymbol{\beta}'}(x)$.

We would like to show that HF° is a weak Heegaard invariant. To this end, we need to define isomorphisms induced by α-equivalences, β-equivalences, diffeomorphisms, and (de)stabilizations between isotopy diagrams. We start with α- and β-equivalences.

Lemma 10.21. *Suppose that we are given admissible diagrams* $(\Sigma, \boldsymbol{\alpha}_1, \boldsymbol{\beta}_1)$, $(\Sigma, \boldsymbol{\alpha}_1, \boldsymbol{\beta}_1')$, $(\Sigma, \boldsymbol{\alpha}_2, \boldsymbol{\beta}_2)$, *and* $(\Sigma, \boldsymbol{\alpha}_2, \boldsymbol{\beta}_2')$ *such that* $\boldsymbol{\alpha}_1 \sim \boldsymbol{\alpha}_2$ *and* $\boldsymbol{\beta}_1 \sim \boldsymbol{\beta}_2 \sim \boldsymbol{\beta}_1' \sim \boldsymbol{\beta}_2'$. *Then the following diagram is commutative:*

$$\begin{array}{ccc} HF^\circ(\Sigma, \boldsymbol{\alpha}_1, \boldsymbol{\beta}_1) & \xrightarrow{\Phi^{\boldsymbol{\alpha}_1}_{\boldsymbol{\beta}_1\to\boldsymbol{\beta}_1'}} & HF^\circ(\Sigma, \boldsymbol{\alpha}_1, \boldsymbol{\beta}_1') \\ \downarrow{\scriptstyle \Phi^{\boldsymbol{\alpha}_1\to\boldsymbol{\alpha}_2}_{\boldsymbol{\beta}_1\to\boldsymbol{\beta}_2}} & & \downarrow{\scriptstyle \Phi^{\boldsymbol{\alpha}_1\to\boldsymbol{\alpha}_2}_{\boldsymbol{\beta}_1'\to\boldsymbol{\beta}_2'}} \\ HF^\circ(\Sigma, \boldsymbol{\alpha}_2, \boldsymbol{\beta}_2) & \xrightarrow{\Phi^{\boldsymbol{\alpha}_2}_{\boldsymbol{\beta}_2\to\boldsymbol{\beta}_2'}} & HF^\circ(\Sigma, \boldsymbol{\alpha}_2, \boldsymbol{\beta}_2'). \end{array}$$

Proof. By equation (10.15),

$$\Phi^{\boldsymbol{\alpha}_1\to\boldsymbol{\alpha}_2}_{\boldsymbol{\beta}_1'\to\boldsymbol{\beta}_2'} \circ \Phi^{\boldsymbol{\alpha}_1}_{\boldsymbol{\beta}_1\to\boldsymbol{\beta}_1'} = \Phi^{\boldsymbol{\alpha}_1\to\boldsymbol{\alpha}_2}_{\boldsymbol{\beta}_1\to\boldsymbol{\beta}_2'} = \Phi^{\boldsymbol{\alpha}_2}_{\boldsymbol{\beta}_2\to\boldsymbol{\beta}_2'} \circ \Phi^{\boldsymbol{\alpha}_1\to\boldsymbol{\alpha}_2}_{\boldsymbol{\beta}_1\to\boldsymbol{\beta}_2}. \qquad \square$$

Definition 10.22. Suppose that the isotopy diagrams $H = (\Sigma, A, B)$ and $H' = (\Sigma, A, B')$ are β-equivalent. Pick admissible representatives $(\Sigma, \boldsymbol{\alpha}, \boldsymbol{\beta})$ and $(\Sigma, \boldsymbol{\alpha}, \boldsymbol{\beta}')$ of H and H', respectively (this is possible by Lemma 10.5). By Lemma 10.21, the isomorphisms $\Phi^{\boldsymbol{\alpha}}_{\boldsymbol{\beta}\to\boldsymbol{\beta}'}$ descend to the direct limit, giving an isomorphism

$$\Phi^A_{B\to B'} \colon HF^\circ(H) \to HF^\circ(H').$$

For α-equivalent diagrams (Σ, A, B) and (Σ, A', B), we define the isomorphism $\Phi^{A\to A'}_B$ analogously.

REMARK 10.23. Note that the isomorphisms Ozsváth and Szabó [**28**, p. 344] associate to α- and β-equivalences are defined by composing continuation maps Γ and triangle maps. The isomorphisms we define in Definition 10.22 only involve triangle maps, hence are better suited to computations, but agree with the isomorphisms of Ozsváth and Szabó by Lemma 10.7.

Next, we go on to define isomorphisms induced by diffeomorphisms.

DEFINITION 10.24. Let $(\Sigma, \boldsymbol{\alpha}, \boldsymbol{\beta})$ be an admissible diagram and $d\colon \Sigma \to \Sigma'$ a diffeomorphism. We write $\boldsymbol{\alpha}' = d(\boldsymbol{\alpha})$ and $\boldsymbol{\beta}' = d(\boldsymbol{\beta})$. Then d induces an isomorphism

$$d_*\colon HF^\circ(\Sigma, \boldsymbol{\alpha}, \boldsymbol{\beta}) \to HF^\circ(\Sigma', \boldsymbol{\alpha}', \boldsymbol{\beta}'),$$

as follows. Let $k = |\boldsymbol{\alpha}| = |\boldsymbol{\beta}|$. Choose a complex structure $\mathfrak{j}$ on Σ and a perturbation J_s of $\mathrm{Sym}^k(\mathfrak{j})$ on $\mathrm{Sym}^k(\Sigma)$. Pushing $\mathfrak{j}$ and J_s forward along d, we get a complex structure $\mathfrak{j}'$ on Σ' and a perturbation J_s' of $\mathrm{Sym}^k(\mathfrak{j}')$ on $\mathrm{Sym}^k(\Sigma')$. Clearly, d induces an isomorphism

$$d_{J_s,J_s'}\colon HF^\circ_{J_s}(\Sigma, \boldsymbol{\alpha}, \boldsymbol{\beta}) \to HF^\circ_{J_s'}(\Sigma', \boldsymbol{\alpha}', \boldsymbol{\beta}').$$

Since the maps $d_{J_s,J_s'}$ commute with the isomorphisms $\Phi_{J_s \to \overline{J_s}}$ of Lemma 10.1, these diffeomorphism maps descend to a map d_* on the direct limit $HF^\circ(\Sigma, \boldsymbol{\alpha}, \boldsymbol{\beta})$.

LEMMA 10.25. *The maps $\Psi^{\boldsymbol{\alpha}}_{\boldsymbol{\beta}\to\boldsymbol{\beta}'}$ commute with the diffeomorphism maps d_* defined above. More precisely, suppose that $(\Sigma, \boldsymbol{\alpha}, \boldsymbol{\beta}, \boldsymbol{\beta}')$ is an admissible triple. Let $d\colon \Sigma \to \overline{\Sigma}$ be a diffeomorphism, and write $\overline{\boldsymbol{\alpha}} = d(\boldsymbol{\alpha})$, $\overline{\boldsymbol{\beta}} = d(\boldsymbol{\beta})$, and $\overline{\boldsymbol{\beta}}' = d(\boldsymbol{\beta}')$. Then we have a commutative rectangle*

$$\begin{array}{ccc} HF^\circ(\Sigma, \boldsymbol{\alpha}, \boldsymbol{\beta}) & \xrightarrow{\Psi^{\boldsymbol{\alpha}}_{\boldsymbol{\beta}\to\boldsymbol{\beta}'}} & HF^\circ(\Sigma, \boldsymbol{\alpha}, \boldsymbol{\beta}') \\ \big\downarrow{\scriptstyle d_*} & & \big\downarrow{\scriptstyle d_*} \\ HF^\circ\left(\overline{\Sigma}, \overline{\boldsymbol{\alpha}}, \overline{\boldsymbol{\beta}}\right) & \xrightarrow{\Psi^{\overline{\boldsymbol{\alpha}}}_{\overline{\boldsymbol{\beta}}\to\overline{\boldsymbol{\beta}}'}} & HF^\circ\left(\overline{\Sigma}, \overline{\boldsymbol{\alpha}}, \overline{\boldsymbol{\beta}}'\right). \end{array}$$

An analogous result holds for the maps $\Psi^{\boldsymbol{\alpha}\to\boldsymbol{\alpha}'}_{\boldsymbol{\beta}}$.

PROOF. If we choose corresponding complex structures and perturbations for Σ and $\overline{\Sigma}$, the statement becomes a tautology. Indeed, $\mathrm{Sym}^k(d)$ is a diffeomorphism between $\mathrm{Sym}^k(\Sigma)$ and $\mathrm{Sym}^k(\overline{\Sigma})$ that takes the triple $(\mathbb{T}_\alpha, \mathbb{T}_\beta, \mathbb{T}_{\beta'})$ to the triple $(\mathbb{T}_{\overline{\alpha}}, \mathbb{T}_{\overline{\beta}}, \mathbb{T}_{\overline{\beta}'})$, and matches up the complex structures and perturbations. Hence the triangle maps $\Psi^{\boldsymbol{\alpha}}_{\boldsymbol{\beta}\to\boldsymbol{\beta}'}$ and $\Psi^{\overline{\boldsymbol{\alpha}}}_{\overline{\boldsymbol{\beta}}\to\overline{\boldsymbol{\beta}}'}$ are conjugate along d_*. □

It follows from Lemma 10.25 that the diffeomorphism maps and the canonical isomorphisms $\Phi^{\boldsymbol{\alpha}\to\boldsymbol{\alpha}'}_{\boldsymbol{\beta}\to\boldsymbol{\beta}'}$ for admissible diagrams $(\Sigma, \boldsymbol{\alpha}, \boldsymbol{\beta})$ and $(\Sigma, \boldsymbol{\alpha}', \boldsymbol{\beta}')$ such that $\boldsymbol{\alpha} \sim \boldsymbol{\alpha}'$ and $\boldsymbol{\beta} \sim \boldsymbol{\beta}'$ also commute, as $\Phi^{\boldsymbol{\alpha}\to\boldsymbol{\alpha}'}_{\boldsymbol{\beta}\to\boldsymbol{\beta}'}$ can be written as a composition of triangle maps. Hence, if H and H' are isotopy diagrams and $d\colon H \to H'$ is a diffeomorphism, then d descents to a map of direct limits

$$d_*\colon HF^\circ(H) \to HF^\circ(H').$$

Finally, we define maps induced by stabilizations. We proceed as Ozsváth and Szabó [**27**, Section 10], [**28**, p. 346]. Suppose that $\mathcal{H}' = (\Sigma', \boldsymbol{\alpha}', \boldsymbol{\beta}')$ is a stabilization of the admissible diagram $\mathcal{H} = (\Sigma, \boldsymbol{\alpha}, \boldsymbol{\beta})$. Then, for suitable almost-complex structures, there is an isomorphism of chain complexes

$$\sigma_{\mathcal{H}\to\mathcal{H}'} \colon CF^\circ(\Sigma, \boldsymbol{\alpha}, \boldsymbol{\beta}) \to CF^\circ(\Sigma', \boldsymbol{\alpha}', \boldsymbol{\beta}'),$$

as defined by Ozsváth and Szabó [**27**, Theorems 10.1 and 10.2]. If $\boldsymbol{\alpha}' = \boldsymbol{\alpha} \cup \{\alpha\}$, $\boldsymbol{\beta}' = \boldsymbol{\beta} \cup \{\beta\}$, and $\alpha \cap \beta = \{c\}$, then $\sigma_{\mathcal{H}\to\mathcal{H}'}$ maps the generator $\mathbf{x} \in \mathbb{T}_\alpha \cap \mathbb{T}_\beta$ to $\mathbf{x} \times \{c\} \in \mathbb{T}_{\alpha'} \cap \mathbb{T}_{\beta'}$. This induces an isomorphism on homology.

Before stating the next lemma, we introduce some notation. If $\mathcal{H}_1 = (\Sigma, \boldsymbol{\alpha}_1, \boldsymbol{\beta}_1)$ and $\mathcal{H}_2 = (\Sigma, \boldsymbol{\alpha}_2, \boldsymbol{\beta}_2)$ are admissible diagrams such that $\boldsymbol{\alpha}_1 \sim \boldsymbol{\alpha}_2$ and $\boldsymbol{\beta}_1 \sim \boldsymbol{\beta}_2$, then we denote $\Phi^{\boldsymbol{\alpha}_1\to\boldsymbol{\alpha}_2}_{\boldsymbol{\beta}_1\to\boldsymbol{\beta}_2}$ by $\Phi_{\mathcal{H}_1\to\mathcal{H}_2}$.

LEMMA 10.26. *The stabilization maps $\sigma_{\mathcal{H}\to\mathcal{H}'}$ commute with the maps $\Phi_{\mathcal{H}_1\to\mathcal{H}_2}$, in the following sense: Let $\mathcal{H}_1 = (\Sigma, \boldsymbol{\alpha}_1, \boldsymbol{\beta}_1)$ and $\mathcal{H}_2 = (\Sigma, \boldsymbol{\alpha}_2, \boldsymbol{\beta}_2)$ be two admissible Heegaard diagrams such that $\boldsymbol{\alpha}_1 \sim \boldsymbol{\alpha}_2$ and $\boldsymbol{\beta}_1 \sim \boldsymbol{\beta}_2$. If $\mathcal{H}_1' = (\Sigma', \boldsymbol{\alpha}_1', \boldsymbol{\beta}_1')$ and $\mathcal{H}_2' = (\Sigma', \boldsymbol{\alpha}_2', \boldsymbol{\beta}_2')$ are stabilizations of $\mathcal{H}_1$ and $\mathcal{H}_2$, respectively, then $\boldsymbol{\alpha}_1' \sim \boldsymbol{\alpha}_2'$, $\boldsymbol{\beta}_1' \sim \boldsymbol{\beta}_2'$, and*

$$\sigma_{\mathcal{H}_2\to\mathcal{H}_2'} \circ \Phi_{\mathcal{H}_1\to\mathcal{H}_2} = \Phi_{\mathcal{H}_1'\to\mathcal{H}_2'} \circ \sigma_{\mathcal{H}_1\to\mathcal{H}_1'}.$$

PROOF. This is verified in [**28**, Lemma 2.15]. Note that the continuation maps in that proof agree with our triangle maps by Lemma 10.7. □

DEFINITION 10.27. Given isotopy diagrams H and H' such that H' is a stabilization of H, we define an isomorphism

$$\sigma_{H\to H'} \colon HF^\circ(H) \to HF^\circ(H')$$

as follows. By definition, there are diagrams $\mathcal{H}$ and $\mathcal{H}'$ representing H and H', respectively, such that $\mathcal{H}'$ is a stabilization of $\mathcal{H}$. There are canonical isomorphisms $i_{\mathcal{H}} \colon HF^\circ(\mathcal{H}) \to HF^\circ(H)$ and $i_{\mathcal{H}'} \colon HF^\circ(\mathcal{H}') \to HF^\circ(H')$ coming from the colimit construction. We define $d_{H\to H'}$ as $i_{\mathcal{H}'} \circ \sigma_{\mathcal{H}\to\mathcal{H}'} \circ i_{\mathcal{H}}^{-1}$. This is independent of the choice of $\mathcal{H}$ and $\mathcal{H}'$ by Lemma 10.26, together with the observation that for any two diagrams $\mathcal{H}_1$ and $\mathcal{H}_2$ representing the same isotopy diagram, $i_{\mathcal{H}_2}^{-1} \circ i_{\mathcal{H}_1} = \Phi_{\mathcal{H}_1\to\mathcal{H}_2}$. If H' is obtained from H by a destabilization, then we set $\sigma_{H\to H'} = (\sigma_{H'\to H})^{-1}$.

Having constructed $HF^\circ(H)$ for any isotopy diagram H (in the class of diagrams for which $HF^\circ(H)$ is defined), and isomorphisms induced by α-equivalences, β-equivalences, diffeomorphisms, stabilizations, and destabilizations, we have proved that HF° is a weak Heegaard invariant. This reproves Theorem 2.26, Theorem 2.27, and Theorem 2.28. However, note that we have already used the invariance of Heegaard Floer homology up to isomorphism for the manifolds $M(R_+, k)$ in the proof of Lemma 10.2, where we constructed the element $\Theta_{\beta,\gamma}$ for $\beta \sim \gamma$. We could have avoided this by imitating the invariance proof of Ozsváth and Szabó [**27**], at the price of making the discussion longer.

Recall that, at the end of Section 2.5, we indicated the necessary checks for obtaining the Spin^c-refinement. If $\mathcal{H}$ is an admissible diagram of the balanced sutured manifold (M, γ) and $\mathfrak{s} \in \text{Spin}^c(M, \gamma)$, then $CF^\circ(\Sigma, \boldsymbol{\alpha}, \boldsymbol{\beta}, \mathfrak{s})$ is generated by those $\mathbf{x} \in \mathbb{T}_\alpha \cap \mathbb{T}_\beta$ for which $\mathfrak{s}_{(M,\gamma)}(\mathbf{x}) = \mathfrak{s}$. It follows from the work of Ozsváth and Szabó [**27**] that the Spin^c-grading is preserved by the isomorphisms $\Phi_{J_s\to J_s'}$, the triangle maps $\Psi^{\boldsymbol{\beta}\to\boldsymbol{\beta}'}_{\boldsymbol{\alpha}}$ for $\boldsymbol{\beta} \sim \boldsymbol{\beta}'$ and $\Psi^{\boldsymbol{\alpha}\to\boldsymbol{\alpha}'}_{\boldsymbol{\beta}}$ for $\boldsymbol{\alpha} \sim \boldsymbol{\alpha}'$ (cf. Lemma 10.4), and the stabilization maps $\sigma_{\mathcal{H}\to\mathcal{H}'}$. Furthermore, given a diagram $\mathcal{H} = (\Sigma, \boldsymbol{\alpha}, \boldsymbol{\beta})$ of (M, γ), a diagram $\mathcal{H}'$ of (M', γ'), and a diffeomorphism $d \colon (M, \gamma) \to (M', \gamma')$ mapping $\mathcal{H}$ to $\mathcal{H}'$, it is straightforward to see that $d_*\left(\mathfrak{s}_{(M,g)}(\mathbf{x})\right) = \mathfrak{s}_{(M',\gamma')}(d(\mathbf{x}))$ for every

$\mathbf{x} \in \mathbb{T}_\alpha \cap \mathbb{T}_\beta$. In particular, if $(M,\gamma) = (M',\gamma')$ and d is isotopic to the identity in (M,γ), then $d_*\colon \mathrm{Spin}^c(M,\gamma) \to \mathrm{Spin}^c(M,\gamma)$ is the identity. The existence of a Spin^c-grading on $HF^\circ(M,\gamma)$ follows once we show that HF° is a strong Heegaard invariant.

10.2. Heegaard Floer homology as a strong Heegaard invariant

In this section, we show that the invariant HF° of isotopy diagrams, together with the maps induced by α-equivalences, β-equivalences, diffeomorphisms, and (de)stabilizations, satisfy the axioms of strong Heegaard invariants listed in Definition 2.32. We postpone the verification of axiom (4), handleswap invariance, to the following section.

We first prove that HF° satisfies axiom (1), functoriality. The α-equivalence and β-equivalence maps $\Phi_B^{A\to A'}$ and $\Phi^A_{B\to B'}$ are functorial by equations (10.15) and (10.18). Functoriality of the diffeomorphism maps d_* follows immediately from the definition. If H' is obtained from H by a stabilization, then the destabilization map $\sigma_{H'\to H} = (\sigma_{H\to H'})^{-1}$, by definition.

Next, we consider axiom (2), commutativity. In Definition 2.29, we defined five different types of distinguished rectangles of the form

$$\begin{array}{ccc} H_1 & \xrightarrow{e} & H_2 \\ \big\downarrow{\scriptstyle f} & & \big\downarrow{\scriptstyle g} \\ H_3 & \xrightarrow{h} & H_4, \end{array}$$

where $H_i = (\Sigma_i, [\boldsymbol{\alpha}_i], [\boldsymbol{\beta}_i])$. For a rectangle of type (1), commutativity follows from equation (10.15). Lemma 10.26 implies commutativity along a rectangle of type (2). Commutativity along a rectangle of type (3) follows from Lemma 10.25.

Now consider a rectangle of type (4). Then there are disjoint disks D_1, $D_2 \subset \Sigma_1$ and punctured tori T_1, $T_2 \subset \Sigma_4$ such that $\Sigma_1 \setminus (D_1 \cup D_2) = \Sigma_4 \setminus (T_1 \cup T_2)$. Let $\boldsymbol{\alpha}_4 \cap \boldsymbol{\beta}_4 \cap T_i = \{c_i\}$ for $i \in \{1,2\}$. Then there are representatives $\mathcal{H}_i = (\Sigma_i, \boldsymbol{\alpha}_i, \boldsymbol{\beta}_i)$ of the isotopy diagrams H_i for $i \in \{1,\dots,4\}$ such that $\boldsymbol{\alpha}_2 \cap \boldsymbol{\beta}_2 \cap T_1 = \{c_1\}$ and $\boldsymbol{\alpha}_3 \cap \boldsymbol{\beta}_3 \cap T_2 = \{c_2\}$, and the four diagrams coincide outside T_1 and T_2. Given a generator $\mathbf{x} \in \mathbb{T}_{\alpha_1} \cap \mathbb{T}_{\beta_1}$,

$$\sigma_{\mathcal{H}_2\to\mathcal{H}_4} \circ \sigma_{\mathcal{H}_1\to\mathcal{H}_2}(\mathbf{x}) = \mathbf{x} \times \{c_1\} \times \{c_2\} = \mathbf{x} \times \{c_2\} \times \{c_1\} = \sigma_{\mathcal{H}_3\to\mathcal{H}_4} \circ \sigma_{\mathcal{H}_1\to\mathcal{H}_3}(\mathbf{x}).$$

So the commutativity already holds on the chain level for an appropriate choice of complex structures. Indeed, if J_s and J_s' are almost complex structures corresponding to different relative neck lengths at ∂T_1 and ∂T_2, then the change of complex structures map $\Phi_{J_s \to J_s'}$ only counts constant curves; see the proof of [**42**, Proposition 4.20].

Finally, for a rectangle of type (5), we can choose a representative $\mathcal{H}_i = (\Sigma_i, \boldsymbol{\alpha}_i, \boldsymbol{\beta}_i)$ of H_i for $i \in \{1,\dots,4\}$ such that $\mathcal{H}_2$ is a stabilization of $\mathcal{H}_1$ and $\mathcal{H}_4$ is a stabilization of $\mathcal{H}_3$; furthermore, $f(\mathcal{H}_1) = \mathcal{H}_3$ and $g(\mathcal{H}_2) = \mathcal{H}_4$. This is possible since, for the stabilization disks $D \subset \Sigma_1$ and $D' \subset \Sigma_3$ and punctured tori $T \subset \Sigma_2$ and $T' \subset \Sigma_4$, the diffeomorphisms satisfy $f(D) = D'$, $g(T) = T'$, and $f|_{\Sigma_1 \setminus D} = g|_{\Sigma_2 \setminus T}$. In particular, if $\boldsymbol{\alpha}_2 \cap \boldsymbol{\beta}_2 \cap T = \{c\}$ and $\boldsymbol{\alpha}_4 \cap \boldsymbol{\beta}_4 \cap T' = \{c'\}$, then $g(c) = c'$. With these choices, for $\mathbf{x} \in \mathbb{T}_{\alpha_1} \cap \mathbb{T}_{\beta_1}$, we have

$$g_* \circ \sigma_{\mathcal{H}_1\to\mathcal{H}_2}(\mathbf{x}) = g(\mathbf{x} \times \{c\}) = g(\mathbf{x}) \times \{g(c)\} = f(\mathbf{x}) \times \{c'\} = \sigma_{\mathcal{H}_3\to\mathcal{H}_4} \circ f_*.$$

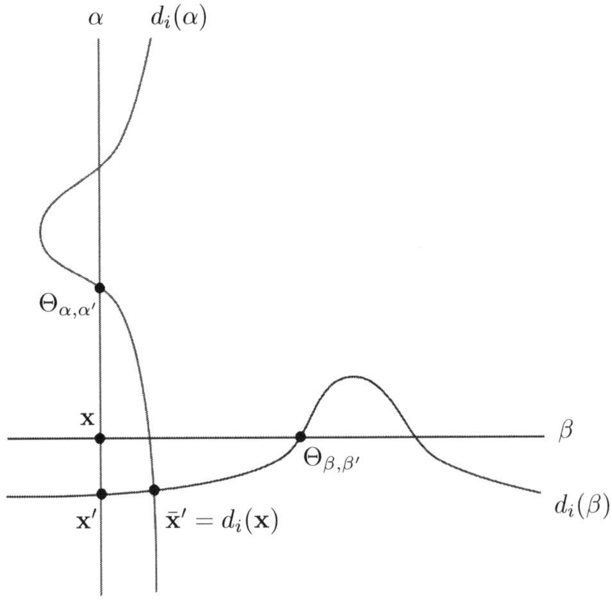

FIGURE 58. A schematic picture illustrating the diffeomorphism d_i.

So we have commutativity on the chain level for an appropriate choice of complex structures (for a sufficiently long neck, as in the proof of [**27**, Theorem 10.2]).

Finally, we verify axiom (3), continuity. This follows from the following result.

PROPOSITION 10.28. *Let $(\Sigma, \boldsymbol{\alpha}, \boldsymbol{\beta})$ be an admissible diagram. Suppose that $d\colon \Sigma \to \Sigma$ is a diffeomorphism isotopic to Id_Σ, and let $\boldsymbol{\alpha}' = d(\boldsymbol{\alpha})$ and $\boldsymbol{\beta}' = d(\boldsymbol{\beta})$. Then*

$$d_* = \Phi^{\boldsymbol{\alpha}\to\boldsymbol{\alpha}'}_{\boldsymbol{\beta}\to\boldsymbol{\beta}'} : HF^\circ(\Sigma, \boldsymbol{\alpha}, \boldsymbol{\beta}) \to HF^\circ(\Sigma, \boldsymbol{\alpha}', \boldsymbol{\beta}'). \tag{10.29}$$

PROOF. Since $d\colon \Sigma \to \Sigma$ is isotopic to the identity, there are Heegaard diagrams $\mathcal{H}_i = (\Sigma, \boldsymbol{\alpha}_i, \boldsymbol{\beta}_i)$ for $i \in \{0, \dots, n\}$, and diffeomorphisms $d_i\colon \mathcal{H}_{i-1} \to \mathcal{H}_i$ for $i \in \{1, \dots, n\}$, such that

- $\mathcal{H}_0 = (\Sigma, \boldsymbol{\alpha}, \boldsymbol{\beta})$ and $\mathcal{H}_n = (\Sigma, \boldsymbol{\alpha}', \boldsymbol{\beta}')$,
- every d_i is Hamiltonian isotopic to Id_Σ for some symplectic form ω_i on Σ,
- $d = d_n \circ \cdots \circ d_1$,
- $|\alpha \cap d_i(\alpha)| = 2$ for every $\alpha \in \boldsymbol{\alpha}_{i-1}$, and
- $|\beta \cap d_i(\beta)| = 2$ for every $\beta \in \boldsymbol{\beta}_{i-1}$.

See Figure 58 for a schematic picture of d_i.

By equation (10.15) and the functoriality of the diffeomorphism maps, it suffices to prove the statement for each d_i. So suppose that $d\colon (\Sigma, \boldsymbol{\alpha}, \boldsymbol{\beta}) \to (\Sigma, \boldsymbol{\alpha}', \boldsymbol{\beta}')$ is a diffeomorphism that is Hamiltonian isotopic to Id_Σ for some symplectic form ω on Σ, and such that $|\alpha \cap d(\alpha)| = 2$ and $|\beta \cap d(\beta)| = 2$ for every $\alpha \in \boldsymbol{\alpha}$ and $\beta \in \boldsymbol{\beta}$. Let $\{ d_t : t \in \mathbb{R} \}$ be a Hamiltonian isotopy supported in $I = [0, 1]$, where $d_0 = \mathrm{Id}_\Sigma$ and $d_1 = d$.

Let $\mathfrak{j}$ be an almost complex structure on Σ, and J_s a perturbation of $\mathrm{Sym}^k(\mathfrak{j})$, where $k = |\boldsymbol{\alpha}| = |\boldsymbol{\beta}|$. For every $t \in I$, let $\mathfrak{j}_t = (d_t)_*(\mathfrak{j})$ and $J_{s,t} = \mathrm{Sym}^k(d_t)_*(J_s)$.

We write $\mathfrak{j}' = \mathfrak{j}_1$ and $J'_s = J_{s,1}$. As in Lemma 10.1 and [**27**, p. 1080], the family $J_{s,t}$ induces an isomorphism $\Phi_{J_s \to J'_s}$ by counting Maslov index zero Whitney disks $u\colon I \times \mathbb{R} \to \mathrm{Sym}^k(\Sigma)$ satisfying $du/ds + J_{s,t}(du/dt) = 0$, where s is the coordinate on I and t is the coordinate on $\mathbb{R}$.

For every $t \in \mathbb{R}$, let $\boldsymbol{\alpha}_t = d_t(\boldsymbol{\alpha})$ and $\boldsymbol{\beta}_t = d_t(\boldsymbol{\beta})$. As defined by Ozsváth and Szabó [**28**, p. 344] (also see [**27**, p. 1086]), the Hamiltonian isotopy induces an isomorphism

$$\Gamma^{\boldsymbol{\alpha} \to \boldsymbol{\alpha}'}_{\boldsymbol{\beta} \to \boldsymbol{\beta}'} \colon HF^\circ_{J_s}(\Sigma, \boldsymbol{\alpha}, \boldsymbol{\beta}) \to HF^\circ_{J_s}(\Sigma, \boldsymbol{\alpha}', \boldsymbol{\beta}')$$

by counting index zero J_s-holomorphic disks with boundary condition $u(1 + it) \in \mathbb{T}_{\alpha_t}$ and $u(0 + it) \in \mathbb{T}_{\beta_t}$ for every $t \in \mathbb{R}$ and connecting some $\mathbf{x} \in \mathbb{T}_\alpha \cap \mathbb{T}_\beta$ and $\mathbf{y} \in \mathbb{T}_{\alpha'} \cap \mathbb{T}_{\beta'}$. As in the proof of [**27**, Lemma 9.6], the quadruple diagram $(\Sigma, \boldsymbol{\alpha}, \boldsymbol{\alpha}', \boldsymbol{\beta}, \boldsymbol{\beta}')$ is admissible. Hence, by [**28**, Lemma 2.12] and Lemma 10.7, we have

$$\Gamma^{\boldsymbol{\alpha} \to \boldsymbol{\alpha}'}_{\boldsymbol{\beta} \to \boldsymbol{\beta}'} = \Gamma^{\boldsymbol{\alpha}'}_{\boldsymbol{\beta} \to \boldsymbol{\beta}'} \circ \Gamma^{\boldsymbol{\alpha} \to \boldsymbol{\alpha}'}_{\boldsymbol{\beta}} = \Psi^{\boldsymbol{\alpha}'}_{\boldsymbol{\beta} \to \boldsymbol{\beta}'} \circ \Psi^{\boldsymbol{\alpha} \to \boldsymbol{\alpha}'}_{\boldsymbol{\beta}} = \Phi^{\boldsymbol{\alpha} \to \boldsymbol{\alpha}'}_{\boldsymbol{\beta} \to \boldsymbol{\beta}'}.$$

So equation (10.29) follows once we show that $d_* = \Gamma^{\boldsymbol{\alpha} \to \boldsymbol{\alpha}'}_{\boldsymbol{\beta} \to \boldsymbol{\beta}'}$. By Definition 10.24, we have $d_* = \Phi_{J'_s \to J_s} \circ d_{J_s, J'_s}$. As $\Phi^{-1}_{J'_s \to J_s} = \Phi_{J_s \to J'_s}$, it suffices to show that

$$d_{J_s, J'_s} = \Phi_{J_s \to J'_s} \circ \Gamma^{\boldsymbol{\alpha} \to \boldsymbol{\alpha}'}_{\boldsymbol{\beta} \to \boldsymbol{\beta}'}.$$

We now express d_{J_s, J'_s} as a continuation map as well. We defined

$$\Gamma_{d_t} \colon HF^\circ_{J_s}(\Sigma, \boldsymbol{\alpha}, \boldsymbol{\beta}) \to HF^\circ_{J'_s}(\Sigma, \boldsymbol{\alpha}', \boldsymbol{\beta}')$$

by counting index zero maps $u\colon I \times \mathbb{R} \to \mathrm{Sym}^k(\Sigma)$ connecting some $\mathbf{x} \in \mathbb{T}_\alpha \cap \mathbb{T}_\beta$ and $\mathbf{y} \in \mathbb{T}_{\alpha'} \cap \mathbb{T}_{\beta'}$ that satisfy $du/ds + J_{s,t}(du/dt) = 0$ and the boundary conditions $u(1 + it) \in \mathbb{T}_{\alpha_t}$ and $u(0 + it) \in \mathbb{T}_{\beta_t}$ for every $t \in \mathbb{R}$. For every such u, let $v(s,t) = \mathrm{Sym}^k(d_t)^{-1}(u(s,t))$. Then v is an index zero J_s-holomorphic disk by the definition of $J_{s,t}$, and satisfies the boundary conditions $v(1 + it) \in \mathbb{T}_\alpha$ and $v(0 + it) \in \mathbb{T}_\beta$ for every $t \in \mathbb{R}$, and is hence a constant disk with image some $\mathbf{x} \in \mathbb{T}_\alpha \cap \mathbb{T}_\beta$. This implies that, on the chain level, Γ_{d_t} maps every $\mathbf{x} \in \mathbb{T}_\alpha \cap \mathbb{T}_\beta$ to $d(\mathbf{x})$, and hence agrees with d_{J_s, J'_s}.

The last step is showing that

$$\Gamma_{d_t} = \Phi_{J_s \to J'_s} \circ \Gamma^{\boldsymbol{\alpha} \to \boldsymbol{\alpha}'}_{\boldsymbol{\beta} \to \boldsymbol{\beta}'}. \tag{10.30}$$

For this, we use techniques analogous to those developed by Ozsváth and Szabó [**27**, pp. 1081, 1089] to prove that the maps $\Phi_{J_s \to J'_s}$ and $\Gamma^{\boldsymbol{\alpha} \to \boldsymbol{\alpha}'}_{\boldsymbol{\beta}}$ are invertible. First, we give some motivation. Consider the product fibration

$$\pi\colon (I \times I) \times \mathrm{Sym}^k(\Sigma) \to I \times I.$$

For every $(t_1, t_2) \in I \times I$, we consider the totally real submanifolds $\mathbb{T}_{\alpha_{t_1}}$ and $\mathbb{T}_{\beta_{t_1}}$ of the fiber $\pi^{-1}(t_1, t_2)$ endowed with the almost complex structure $\mathrm{Sym}^k(\mathfrak{j}_{t_2})$ and perturbation J_{s,t_2}. Then Γ_{d_t} is the continuation map along the path $h_0(t) = (t,t)$ in $I \times I$ from $(0,0)$ to $(1,1)$, while $\Phi_{J_s \to J'_s} \circ \Gamma^{\boldsymbol{\alpha} \to \boldsymbol{\alpha}'}_{\boldsymbol{\beta} \to \boldsymbol{\beta}'}$ is the continuation map along the path $h_1(t)$ that is $(2t, 0)$ for $t \in [0, 1/2]$, and is $h_1(t) = (1, 2t)$ for $t \in [1/2, 1]$. Since h_0 and h_1 are homotopic in $I \times I$ fixing the endpoints, they induce chain homotopic chain maps, as we will now show.

Let $h_\tau(t) = (h_{\tau,1}(t), h_{\tau,2}(t))$ for $\tau \in I$ be a homotopy from h_0 to h_1 in $I \times I$ fixing the endpoints. Then $h_\tau(t)$ induces a chain homotopy between the two sides of equation (10.30), as follows. For $(t, \tau) \in I \times I$, we write $\boldsymbol{\alpha}_{t,\tau} = \boldsymbol{\alpha}_{h_{\tau,1}(t)}$ and

$\boldsymbol{\beta}_{t,\tau} = \boldsymbol{\beta}_{h_{\tau,1}(t)}$. Fix $\tau \in I$, and let $\mathbf{x} \in \mathbb{T}_\alpha \cap \mathbb{T}_\beta$ and $\mathbf{y} \in \mathbb{T}_{\alpha'} \cap \mathbb{T}_{\beta'}$. Then $\pi_2^\tau(\mathbf{x}, \mathbf{y})$ is the set of homotopy classes of maps $u\colon I \times \mathbb{R} \to \mathrm{Sym}^k(\Sigma)$ with boundary conditions $u(1+it) \in \mathbb{T}_{\alpha_{t,\tau}}$ and $u(0+it) \in \mathbb{T}_{\beta_{t,\tau}}$ for every $t \in \mathbb{R}$, and such that $\lim_{t\to-\infty} u(s+it) = \mathbf{x}$ and $\lim_{t\to\infty} u(s+it) = \mathbf{y}$.

For $\phi \in \pi_2^\tau(\mathbf{x}, \mathbf{y})$, let $\mathcal{M}_\tau(\phi)$ be the moduli space of maps u representing ϕ such that

$$\frac{du}{ds} + J_{s,h_{\tau,2}(t)} \frac{du}{dt} = 0.$$

For every $\tau \in I$, the homotopy h induces a canonical identification between $\pi_2^0(\mathbf{x}, \mathbf{y})$ and $\pi_2^\tau(\mathbf{x}, \mathbf{y})$. Using these identifications, for every $\phi \in \pi_2^0(\mathbf{x}, \mathbf{y})$, we define the moduli space

$$\mathcal{M}^h(\phi) = \bigcup_{\tau \in I} \mathcal{M}_\tau(\phi) \times \{\tau\} = \left\{ (u, \tau) \in C^\infty\left(I \times \mathbb{R}, \mathrm{Sym}^k(\Sigma)\right) \times I : \, u \in \mathcal{M}_\tau(\phi) \right\}.$$

Then $\dim\left(\mathcal{M}^h(\phi)\right) = \mu(\phi) + 1$. For $\mathbf{x} \in \mathbb{T}_\alpha \cap \mathbb{T}_\beta$, let

$$H^h(\mathbf{x}) = \sum_{\mathbf{y} \in \mathbb{T}_{\alpha'} \cap \mathbb{T}_{\beta'}} \sum_{\substack{\phi \in \pi_2^0(\mathbf{x},\mathbf{y}) \\ \mu(\phi) = -1}} \left(\left| \mathcal{M}^h(\phi) \right| \mod 2 \right) \mathbf{y}$$

in case of $\widehat{HF}$ or SFH, and replacing $\mathbf{x}$ with $[\mathbf{x}, i]$ and $\mathbf{y}$ with $[\mathbf{y}, i - n_z(\phi)]$ in case of HF^+, HF^-, and HF^∞.

To see that H^h gives a chain homotopy between Γ_{d_t} and $\Phi_{J_s \to J_s'} \circ \Gamma^{\boldsymbol{\alpha} \to \boldsymbol{\alpha}'}_{\boldsymbol{\beta} \to \boldsymbol{\beta}'}$, we consider the ends of the moduli spaces $\mathcal{M}^h(\psi)$ for $\psi \in \pi_2^0(\mathbf{x}, \mathbf{y})$ such that $\mu(\psi) = 0$. There are three types of ends: The ends at $\tau = 0$ contribute to the map Γ_{d_t}. The ends at $\tau = 1$ contribute to $\Phi_{J_s \to J_s'} \circ \Gamma^{\boldsymbol{\alpha} \to \boldsymbol{\alpha}'}_{\boldsymbol{\beta} \to \boldsymbol{\beta}'}$; to see that continuation maps are natural under juxtaposition of the paths involved, the proof of [**28**, Lemma 2.12] generalizes readily. Finally, broken strips contribute to $H^h \circ \partial + \partial' \circ H^h$, where ∂ is the boundary maps for $\mathbb{T}_\alpha$, $\mathbb{T}_\beta$, and $(\mathfrak{j}, J_s)$, while ∂' is the boundary map for $\mathbb{T}_\alpha'$, $\mathbb{T}_\beta'$, and $(\mathfrak{j}', J_s')$. Hence, on the chain level,

$$\Gamma_{d_t} + \Phi_{J_s \to J_s'} \circ \Gamma^{\boldsymbol{\alpha} \to \boldsymbol{\alpha}'}_{\boldsymbol{\beta} \to \boldsymbol{\beta}'} \equiv H^h \circ \partial + \partial' \circ H^h \mod 2,$$

and equation (10.30) follows. □

10.3. Simple handleswap invariance of Heegaard Floer homology

In this section, we prove simple handleswap invariance of Heegaard Floer homology. The technical details of our proof are modeled on the proof of stabilization invariance due to Lipshitz [**19**], and invariance of HF^+ under adding extra basepoints due to Ozsváth and Szabó [**29**]. The main analytical input is due to Lipshitz [**19**] and Lipshitz, Ozsváth, and Thurston [**22**].

Let

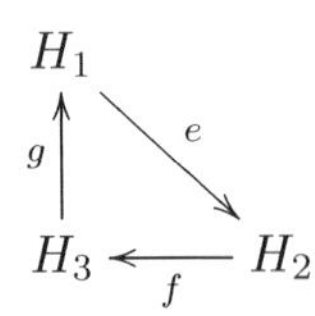

be a simple handleswap, as in Definition 2.31, where $H_i = (\Sigma \# \Sigma_0, \boldsymbol{\alpha}_i, \boldsymbol{\beta}_i)$ are admissible diagrams for $i \in \{1, 2, 3\}$, the edge e is an α-equivalence, f is a β-equivalence, and g is a diffeomorphism. Note that $\boldsymbol{\beta}_1 = \boldsymbol{\beta}_2$ and $\boldsymbol{\alpha}_2 = \boldsymbol{\alpha}_3$. We choose representatives of the isotopy diagrams H_i that violate condition (4) of Definition 2.31, in that we require $\boldsymbol{\alpha}_1 \setminus \{\alpha_1\}$ and $\boldsymbol{\alpha}_2 \setminus \{\alpha_1'\}$, and similarly, $\boldsymbol{\beta}_2 \setminus \{\beta_2\}$ and $\boldsymbol{\beta}_3 \setminus \{\beta_2'\}$ to be small Hamiltonian isotopic translates of each other, and adjust the diffeomorphism g accordingly. This will allow us to compute the isomorphisms induced by e and f via a single triangle map Ψ, as opposed to the maps Φ that are compositions of two triangle maps. Changing g by an isotopy does not change the map $g_*\colon HF^\circ(H_3) \to HF^\circ(H_1)$ by the Continuity Axiom, Proposition 10.28.

The α- and β-equivalence maps corresponding to the arrows e and f can be computed with the triangle maps $\Phi_e := \Psi^{\boldsymbol{\alpha}_1 \to \boldsymbol{\alpha}_2}_{\boldsymbol{\beta}_1}$ and $\Phi_f := \Psi^{\boldsymbol{\alpha}_2}_{\boldsymbol{\beta}_2 \to \boldsymbol{\beta}_3}$. The map corresponding to the arrow g is the map induced by a diffeomorphism, which we also denote by g. Simple handleswap invariance amounts to proving the following:

THEOREM 10.31. *The induced maps g_*, Φ_e, and Φ_f satisfy*

$$g_* \circ \Phi_f \circ \Phi_e = \mathrm{Id}_{HF^\circ(H_1)}.$$

To prove this, we will explicitly count triangles by a neck stretching argument. Let $\mathcal{T}_0 = (\Sigma_0, \boldsymbol{\alpha}_0', \boldsymbol{\alpha}_0, \boldsymbol{\beta}_0)$ be the Heegaard triple shown in Figure 59, where Σ_0 is a surface of genus two, $p_0 \in \Sigma_0 \setminus (\boldsymbol{\alpha}_0' \cup \boldsymbol{\alpha}_0 \cup \boldsymbol{\beta}_0)$ is a distinguished point, $\boldsymbol{\alpha}_0' = \{\alpha_1', \alpha_2'\}$, $\boldsymbol{\alpha}_0 = \{\alpha_1, \alpha_2\}$, and $\boldsymbol{\beta}_0 = \{\beta_1, \beta_2\}$. Suppose that $\mathcal{T} = (\Sigma, \boldsymbol{\alpha}', \boldsymbol{\alpha}, \boldsymbol{\beta})$ is a Heegaard triple with a distinguished point $p \in \Sigma \setminus (\boldsymbol{\alpha}' \cup \boldsymbol{\alpha} \cup \boldsymbol{\beta})$. We consider the Heegaard triple

$$\mathcal{T} \# \mathcal{T}_0 = (\Sigma \# \Sigma_0, \boldsymbol{\alpha}' \cup \boldsymbol{\alpha}_0', \boldsymbol{\alpha} \cup \boldsymbol{\alpha}_0, \boldsymbol{\beta} \cup \boldsymbol{\beta}_0),$$

where the connected sum is taken at $p \in \Sigma$ and $p_0 \in \Sigma_0$. Write $\mathbb{T}_{\alpha_0} \cap \mathbb{T}_{\beta_0} = \{\mathbf{a}\}$, $\mathbb{T}_{\alpha_0'} \cap \mathbb{T}_{\beta_0} = \{\mathbf{b}\}$, and

$$\mathbb{T}_{\alpha_0'} \cap \mathbb{T}_{\alpha_0} = \{\, \theta_1^+\theta_2^+, \theta_1^+\theta_2^-, \theta_1^-\theta_2^+, \theta_1^-\theta_2^- \,\},$$

where the intersection points $\theta_1^\pm \in \alpha_1' \cap \alpha_1$ and $\theta_2^\pm \in \alpha_2' \cap \alpha_2$ are marked in Figure 59. We will write Θ for $\theta_1^+\theta_2^+$.

The main technical result needed to prove Theorem 10.31 is the following count of holomorphic triangles:

PROPOSITION 10.32. *Suppose that $\mathcal{T} = (\Sigma, \boldsymbol{\alpha}', \boldsymbol{\alpha}, \boldsymbol{\beta})$ is an admissible Heegaard triple, and let $\mathcal{T} \# \mathcal{T}_0 = (\Sigma \# \Sigma_0, \boldsymbol{\alpha}' \cup \boldsymbol{\alpha}_0', \boldsymbol{\alpha} \cup \boldsymbol{\alpha}_0, \boldsymbol{\beta} \cup \boldsymbol{\beta}_0)$ be as above. For a sufficiently stretched and appropriately generic almost complex structure (specified precisely in Subsection 10.3.1), the triangle map $F^\circ_{\mathcal{T}\#\mathcal{T}_0}$ associated to the triple $\mathcal{T}\#\mathcal{T}_0$ is given by*

$$(\mathbf{x} \times \Theta) \otimes (\mathbf{y} \times \mathbf{a}) \mapsto F^\circ_{\mathcal{T}}(\mathbf{x} \otimes \mathbf{y}) \times \mathbf{b}$$

for every $\mathbf{x} \in \mathbb{T}_{\alpha'} \cap \mathbb{T}_\alpha$ and $\mathbf{y} \in \mathbb{T}_\alpha \cap \mathbb{T}_\beta$.

We will prove Proposition 10.32 in Subsection 10.3.1. The same proof can easily be adapted to show the following. Let $\mathcal{T}' = (\Sigma, \boldsymbol{\alpha}', \boldsymbol{\beta}, \boldsymbol{\beta}')$ denote a Heegaard triple with a distinguished point $p \in \Sigma \setminus (\boldsymbol{\alpha}' \cup \boldsymbol{\beta} \cup \boldsymbol{\beta}')$. Furthermore, let $\mathcal{T}_0' = (\Sigma_0, \boldsymbol{\alpha}_0', \boldsymbol{\beta}_0, \boldsymbol{\beta}_0')$ denote the Heegaard triple shown in Figure 60, where Σ_0 is a surface of genus two, $p_0 \in \Sigma_0 \setminus (\boldsymbol{\alpha}_0' \cup \boldsymbol{\beta}_0 \cup \boldsymbol{\beta}_0')$ is a distinguished point, $\boldsymbol{\alpha}_0' = \{\alpha_1', \alpha_2'\}$, $\boldsymbol{\beta}_0 = \{\beta_1, \beta_2\}$, and $\boldsymbol{\beta}_0' = \{\beta_1', \beta_2'\}$. We write $\mathbb{T}_{\alpha_0'} \cap \mathbb{T}_{\beta_0} = \{\mathbf{b}\}$, $\mathbb{T}_{\alpha_0'} \cap \mathbb{T}_{\beta_0'} = \{\mathbf{c}\}$, and let Θ' denote the top-graded intersection point of $\mathbb{T}_{\beta_0} \cap \mathbb{T}_{\beta_0'}$. We consider the Heegaard triple

$$\mathcal{T}' \# \mathcal{T}_0' = (\Sigma \# \Sigma_0, \boldsymbol{\alpha}' \cup \boldsymbol{\alpha}_0', \boldsymbol{\alpha} \cup \boldsymbol{\alpha}_0, \boldsymbol{\beta}' \cup \boldsymbol{\beta}_0'),$$

where the connected sum is taken at $p \in \Sigma$ and $p_0 \in \Sigma_0$.

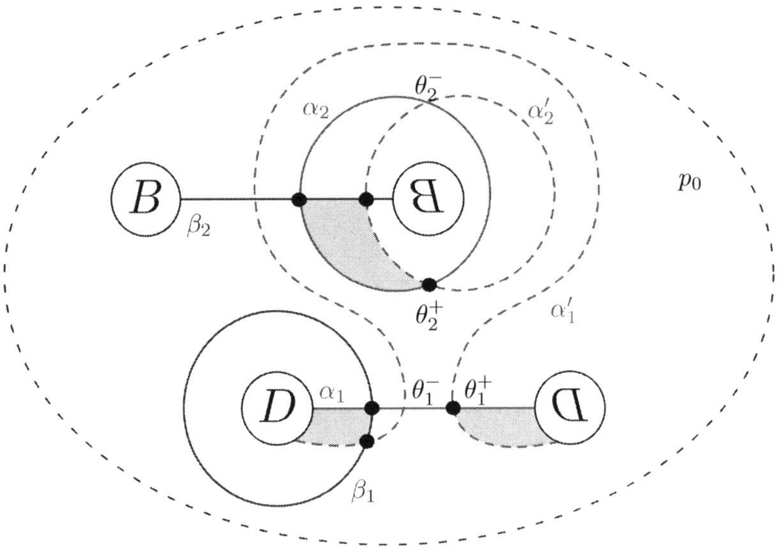

FIGURE 59. The Heegaard triple $(\Sigma_0, \boldsymbol{\alpha}_0', \boldsymbol{\alpha}_0, \boldsymbol{\beta}_0, p_0)$ we will use for computing the α-equivalence map Φ_e. Shaded in gray is a Maslov index zero triangle in $\pi_2(\Theta, \mathbf{a}, \mathbf{b})$.

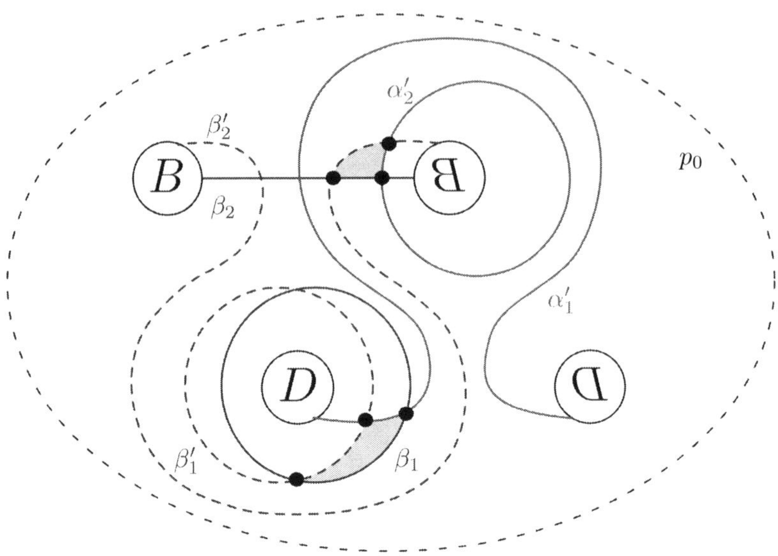

FIGURE 60. The Heegaard triple $(\Sigma_0, \boldsymbol{\alpha}_0', \boldsymbol{\beta}_0, \boldsymbol{\beta}_0', p_0)$ we will use for computing the β-equivalence map Φ_f. Shaded in gray is a Maslov index zero triangle in $\pi_2(\mathbf{b}, \Theta', \mathbf{c})$.

PROPOSITION 10.33. *Suppose that $\mathcal{T}' = (\Sigma, \boldsymbol{\alpha}', \boldsymbol{\beta}, \boldsymbol{\beta}')$ is an admissible Heegaard triple, and let $\mathcal{T}'\#\mathcal{T}_0' = (\Sigma\#\Sigma_0, \boldsymbol{\alpha}' \cup \boldsymbol{\alpha}_0', \boldsymbol{\alpha} \cup \boldsymbol{\alpha}_0, \boldsymbol{\beta}' \cup \boldsymbol{\beta}_0')$ be as above. For a sufficiently stretched and appropriately generic almost complex structure (specified precisely in Subsection* 10.3.1*), the triangle map $F^\circ_{\mathcal{T}'}$ associated to the triple $\mathcal{T}'\#\mathcal{T}_0'$*

is given by

$$(\mathbf{x} \times \mathbf{b}) \otimes (\mathbf{y} \times \Theta') \mapsto F^{\circ}_{\mathcal{T}'}(\mathbf{x} \otimes \mathbf{y}) \times \mathbf{c}$$

for every $\mathbf{x} \in \mathbb{T}_{\alpha'} \cap \mathbb{T}_{\beta}$ *and* $\mathbf{y} \in \mathbb{T}_{\beta} \cap \mathbb{T}_{\beta'}$.

REMARK 10.34. In Propositions 10.32 and 10.33, our notation might suggest that $\boldsymbol{\alpha} \sim \boldsymbol{\alpha}'$ and $\boldsymbol{\beta} \sim \boldsymbol{\beta}'$. However, they hold for arbitrary admissible triples $\mathcal{T}$ and $\mathcal{T}'$. We use this notation since we will only need to apply Propositions 10.32 and 10.33 when the curves $\boldsymbol{\alpha}$ are isotopic to $\boldsymbol{\alpha}'$ and the curves $\boldsymbol{\beta}$ are isotopic to $\boldsymbol{\beta}'$.

We now use the above counts of holomorphic triangles to prove invariance under simple handleswaps:

PROOF OF THEOREM 10.31. Consider the diagram $H_1 = (\Sigma \# \Sigma_0, \boldsymbol{\alpha}_1, \boldsymbol{\beta}_1)$. Recall that Σ_0 is a genus two surface. Let $\boldsymbol{\alpha}_0 = \boldsymbol{\alpha}_1 \cap \Sigma_0 = \{\alpha_1, \alpha_2\}$ and $\boldsymbol{\beta}_0 = \boldsymbol{\beta}_1 \cap \Sigma_0 = \{\beta_1, \beta_2\}$. The diagram $H_0 = (\Sigma_0, \boldsymbol{\alpha}_0, \boldsymbol{\beta}_0)$ is shown in Figures 4 and 59. Furthermore, let $H = (\Sigma, \boldsymbol{\alpha}, \boldsymbol{\beta})$, where $\boldsymbol{\alpha} = \boldsymbol{\alpha}_1 \cap \Sigma$ and $\boldsymbol{\beta} = \boldsymbol{\beta}_1 \cap \Sigma$. By Definition 2.31, we can write

$$H_1 = H \# H_0 = (\Sigma \# \Sigma_0, \boldsymbol{\alpha} \cup \boldsymbol{\alpha}_0, \boldsymbol{\beta} \cup \boldsymbol{\beta}_0).$$

Then H is admissible since H_1 is admissible.

The α-equivalence e corresponds to handlesliding α_1 over α_2. Let $\boldsymbol{\alpha}_0' = \{\alpha_1', \alpha_2'\}$ be as in Figure 59. The curve α_2' is a small Hamiltonian translate of α_2, and α_1' is the result of handlesliding α_1 over α_2. Furthermore, let $\boldsymbol{\alpha}'$ be a small Hamiltonian translate of $\boldsymbol{\alpha}$ on Σ. Then

$$H_2 = (\Sigma \# \Sigma_0, \boldsymbol{\alpha}' \cup \boldsymbol{\alpha}_0', \boldsymbol{\beta} \cup \boldsymbol{\beta}_0),$$

and $\Phi_e = \Psi^{\boldsymbol{\alpha} \cup \boldsymbol{\alpha}_0 \to \boldsymbol{\alpha}' \cup \boldsymbol{\alpha}_0'}_{\boldsymbol{\beta} \cup \boldsymbol{\beta}_0}$. As before, we write $\mathbb{T}_{\alpha_0} \cap \mathbb{T}_{\beta_0} = \{\mathbf{a}\}$ and $\mathbb{T}_{\alpha_0'} \cap \mathbb{T}_{\beta_0} = \{\mathbf{b}\}$. By Proposition 10.32, for an appropriate choice of almost complex structure, we have

$$\begin{aligned}\Psi^{\boldsymbol{\alpha} \cup \boldsymbol{\alpha}_0 \to \boldsymbol{\alpha}' \cup \boldsymbol{\alpha}_0'}_{\boldsymbol{\beta} \cup \boldsymbol{\beta}_0}(\mathbf{y} \times \mathbf{a}) &= F_{\alpha' \cup \alpha_0', \alpha \cup \alpha_0, \beta \cup \beta_0}((\Theta_{\alpha', \alpha} \times \Theta) \otimes (\mathbf{y} \times \mathbf{a})) = \\ F_{\alpha', \alpha, \beta}(\Theta_{\alpha', \alpha} \otimes \mathbf{y}) \times \mathbf{b} &= \Psi^{\boldsymbol{\alpha} \to \boldsymbol{\alpha}'}_{\boldsymbol{\beta}}(\mathbf{y}) \times \mathbf{b}.\end{aligned} \tag{10.35}$$

The β-equivalence map Φ_f can be computed similarly. Let $\boldsymbol{\beta}_0' = \{\beta_1', \beta_2'\}$, where β_1' is a Hamiltonian translate of β_1 and β_2' is obtained by handlesliding β_2 over β_1; see Figure 60. We write $\mathbb{T}_{\alpha_0'} \cap \mathbb{T}_{\beta_0'} = \{\mathbf{c}\}$. By Proposition 10.33, we have

$$\Psi^{\boldsymbol{\alpha}' \cup \boldsymbol{\alpha}_0'}_{\boldsymbol{\beta} \cup \boldsymbol{\beta}_0 \to \boldsymbol{\beta}' \cup \boldsymbol{\beta}_0'}(\mathbf{x} \times \mathbf{b}) = \Psi^{\boldsymbol{\alpha}'}_{\boldsymbol{\beta} \to \boldsymbol{\beta}'}(\mathbf{x}) \times \mathbf{c}. \tag{10.36}$$

Outside Σ_0, we can take g to be a diffeomorphism isotopic to the identity and mapping $\boldsymbol{\alpha}'$ to $\boldsymbol{\alpha}$ and $\boldsymbol{\beta}'$ to $\boldsymbol{\beta}$. By construction, we have that

$$g_*(\mathbf{x} \times \mathbf{c}) = (g|_\Sigma)_*(\mathbf{x}) \times \mathbf{a}. \tag{10.37}$$

Combining Equations (10.35), (10.36), and (10.37), we see that

$$\left(g_* \circ \Psi^{\boldsymbol{\alpha}' \cup \boldsymbol{\alpha}_0'}_{\boldsymbol{\beta} \cup \boldsymbol{\beta}_0 \to \boldsymbol{\beta}' \cup \boldsymbol{\beta}_0'} \circ \Psi^{\boldsymbol{\alpha} \cup \boldsymbol{\alpha}_0 \to \boldsymbol{\alpha}' \cup \boldsymbol{\alpha}_0'}_{\boldsymbol{\beta} \cup \boldsymbol{\beta}_0}\right)(\mathbf{x} \times \mathbf{a}) = \left((g|_\Sigma)_* \circ \Psi^{\boldsymbol{\alpha}'}_{\boldsymbol{\beta} \to \boldsymbol{\beta}'} \circ \Psi^{\boldsymbol{\alpha} \to \boldsymbol{\alpha}'}_{\boldsymbol{\beta}}\right)(\mathbf{x}) \times \mathbf{a}.$$

By Proposition 10.28, we have

$$(g|_\Sigma)_* \circ \Psi^{\boldsymbol{\alpha}'}_{\boldsymbol{\beta} \to \boldsymbol{\beta}'} \circ \Psi^{\boldsymbol{\alpha} \to \boldsymbol{\alpha}'}_{\boldsymbol{\beta}} \simeq \mathrm{Id}_{CF(H)}.$$

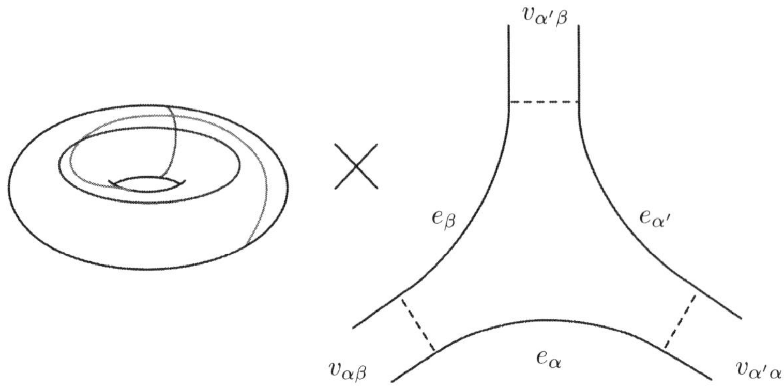

FIGURE 61. An example of a 4-manifold $\Sigma \times \Delta$ used to define the triangle maps in the cylindrical formulation.

If ∂_H denotes the differential on $CF(H)$, then stabilization invariance (see [**27**, Section 10] and [**19**, Section 12]) implies that the differential on $CF(H_1)$ is $\partial_H \otimes \mathrm{Id}_{CF(H_0)}$ for a sufficiently stretched almost complex structure. Hence

$$g_* \circ \Phi_f \circ \Phi_e = \left((g|_\Sigma)_* \circ \Psi^{\boldsymbol{\alpha}'}_{\boldsymbol{\beta}\to\boldsymbol{\beta}'} \circ \Psi^{\boldsymbol{\alpha}\to\boldsymbol{\alpha}'}_{\boldsymbol{\beta}}\right) \otimes \mathrm{Id}_{CF(H_0)} \simeq \mathrm{Id}_{CF(H)} \otimes \mathrm{Id}_{CF(H_0)},$$

completing the proof. □

10.3.1. Moduli spaces, compactness, transversality, and gluing results. In order to prove Propositions 10.32 and 10.33, we describe the limiting behavior of pseudo-holomorphic curves as one stretches the connected sum tube between Σ and Σ_0. In principle, it is possible to describe the behavior of holomorphic triangles appearing in $\mathrm{Sym}^{g(\Sigma)+g(\Sigma_0)}(\Sigma\#\Sigma_0)$ as one stretches the almost complex structure (see, for example, Ozsváth and Szabó [**27**, Section 10]), but it is easier to use the cylindrical reformulation of Heegaard Floer homology due to Lipshitz [**19**]. Most of the material in this section is an elaboration on Lipshitz's proof of stabilization invariance in [**19**, Section 12].

Lipshitz showed that the Heegaard Floer chain complexes can also be defined by counting holomorphic curves in $\Sigma \times [0,1] \times \mathbb{R}$. We endow this manifold with the split symplectic form $\omega = dA + ds \wedge dt$ for an area form dA on Σ, where s is the coordinate on $[0,1]$ and t is the coordinate on $\mathbb{R}$. The holomorphic triangle maps described by Ozsváth and Szabó are replaced in his theory with counts of pseudo-holomorphic curves mapping into $\Sigma \times \Delta$, where Δ is the subset of the complex plane shown in Figure 61, viewed as having three cylindrical ends of the form $[0,1] \times [0,\infty)$ or $[0,1] \times (-\infty, 0]$.

To prove stabilization invariance, Lipshitz [**19**, p. 959] considers almost complex structures J on $\Sigma \times [0,1] \times \mathbb{R}$ that satisfy the following five axioms:

($J1$) J is tamed by ω.
($J2$) J is split in a cylindrical neighborhood of $P \times [0,1] \times \mathbb{R}$, where $P \subset \Sigma \setminus (\boldsymbol{\alpha} \cup \boldsymbol{\beta})$ is a finite collection of points with at least one point in each component of $\Sigma \setminus (\boldsymbol{\alpha} \cup \boldsymbol{\beta})$.
($J3$) J is translation invariant in the $\mathbb{R}$-factor.

($J4$) $J(\partial/\partial s) = \partial/\partial t$.

($J5'$) There is a 2-plane distribution ξ on $\Sigma\times[0,1]\times\{0\}$ such that the restriction of ω to ξ is non-degenerate, J preserves ξ, and the restriction of J to ξ is compatible with ω. We further assume that ξ is tangent to Σ near $(\boldsymbol{\alpha}\cup\boldsymbol{\beta})\times[0,1]\times\{0\}$ and near $\Sigma\times\{0,1\}\times\{0\}$.

Lipshitz often considers an axiom ($J5$) that is more restrictive than ($J5'$), which requires that the 2-planes $T(\Sigma\times\{(s,t)\})$ are complex lines inside $\Sigma\times[0,1]\times\mathbb{R}$ for $(s,t)\in[0,1]\times\mathbb{R}$. Axiom ($J5'$) is used to ensure that certain curves achieve transversality in the proof of stabilization invariance.

For the proof of Proposition 10.32, we will consider almost complex structures J on $\Sigma\times\Delta$ that satisfy the following:

($J'1'$) J is tamed by the split symplectic form on $\Sigma\times\Delta$.

($J'2'$) There is a finite collection of points $P\subset\Sigma\setminus(\boldsymbol{\alpha}'\cup\boldsymbol{\alpha}\cup\boldsymbol{\beta})$, with at least one point in each component of $\Sigma\setminus(\boldsymbol{\alpha}'\cup\boldsymbol{\alpha}\cup\boldsymbol{\beta})$, such that the almost complex structure is split on a product neighborhood of $P\times\Delta$; i.e., $J = \mathrm{j}_\Sigma\times\mathrm{j}_\Delta$.

($J'3'$) Near the cylindrical ends of Δ, the almost complex structure J agrees with cylindrical almost complex structures on $\Sigma\times[0,1]\times\mathbb{R}$ satisfying condition ($J5'$) above.

($J'4'$) The 2-planes $T_d(\{p\}\times\Delta)$ are complex lines of J for all $(p,d)\in\Sigma\times\Delta$.

($J'5'$) The 2-planes $T_p(\Sigma\times\{d\})$ are complex lines of J for (p,d) near $(\boldsymbol{\alpha}'\cup\boldsymbol{\alpha}\cup\boldsymbol{\beta})\times\Delta$ and for $(p,d)\in\Sigma\times U$, where $U\subset\Delta$ is an open subset containing $\partial\Delta\setminus\{v_{\alpha'\alpha}, v_{\alpha\beta}, v_{\alpha'\beta}\}$.

Lipshitz considers axioms ($J'1$)–($J'4$) in [**19**, Section 10.2] for the triangle maps. Our axioms ($J'1'$) and ($J'2'$) coincide with ($J'1$) and ($J'2$), respectively, but ($J'3'$)–($J'5'$) together are more generic than the axioms ($J'3$) and ($J'4$), and are analogous to the axiom ($J5'$) that Lipshitz uses for holomorphic strips.

Note that the projections $\pi_\Delta\colon\Sigma\times\Delta\to\Delta$ and $\pi_\Sigma\colon\Sigma\times\Delta\to\Sigma$ are not necessarily holomorphic maps for almost complex structures satisfying the above axioms. Nonetheless, if $u\colon S\to\Sigma\times\Delta$ is a J-holomorphic curve, then the map $\pi_\Sigma\circ u$ is either constant, or an open map:

LEMMA 10.38. *Let J be an almost complex structure on $\Sigma\times\Delta$ that satisfies axioms ($J'1'$)–($J'5'$). If $u\colon S\to\Sigma\times\Delta$ is J-holomorphic and $\pi_\Sigma\circ u$ is nonconstant on a component S_0 of S, then $\pi_\Sigma\circ u|_{S_0}$ is an open map. In fact, there are coordinates near any critical point of $\pi_\Sigma\circ u|_{S_0}$ where $\pi_\Sigma\circ u$ takes the form $z\mapsto z^k$ for some $k>0$.*

PROOF. By condition ($J'4'$), the fiber bundle $\pi_\Sigma\colon\Sigma\times\Delta\to\Sigma$ has holomorphic fibers. The hypotheses of [**19**, Lemma 3.1] are satisfied, immediately yielding the result. □

In the cylindrical setting, a (possibly nodal) holomorphic triangle in the triple diagram $(\Sigma,\boldsymbol{\alpha}',\boldsymbol{\alpha},\boldsymbol{\beta})$ with $d=|\boldsymbol{\alpha}'|=|\boldsymbol{\alpha}|=|\boldsymbol{\beta}|$ is a map $u\colon S\to\Sigma\times\Delta$ that satisfies the following:

($M1$) (S,j) is a (possibly nodal) Riemann surface with boundary and $3d$ punctures on ∂S.

($M2$) u is locally nonconstant and (j,J)-holomorphic.

($M3$) $u(\partial S)\subset(\boldsymbol{\alpha}'\times e_{\alpha'})\cup(\boldsymbol{\alpha}\times e_\alpha)\cup(\boldsymbol{\beta}\times e_\beta)$.

($M4$) u has finite energy.

$(M5)$ For each $i \in \{1, \ldots, d\}$ and $\sigma \in \{\alpha', \alpha, \beta\}$, the preimage $u^{-1}(\sigma_i \times e_\sigma)$ consists of exactly one component of the punctured boundary of S.

$(M6)$ As one approaches the punctures along ∂S, the map u converges to a collection of Reeb chords (i.e., intersection points on the Heegaard triple) in the cylindrical ends of $\Sigma \times \Delta$.

For the holomorphic triangle maps, we additionally impose the following axioms:

$(M7)$ $\pi_\Delta \circ u$ is nonconstant on each component of S.

$(M8)$ S is smooth (i.e., not nodal), and the map $u\colon S \to \Sigma \times \Delta$ is an embedding.

If ψ is a homology class of triangles on a Heegaard triple $\mathcal{T}$, we write $\mathcal{M}(\psi)$ for the moduli space of holomorphic maps $u\colon S \to \Sigma \times \Delta$ that satisfy $(M1)$–$(M6)$. If S is a decorated surface (possibly with nodes), then we write $\mathcal{M}(\psi, S)$ for the subset of $\mathcal{M}(\psi)$ consisting of holomorphic curves with underlying source S. We will see that, if ψ is a Maslov index 0 homology class of triangles, then, for a generic choice of almost complex structure J satisfying $(J'1')$–$(J'5')$, any J-holomorphic map $u\colon S \to \Sigma \times \Delta$ representing ψ, and satisfying $(M1)$–$(M6)$ also satisfies $(M7)$ and $(M8)$; see Equation (10.47) below and the surrounding discussion.

Maps $u\colon S \to \Sigma \times [0,1] \times \mathbb{R}$ that satisfy the obvious analogs of $(M1)$–$(M6)$ are called holomorphic strips. We will need to consider compactifications of spaces of holomorphic triangles. To this end, we make the following definition:

DEFINITION 10.39. If $\mathcal{T} = (\Sigma, \boldsymbol{\alpha}', \boldsymbol{\alpha}, \boldsymbol{\beta})$ is a triple diagram, a *broken holomorphic triangle* on $\mathcal{T}$ representing a homology class ψ of triangles is a collection of (possibly nodal, non-embedded) (j, J)-holomorphic curves $(u_1, v_1, \ldots, v_n, w_1, \ldots, w_m)$ such that the following hold:

$(BT1)$ u_1 maps into $\Sigma \times \Delta$, and satisfies $(M1)$ and $(M3)$–$(M6)$.

$(BT2)$ $v_1, \ldots, v_n$ map into $\Sigma \times [0,1] \times \mathbb{R}$, satisfy the analogs of $(M1)$ and $(M3)$–$(M6)$, and represent homology classes of disks in the diagrams $(\Sigma, \boldsymbol{\alpha}', \boldsymbol{\alpha})$, $(\Sigma, \boldsymbol{\alpha}', \boldsymbol{\beta})$, and $(\Sigma, \boldsymbol{\alpha}, \boldsymbol{\beta})$.

$(BT3)$ $w_1, \ldots, w_m$ map into $\Sigma \times \Delta$ and $\Sigma \times [0,1] \times \mathbb{R}$. For every $i \in \{1, \ldots, m\}$, the curve w_i has d boundary components, each with a single puncture, and they all map onto one set of attaching curves (boundary degenerations).

$(BT4)$ The total homology class of the curves $u_1, v_1, \ldots, v_n, w_1, \ldots, w_m$ is equal to ψ.

The appropriate notion of convergence of holomorphic triangles to a broken holomorphic triangle is somewhat involved; see **[6]** and **[1]**. Roughly speaking, the source curve can degenerate along arcs connecting boundary components, or simple closed curves along the interior, as in the Deligne-Mumford compactification of stable curves. Additionally, different parts of the source curve can be sent off to ∞ in the cylindrical ends, resulting in a holomorphic building, with one story in $\Sigma \times \Delta$, and all other stories in one of the three cylindrical ends of $\Sigma \times \Delta$. Finally, constant components might appear.

In our definition of a broken holomorphic triangle, for notational simplicity, we have omitted the relative ordering of different stories of the resulting holomorphic building. For the purposes of this section, this additional structure will not be important, as we will only encounter broken triangles with one or two stories.

PROPOSITION 10.40. *Let J be an almost complex structure on $\Sigma \times \Delta$ satisfying $(J'1')$–$(J'5')$. If u_i is a sequence of J-holomorphic triangles satisfying $(M1)$–$(M6)$*

in $\mathcal{M}(\psi)$ for a homology class ψ, then there is a subsequence that converges to a broken holomorphic triangle.

The above result follows by adapting the arguments of Lipshitz [**19**, Section 7] to holomorphic triangles, and follows from standard results about compactness in symplectic field theory; see Bourgeois, Eliashberg, Hofer, Wysocki, and Zehnder [**6**] and Abbas [**1**]. For analogous results in a closely related context, see Lipshitz, Ozsváth, and Thurston [**22**, Section 5.4].

We need an additional compactness result that concerns the neck stretching procedure. Suppose we are given almost complex structures J and J_0 on $\Sigma \times \Delta$ and $\Sigma_0 \times \Delta$ that are each split on sets $D \times \Delta$ and $D_0 \times \Delta$, for disks $D \subset \Sigma$ and $D_0 \subset \Sigma_0$ containing the points p and p_0, respectively. We can form an almost complex structure $J(T)$ on $(\Sigma \# \Sigma_0) \times \Delta$ by inserting a connected sum tube of length T between Σ and Σ_0. We need to understand how a sequence of $J(T)$-holomorphic triangles behaves as $T \to \infty$.

PROPOSITION 10.41. *If u_{T_i} is a sequence of holomorphic triangles on $(\Sigma \# \Sigma_0) \times \Delta$ for a sequence of almost complex structures $J(T_i)$ with $T_i \to \infty$, then a subsequence can be extracted that converges to a triple (U, V, U_0), where U is a broken holomorphic triangle in $\Sigma \times \Delta$ and U_0 is a broken holomorphic triangle in $\Sigma_0 \times \Delta$. Furthermore, V is a collection of holomorphic curves mapping into the neck regions $S^1 \times \mathbb{R} \times \Delta$ or $S^1 \times \mathbb{R} \times [0,1] \times \mathbb{R}$ that are asymptotic to possibly multiply covered Reeb orbits of the form $S^1 \times \{d\}$ in $S^1 \times \Delta$ or $S^1 \times [0,1] \times \mathbb{R}$ for $d \in \Delta$ or $d \in [0,1] \times \mathbb{R}$, respectively.*

The splitting process is described in detail by Lipshitz [**19**, Appendix A], and follows from adapting [**19**, Proposition 12.4 or Sublemma A.12] to the setting of holomorphic triangles. See also Ozsváth and Szabó [**29**, Section 5.1] and Lipshitz, Ozsváth, and Thurston [**22**, Proposition 5.24]. The process of splitting a symplectic manifold along a hypersurface is considered in more generality in [**6**, Section 9].

We note that one could enhance the previous proposition to keep track of the *level splittings* (see [**19**, p. 993]) of the curves appearing in the limit, in analogy with the notion of holomorphic twin tower appearing in [**6**] and [**19**, Proposition 12.4]. We could also keep track of how curves in U and U_0 that are in the same level splitting match up across the connected sum point, somewhat analogous to the notion of a holomorphic comb appearing in [**22**], or the more complicated limits of holomorphic polygonal combs appearing in the setting of bordered Heegaard Floer homology in [**21**, Section 5.6]. There is a simple reason why we omit this data from our definition: In our proof of handleswap invariance, we will reduce to the case when the broken triangles on $\Sigma \times \Delta$ and $\Sigma_0 \times \Delta$ each consist of a single holomorphic triangle satisfying $(M1)$–$(M8)$.

Consider the triple diagram $(\Sigma, \boldsymbol{\alpha}', \boldsymbol{\alpha}, \boldsymbol{\beta})$, and fix a point $p \in \Sigma \setminus (\boldsymbol{\alpha}' \cup \boldsymbol{\alpha} \cup \boldsymbol{\beta})$. Using Lemma 10.38, if $u \colon S \to \Sigma \times \Delta$ is a holomorphic curve satisfying $(M1)$–$(M6)$, then condition $(J'4')$ ensures that $\{p\} \times \Delta$ and the image of u intersect at only finitely many points, and all intersections are positive. We write $(\pi_\Sigma \circ u)^{-1}(p) = (x_1, \dots, x_{n_p(u)}) \in \mathrm{Sym}^{n_p(u)}(S)$, where x_i appears with multiplicity m if it is a branch point of $\pi_\Sigma \circ u$ of order m, and define

$$\rho^p(u) = \big(\pi_\Delta \circ u(x_1), \dots, \pi_\Delta \circ u(x_{n_p(u)})\big) \in \mathrm{Sym}^{n_p(u)}(\Delta). \tag{10.42}$$

Note that this definition makes sense even if u is nodal: If a node x of the source is mapped to $\{p\} \times \Delta$, then $\pi_\Delta \circ u(x)$ appears with multiplicity $m_1 + m_2$ in $\rho^p(u)$, where m_1 and m_2 are the multiplicities of branching of $\pi_\Sigma \circ u$ along the two sheets.

DEFINITION 10.43. Suppose ψ is a homology class of triangles and $n_p(\psi) = k$. Given a subset $X \subset \mathrm{Sym}^k(\Delta)$, we define $\mathcal{M}(\psi, X)$ to be the moduli space of holomorphic curves in the homology class ψ that satisfy $(M1)$–$(M6)$ and match X at the marked point p:

$$\mathcal{M}(\psi, X) = \{\, u \in \mathcal{M}(\psi) : \rho^p(u) \in X \,\}.$$

Similarly, if S is a decorated source curve (smooth or nodal), then we define the moduli space $\mathcal{M}(\psi, S, X)$ to be the moduli space of holomorphic curves in the homology class ψ with source curve S that match X at the marked point p.

DEFINITION 10.44. The *fat diagonal* $\mathrm{Diag}^k(\Delta)$ in $\mathrm{Sym}^k(\Delta)$ is the subset consisting of elements of $\mathrm{Sym}^k(\Delta)$ with at least one repeated entry.

DEFINITION 10.45. We say a holomorphic curve $u : S \to \Sigma \times \Delta$ with connected source S is *somewhere injective* if there is a point $z \in S$ such that $du(z) \neq 0$ and $u^{-1}(u(z)) = \{z\}$. Such a point $z \in S$ satisfying the above condition is an *injective point.*

LEMMA 10.46. *Let $(\Sigma, \boldsymbol{\alpha}', \boldsymbol{\alpha}, \boldsymbol{\beta})$ be a triple diagram, and fix a point $p \in \Sigma \backslash (\boldsymbol{\alpha}' \cup \boldsymbol{\alpha} \cup \boldsymbol{\beta})$. Suppose that $\mathbf{d} \in \mathrm{Sym}^k(\Delta)$ is an element that is not in the fat diagonal. Let $u\colon S \to \Sigma \times \Delta$ be a holomorphic curve satisfying $(M1)$–$(M6)$ for an almost complex structure satisfying $(J'1')$–$(J'5')$ on Σ such that $\rho^p(u) = \mathbf{d}$. Then each component of u is somewhere injective.*

PROOF. A modification of [**19**, Lemma 3.3] shows that any component of u that is asymptotic at a puncture to a Reeb chord in an end of $\Sigma \times \Delta$ is somewhere injective. Thus the only components left to consider are ones that have no boundary components; i.e., closed components mapping into the interior of $\Sigma \times \Delta$. By assumption, $(\pi_\Sigma \circ u)^{-1}(p)$ contains exactly $|\mathbf{d}|$ points. Using condition $(J'4')$, the fact that u has no constant components, positivity of intersections of pseudo-holomorphic curves, and the assumption that the entries of $\mathbf{d}$ are distinct, we see that $(\pi_\Sigma \circ u)^{-1}(q)$ contains $|\mathbf{d}|$ points for any q close to p. This implies that du does not vanish at any of the points in $(\pi_\Sigma \circ u)^{-1}(p)$, and each point $x \in (\pi_\Sigma \circ u)^{-1}(p)$ has the property that $u(z) \neq u(x)$ for any $z \neq x$ in the source. Since each closed component must have nontrivial intersection with $\{p\} \times \Delta$ by Lemma 10.38, we conclude that each component of u is somewhere injective. □

We now briefly discuss the index of holomorphic curves in the cylindrical setting. If ψ is a homology class of triangles, then one can define the Maslov index of ψ, denoted $\mu(\psi)$, in analogy to the original setting of Heegaard Floer homology. Similarly, if $u\colon S \to \Sigma \times \Delta$ is a holomorphic triangle representing ψ, one can consider the Fredholm index of the linearized $\bar{\partial}$ operator at u, for which we write $\mathrm{ind}(\psi, S)$. The Fredholm index gives the expected dimension of the moduli space $\mathcal{M}(\psi, S)$. According to Lipshitz [**19**, Section 4], the Maslov index $\mu(\psi)$ agrees with the Fredholm index $\mathrm{ind}(\psi, S)$ at a holomorphic disk or triangle u that has smooth source and is embedded. For a curve u that is immersed, adapting [**22**, Proposition 5.69], we instead have

$$\mathrm{ind}(\psi, S) = \mu(\psi) - 2\,\mathrm{sing}(u), \tag{10.47}$$

where $\operatorname{sing}(u)$ denotes the order of singularity of u. The quantity $\operatorname{sing}(u)$ is non-negative for u holomorphic, and is zero if and only if u is embedded. A double point along the interior of S contributes $+1$ to $\operatorname{sing}(u)$. A double point along the boundary of S contributes $+\frac{1}{2}$ to $\operatorname{sing}(u)$. See also [**20**, Proposition 4.2'].

We state the following transversality result, and sketch the proof:

PROPOSITION 10.48. *Let $(\Sigma, \boldsymbol{\alpha}', \boldsymbol{\alpha}, \boldsymbol{\beta})$ be a triple diagram, and fix a point $p \in \Sigma \setminus (\boldsymbol{\alpha}' \cup \boldsymbol{\alpha} \cup \boldsymbol{\beta})$. Suppose $X \subset \operatorname{Sym}^k(\Delta)$ for some $k \in \mathbb{N}$ is a non-empty submanifold that does not intersect the fat diagonal. Furthermore, suppose that, for every $x \in X$, the k-tuple x has no coordinate in the open set $U \subset \Delta$ from $(J'5')$ that contains $\partial\Delta \setminus \{v_{\alpha'\alpha}, v_{\alpha\beta}, v_{\alpha'\beta}\}$. Then, for a generic choice of almost complex structure J, the set $\mathcal{M}(\psi, S, X)$ is a smooth manifold of dimension*

$$\operatorname{ind}(\psi, S) - \operatorname{codim}(X),$$

where $\operatorname{ind}(\psi, S)$ denotes the Fredholm index.

When $X = \operatorname{Sym}^k(\Delta)$, the same statement holds near any curve u that has no component T on which $\pi_\Delta \circ u|_T$ is constant and has image in U, and such that all components of u are somewhere injective.

SKETCH OF PROOF. We first note that if X does not intersect the fat diagonal in $\operatorname{Sym}^k(\Delta)$, then, by Lemma 10.46, each curve u that satisfies $\rho^p(u) \in X$ for some $p \in \Sigma \setminus (\boldsymbol{\alpha}' \cup \boldsymbol{\alpha} \cup \boldsymbol{\beta})$ is automatically somewhere injective with no assumption on the almost complex structure being generic.

Proposition 10.48 follows by modifying [**19**, Proposition 3.7] to take into account the map $\rho^p \colon \mathcal{M}(\psi, S) \to \operatorname{Sym}^k(\Delta)$. The required changes to the proof of [**19**, Proposition 3.7] can be adapted from [**23**, Proposition 3.4.2], as we briefly explain. One considers the universal moduli space $\mathcal{B}^\ell(\psi, S)$ of triples (u, j, J), where j is a complex structure on S, while J is a C^ℓ almost complex structure on $\Sigma \times \Delta$ satisfying $(J'1')$–$(J'5')$, and u is a (j, J)-holomorphic map from S to $\Sigma \times \Delta$ with each component of u somewhere injective. By Lemma 10.38, around a critical point, $\pi_\Sigma \circ u$ is of the form $z \mapsto z^n$ for some $n > 0$, hence the set of critical points of $\pi_\Sigma \circ u$ is discrete. As the set of injective points of u is open and dense, we can find an injective point of u in each component of S where $d(\pi_\Sigma \circ u)$ is non-zero. Adapting the proof of [**19**, Proposition 3.7], one can use this to show that $D\bar{\partial}$ is surjective at (u, j, J), and hence that $\mathcal{B}^\ell(\psi, S)$ is a Banach manifold.

The next step is to show that the universal evaluation map

$$\rho^p \colon \mathcal{B}^\ell(\psi, S) \to \operatorname{Sym}^k(\Delta)$$

is a submersion at triples (u, j, J) where $\rho^p(u)$ is not in the fat diagonal. The fact that the map ρ^p is a submersion when $\rho^p(u)$ is not in the fat diagonal can be proven by modifying [**23**, Proposition 3.4.2]. For almost complex structures satisfying $(J'1')$–$(J'5')$, this fact can also be proven by hand, as we now describe. Note that, since we are using the split symplectic form on $\Sigma \times \Delta$, if ϕ_Δ is a symplectomorphism of Δ, then the map $\operatorname{Id}_\Sigma \times \phi_\Delta$ will be a symplectomorphism of $\Sigma \times \Delta$. Furthermore, $(\operatorname{Id}_\Sigma \times \phi_\Delta)^*(J)$ satisfies $(J'1')$–$(J'5')$ for any J satisfying $(J'1')$–$(J'5')$. Suppose that $\rho^p(u) = \mathbf{d}$, where $\mathbf{d}$ is not in the fat diagonal. At each $d_i \in \mathbf{d}$, pick an arbitrary vector $v_i \in T_{d_i}\Delta$. Pick any Hamiltonian function H on Δ such that the Hamiltonian vector field X_H satisfies

$$(X_H)_{d_i} = v_i.$$

Multiplying H by bump functions, we can assume that H is supported in a neighborhood of the points d_i. Let $\phi_t\colon \Delta\times[-1,1]\to\Delta$ denote the flow of X_H. Then ϕ_t is a symplectomorphism for each t. We define the action of $\mathrm{Id}_\Sigma\times\phi_t$ on $\mathcal{B}^\ell(\psi,S)$ by

$$(\mathrm{Id}_\Sigma\times\phi_t)^*(u,j,J):=\left((\mathrm{Id}_\Sigma\times\phi_t)^{-1}\circ u, j, (\mathrm{Id}_\Sigma\times\phi_t)^*(J)\right).$$

Then

$$\left.\frac{d}{dt}\right|_{t=0}\rho^p\left((\mathrm{Id}_\Sigma\times\phi_t)^*(u,j,J)\right)=(-v_1,\dots,-v_k),$$

which shows that $\rho^p\colon\mathcal{B}^\ell(\psi,S)\to\mathrm{Sym}^k(\Delta)$ is a submersion at (u,j,J).

We define the universal matched moduli space

$$\mathcal{B}^\ell(\psi,S,X)=\{\,(u,j,J)\in\mathcal{B}^\ell(\psi,S):\ \rho^p(u)\in X\}.$$

Near any (u,j,J) where ρ^p is a submersion, this is a Banach manifold of codimension $\mathrm{codim}(X)$ in $\mathcal{B}^\ell(\psi,S)$. We write $\mathcal{J}^\ell$ for the space of C^ℓ almost complex structures on $\Sigma\times\Delta$ that satisfy $(J'1')$–$(J'5')$. If we apply the Sard–Smale theorem to the Fredholm map $\pi\colon\mathcal{B}^\ell(\psi,S,X)\to\mathcal{J}^\ell$, we obtain that a generic choice of $J\in\mathcal{J}^\ell$ is a regular value of π. The Fredholm index of π is $\mathrm{ind}(\psi,S)-\mathrm{codim}(X)$. Hence $\mathcal{M}(\psi,S,X)=\pi^{-1}(J)$ is a smooth manifold of dimension $\mathrm{ind}(\psi,S)-\mathrm{codim}(X)$ for a generic choice of J. An approximation argument yields the statement for C^∞ almost complex structures.

When $X=\mathrm{Sym}^k(\Delta)$, the above proof works given the additional assumptions in the statement of the proposition, and we do not need to use the universal evaluation map. □

REMARK 10.49. The only curves that we cannot achieve transversality for are curves where $\pi_\Delta\circ u$ is constant and takes a value in the set U containing $\partial\Delta\setminus\{v_{\alpha'\alpha},v_{\alpha\beta},v_{\alpha'\beta}\}$, as well as curves with closed components that are multiply covered.

We finally need to discuss gluing results for holomorphic triangles. If ψ is a homology class of triangles on the triple $\mathcal{T}$ and ψ_0 is a homology class of triangles on $\mathcal{T}_0$, then we can take the connected sum $\psi\#\psi_0$ of the two homology classes, and consider the moduli space $\mathcal{M}_{J(T)}(\psi\#\psi_0)$ for large T.

Let p and p_0 be the connected sum points on Σ and Σ_0, respectively. If $u\in\mathcal{M}(\psi)$ and $u_0\in\mathcal{M}(\psi_0)$, we consider the divisors $\rho^p(u)\in\mathrm{Sym}^{n_p(u)}(\Delta)$ and $\rho^{p_0}(u_0)\in\mathrm{Sym}^{n_{p_0}(u_0)}(\Delta)$ defined in equation (10.42). Then we have the following gluing result:

PROPOSITION 10.50. *Let u and u_0 be holomorphic triangles representing homology classes ψ and ψ_0 in $\Sigma\times\Delta$ and $\Sigma_0\times\Delta$ of Maslov indices 0 and $2k$, respectively. Suppose that*

$$\rho^p(u)=\rho^{p_0}(u_0)\in\mathrm{Sym}^k(\Delta)\setminus\mathrm{Diag}^k(\Delta),$$

where $k=n_p(\psi)=n_{p_0}(\psi_0)$, and that the moduli spaces $\mathcal{M}(\psi)$ and $\mathcal{M}(\psi_0,\rho^p(u))$ are transversely cut out near u and u_0, respectively. Then there is a homeomorphism h between a neighborhood of (u,u_0) in the compactified 1-dimensional moduli space

$$\overline{\bigcup_T\mathcal{M}_{J(T)}(\psi\#\psi_0)}$$

(cf. Proposition 10.41) and $[0,1)$ such that $h(u,u_0)=\{0\}$.

The above proposition follows immediately from the work of Lipshitz [**19**, Proposition A.2], together with a computation of the relevant Fredholm indices associated to $\mathcal{M}(\psi)$ and $\mathcal{M}(\psi_0, \rho^p(u))$ from Proposition 10.48. See also Ozsváth and Szabó [**29**, Theorem 5.1] for a similar result. The conditions that $\rho^p(u) \notin \mathrm{Diag}^k(\Delta)$, and that u has Maslov index zero ensure that the Reeb orbits involved are embedded, and that the techniques of Lipshitz [**19**] apply. This is the only case that we need. To relax these conditions, one would need to adapt [**19**, Proposition A.2] to a setting that allowed for multiply-covered Reeb orbits.

10.3.2. Proof of Proposition 10.32. In this section, we prove the main holomorphic triangle count that features in the proof of handleswap invariance. We first prove several simpler computations that will be used in the proof of the main count.

LEMMA 10.51. *Consider the triple diagram $\mathcal{T}_0 = (\Sigma_0, \boldsymbol{\alpha}_0', \boldsymbol{\alpha}_0, \boldsymbol{\beta}_0)$ in Figure 59. If $x \in \mathbb{T}_{\alpha_0'} \cap \mathbb{T}_{\alpha_0}$ and $\psi_0 \in \pi_2(x, \mathbf{a}, \mathbf{b})$, then*

$$\mu(\psi_0) = 2n_{p_0}(\psi_0) + \mu(x, \Theta), \tag{10.52}$$

where $\mu(x, \Theta)$ denotes the relative Maslov grading of x and Θ.

PROOF. We first prove equation (10.52) when $x = \Theta$. This holds when ψ_0 is the Maslov index zero triangle domain shown in Figure 59 as $\mu(\psi_0) = 0$ and $n_{p_0}(\psi_0) = 0$. The formula respects adding a multiple of $[\Sigma_0]$ as well as triply periodic domains. Indeed, every triply periodic domain in $\mathcal{T}_0$ is a sum of doubly periodic domains. There are no doubly periodic domains in $(\Sigma_0, \boldsymbol{\alpha}_0, \boldsymbol{\beta}_0, p_0)$ or $(\Sigma_0, \boldsymbol{\alpha}_0', \boldsymbol{\beta}_0, p_0)$, and every doubly periodic domain in $(\Sigma_0, \boldsymbol{\alpha}_0', \boldsymbol{\alpha}_0, p_0)$ has Maslov index zero.

For a general $x \in \mathbb{T}_{\alpha_0'} \cap \mathbb{T}_{\alpha_0}$, pick a domain $\phi \in \pi_2(x, \Theta)$. Then $\psi_0 - \phi$ represents an element of $\pi_2(\Theta, \mathbf{a}, \mathbf{b})$, and hence by the above, $\mu(\psi_0) = \mu(\phi) + 2n_{p_0}(\psi_0)$. Finally, note that $\mu(\phi) = \mu(x, \Theta)$; this is independent of the choice of ϕ since every periodic domain in $(\Sigma_0, \boldsymbol{\alpha}_0', \boldsymbol{\alpha}_0, p_0)$ has Maslov index zero. □

Before we consider the curves counted by the map $F^\circ_{\mathcal{T}\#\mathcal{T}_0}$ in full generality, we prove two simple lemmas that will be useful later when we perform more complicated counts of holomorphic triangles.

LEMMA 10.53. *The differential on $\widehat{CF}(\Sigma_0, \boldsymbol{\alpha}_0', \boldsymbol{\alpha}_0, p_0)$ vanishes.*

PROOF. Observe that $(\Sigma_0, \boldsymbol{\alpha}_0', \boldsymbol{\alpha}_0)$ represents $(S^1 \times S^2)\#(S^1 \times S^2)$, and hence

$$\dim \widehat{HF}(\Sigma_0, \boldsymbol{\alpha}_0', \boldsymbol{\alpha}_0, p_0) = 4$$

over $\mathbb{F}_2$. But $\widehat{CF}(\Sigma_0, \boldsymbol{\alpha}_0', \boldsymbol{\alpha}_0, p_0)$ is also 4-dimensional, and hence the differential must vanish. □

LEMMA 10.54. *The map*

$$\Psi_{\boldsymbol{\beta}_0}^{\boldsymbol{\alpha}_0 \to \boldsymbol{\alpha}_0'} \colon \widehat{CF}(\Sigma_0, \boldsymbol{\alpha}_0, \boldsymbol{\beta}_0, p_0) \to \widehat{CF}(\Sigma_0, \boldsymbol{\alpha}_0', \boldsymbol{\beta}_0, p_0)$$

satisfies $\Psi_{\boldsymbol{\beta}_0}^{\boldsymbol{\alpha}_0 \to \boldsymbol{\alpha}_0'}(\mathbf{a}) = \mathbf{b}$.

PROOF. By Proposition 10.11, the map $\Psi_{\boldsymbol{\beta}_0}^{\boldsymbol{\alpha}_0 \to \boldsymbol{\alpha}_0'}$ is a quasi-isomorphism of 1-dimensional chain complexes over $\mathbb{F}_2$ with vanishing differentials, and such a map is uniquely determined. □

We remark that Lemma 10.54 implies handleswap invariance in the hat version, or in the case of sutured Floer homology, regardless of the almost complex structure, but only for handleswaps where the region containing p also contains a basepoint or a boundary component of the sutured Heegaard diagram. (As pointed out to us by Sungkyung Kang, general handleswaps can be reduced to these using the Commutativity Axiom.)

For the $+$, $-$, and ∞ versions, and for general simple handleswaps in the hat version, we must consider how moduli spaces of holomorphic curves degenerate as we degenerate the almost complex structure.

PROOF OF PROPOSITION 10.32. If $\psi \in \pi_2(\mathbf{x}, \mathbf{y}, \mathbf{z})$ is a homology class of triangles on $\mathcal{T} = (\Sigma, \boldsymbol{\alpha}', \boldsymbol{\alpha}, \boldsymbol{\beta})$, and $\psi_0 \in \pi_2(x, \mathbf{a}, \mathbf{b})$ is a homology class on $\mathcal{T}_0 = (\Sigma_0, \boldsymbol{\alpha}_0', \boldsymbol{\alpha}_0, \boldsymbol{\beta}_0)$ with $n_p(\psi) = n_{p_0}(\psi_0)$, then we can form the homology class $\psi \# \psi_0$. By Lemma 10.51 and the Maslov index formula of Sarkar [**33**, Theorem 4.1], we have

$$\mu(\psi \# \psi_0) = \mu(\psi) + \mu(\psi_0) - 2n_{p_0}(\psi_0) = \mu(\psi) + \mu(x, \Theta), \tag{10.55}$$

since when we take the connected sum of the two homology classes, we must remove two balls with multiplicity $n_{p_0}(\psi_0) = n_p(\psi)$, and each ball has Euler measure one.

Suppose that

$$\psi \# \psi_0 \in \pi_2(\mathbf{x} \times \Theta, \mathbf{y} \times \mathbf{a}, \mathbf{z} \times \mathbf{b})$$

is a homology class of triangles that has holomorphic representatives u_{T_i} for a sequence of neck lengths T_i approaching ∞. By Proposition 10.41, we can extract a subsequence that converges to a broken triangle U on $(\Sigma, \boldsymbol{\alpha}', \boldsymbol{\alpha}, \boldsymbol{\beta})$ representing ψ, a broken triangle U_0 on $(\Sigma_0, \boldsymbol{\alpha}_0', \boldsymbol{\alpha}_0, \boldsymbol{\beta}_0)$ representing ψ_0, and a collection of holomorphic curves V mapping into the connected sum regions $S^1 \times \mathbb{R} \times \Delta$ or $S^1 \times \mathbb{R} \times [0,1] \times \mathbb{R}$. Supposing that $\mu(\psi \# \psi_0) = 0$, we know from equation (10.55) that

$$\mu(\psi \# \psi_0) = \mu(\psi) + \mu(\Theta, \Theta) = \mu(\psi), \tag{10.56}$$

and hence $\mu([U]) = 0$.

We list the steps of the rest of the proof:

(1) We show that U consists of a single Maslov index zero holomorphic triangle u satisfying $(M1)$–$(M8)$, as well as potentially some constant holomorphic curves.
(2) We show that U_0 consists of a single Maslov index $2k$ triangle u_0 satisfying $(M1)$–$(M8)$ and $\rho^{p_0}(u_0) = \rho^p(u)$, as well as potentially some constant holomorphic curves.
(3) We show that V consists of only a collection of trivial cylinders, and that there are no constant holomorphic components in U or U_0.
(4) We use the fibered product description of the moduli spaces in Proposition 10.50 to yield the relevant counts of curves on $\mathcal{T} \# \mathcal{T}_0$ in terms of the counts on $\mathcal{T}$.

First, let us restrict our attention to the components of U and U_0 that are nonconstant. Write U' and U_0' for the collections obtained by removing components of U and U_0, respectively, that map to a single point. Note that this does not change the homology classes of U and U', hence $\mu([U]) = \mu([U'])$ and $\mu([U_0]) = \mu([U_0'])$.

A priori, the broken triangle U' could consist of curves mapping into both $\Sigma \times \Delta$ and $\Sigma \times [0,1] \times \mathbb{R}$. However, holomorphic curves mapping into $\Sigma \times [0,1] \times \mathbb{R}$ that satisfy the analogs of $(M1)$–$(M6)$ and have nonzero domain on Σ have Fredholm index at least one, by transversality, and since there is an $\mathbb{R}$-action on the moduli space. As equation (10.47) also holds for curves mapping into $\Sigma \times [0,1] \times \mathbb{R}$, we obtain that these curves must also have Maslov index at least one. Similarly, if we choose the almost complex structure J on $\Sigma \times \Delta$ generically, the curves mapping into $\Sigma \times \Delta$ must have nonnegative Fredholm index, and hence also Maslov index by equation (10.47). Since

$$\mu([U']) = \mu([U]) = 0,$$

the limiting curve must map only into $\Sigma \times \Delta$. Clearly, the limiting curve in U' must satisfy $(M1)$–$(M4)$ and $(M6)$. Let us write u for the single holomorphic triangle in U'.

We now show that u satisfies $(M5)$. Degeneration of a source curve along a simple closed curve in the source preserves $(M5)$, though such a degeneration is impossible since u has Maslov index zero, and nodal curves would have negative expected dimension by equation (10.47). Collapsing along an arc connecting two boundary components of different type (e.g., an α boundary component and a β boundary component) would result in strip breaking, which we have already ruled out.

The final possibility is that the source collapses along arcs with endpoints along boundary components of the same type (e.g., two α boundary components). Since $u|_{\partial S}$ is monotone along each arc of the punctured boundary for a holomorphic curve with $u|_{\partial S}$ nonconstant, using the maximum modulus principle, it is not hard to see that some components of the resulting curve must be constant under the map π_Δ and map into $e_{\alpha'} \cup e_\alpha \cup e_\beta$. These correspond to boundary degenerations. The remaining components of the limit satisfy $(M5)$. However, a boundary degeneration increases the Maslov index by at least two. For example, an α boundary degeneration has domain equal to a nonnegative sum of components of $\Sigma \setminus \boldsymbol{\alpha}$, and upon examining the formula of Lipshitz [**19**, Proposition 4.10], splicing in such a class raises the Maslov index by twice the sum of multiplicities. Since $\mu([u]) = 0$, and the remaining curves are transversely cut out, boundary degenerations are thus prohibited. It follows that axiom $(M5)$ holds, as it is satisfied by all the curves u_{T_i}, and we have ruled out a sequence of curves collapsing along any arcs or simple closed curves in the source.

Similarly, $(M8)$ is satisfied because we have ruled out nodal source curves in U'. Condition $(M7)$ holds since holomorphic curves with $\pi_\Delta \circ u$ constant are either constant curves (which we are excluding, at the moment), nonconstant closed surfaces, or boundary degenerations, and our previous discussion rules out the latter two, using Maslov index considerations. This concludes Step (1).

We now proceed to Step (2). If v is a holomorphic triangle on $(\Sigma, \boldsymbol{\alpha}', \boldsymbol{\alpha}, \boldsymbol{\beta})$, and $\rho^p(v)$ contains a point of multiplicity greater than one, then $(\pi_\Sigma \circ v)^{-1}(p)$ either contains a double point of v, or a branch point of $\pi_\Sigma \circ v$. For a curve whose projection to Σ is locally non-constant, the set of critical points of $\pi_\Sigma \circ v$ is discrete by Lemma 10.38 (in fact, since J is split in a neighborhood of p, we could just use elementary complex analysis). In particular, by perturbing the connected sum point p slightly, we can assume that p is not the image of a branch point of $\pi_\Sigma \circ v$ for any Maslov index zero triangle v. Since u is also embedded, we can assume that $\rho^p(u)$ is not contained in the fat diagonal.

Since $\rho^p(u) \notin \mathrm{Diag}^k(\Delta)$, the curves appearing in V (mapping into $S^1 \times \mathbb{R} \times \Delta$ or $S^1 \times \mathbb{R} \times [0,1] \times \mathbb{R}$) are asymptotic to embedded Reeb orbits of the form $S^1 \times \{d\}$ in $S^1 \times \Delta$ or $S^1 \times [0,1] \times \mathbb{R}$ for some d in Δ or $[0,1] \times \mathbb{R}$. However, if v is a curve in V, then $\pi_\Delta \circ v$ must be constant by the maximum modulus principle. In particular, the asymptotics of V agree with the asymptotics of the curve $u \in U'$. Also, since U' consists of a single holomorphic triangle, the curves in V can only map into $S^1 \times \mathbb{R} \times \Delta$.

By Lemma 10.51, we have $\mu(\psi_0) = 2n_{p_0}(\psi_0)$. The broken holomorphic triangle U_0' could *a priori* consist of a genuinely broken holomorphic triangle, with nodal curves, boundary degenerations, and holomorphic strips. The curves in U_0' are obtained by completing the limiting curves in $(\Sigma_0 \setminus \{p_0\}) \times \Delta$. By [**6**, Section 10.2.3], curves in adjacent level splittings must have asymptotics that agree. Since the asymptotics of the curves in V agree with those of u, by the previous paragraph, we conclude that there must be curves in U_0' that satisfy a matching condition with u. The matching condition reads

$$\rho^p(u) = \rho^{p_0}(u_0).$$

Since u satisfies $(M1)$–$(M8)$, the divisor $\rho^p(u)$ is a collection of $n_p(u)$ points (possibly with repetition) in the interior of Δ. Let

$$\mathcal{D} := \{\, \rho^p(u) : u \text{ holomorphic on } \mathcal{T},\ \mu(u) = 0 \,\}.$$

Since $(\Sigma, \boldsymbol{\alpha}', \boldsymbol{\alpha}, \boldsymbol{\beta})$ is admissible and J is generic, the set $\mathcal{D}$ is finite. As described above, by perturbing the connected sum point p slightly, we can assume that $\mathcal{D} \cap \mathrm{Diag}^k(\Delta) = \emptyset$. By Lemma 10.46, this ensures that if u_0 is any (not necessarily embedded) holomorphic curve in $\Sigma_0 \times \Delta$ that satisfies

$$\rho^{p_0}(u_0) \in \mathcal{D},$$

then u_0 is somewhere injective on each component.

Thus, for a generic almost complex structure on $\Sigma_0 \times \Delta$, by Proposition 10.48, the moduli space

$$\mathcal{M}(\psi_0, \mathbf{d}) = \{\, u_0 \in \mathcal{M}(\psi_0) : \rho^{p_0}(u_0) = \mathbf{d} \,\}$$

is a transversely cut out manifold of dimension

$$\mu(\psi_0) - \mathrm{codim}(\{\mathbf{d}\}) = 2n_{p_0}(\psi_0) - 2n_p(u) = 0$$

for every $\mathbf{d} \in \mathcal{D}$. Repeating the argument that we made previously for the curves appearing in $\Sigma \times \Delta$ shows that, for a generic almost complex structure on $\Sigma_0 \times \Delta$, the broken triangle U_0' consists of exactly one curve $u_0 \in \mathcal{M}(\psi_0, \mathbf{d})$ mapping into $\Sigma_0 \times \Delta$, and that u_0 satisfies $(M1)$–$(M8)$. This concludes Step (2).

We now argue that V consists of only trivial cylinders, and that there are no non-trivial constant holomorphic curves in U and U_0. The argument follows by adapting standard arguments (for example [**22**, Lemma 5.57]). For a nontrivial constant holomorphic curve to appear in the limit, it must be stable; i.e., it cannot consist of a sphere with exactly one or two punctures. The original sequence of curves u_{T_i} had Maslov index zero and satisfied $(M1)$–$(M8)$. Writing $\widehat{S}$ for the source curve of u_{T_i} (which we can assume are topologically the same for all i, by passing to a subsequence), by [**19**, Section 10.2], we have

$$\mathrm{ind}\left(\psi \# \psi_0, \widehat{S}\right) = \frac{1}{2}(d+2) - \chi(\widehat{S}) + 2e(\psi \# \psi_0), \tag{10.57}$$

where $d = |\boldsymbol{\alpha}'| = |\boldsymbol{\alpha}| = |\boldsymbol{\beta}|$ is the number of attaching curves of each type in $\mathcal{T}$. Notice too that none of the constant curves in U or U_0 can map to $\{p\} \times \Delta$ or $\{p_0\} \times \Delta$, since the collections U and U_0 were obtained by taking curves in $(\Sigma \setminus \{p\}) \times \Delta$ and $(\Sigma_0 \setminus \{p_0\}) \times \Delta$ that were asymptotic to Reeb orbits in the ends corresponding to the points p and p_0, and then completing over these punctures to get curves in $\Sigma \times \Delta$ and $\Sigma_0 \times \Delta$.

Hence, we can be reconstruct the surface $\widehat{S}$ from the curves in U, V, and U_0 by gluing the source curves along their nodes. First, glue the source curves of the constant holomorphic curves together along any shared nodes. Assuming that this has been done, leaving only components $T_1, \dots, T_n$ of the constant holomorphic curves with no shared nodes, we note that there can only be one node between any T_i and S or S_0, as u and u_0 are embedded and have no nodal singularities. Since none of the constant curves appearing in the limit are once or twice punctured spheres, by stability of the limiting curves, we conclude that each T_i has genus at least one, and has at most one node shared with S or S_0. Hence, gluing the T_i to S and S_0 only raises the genus of S and S_0. Similarly, if V contains curves that are not trivial cylinders, it is easy to see that they must have positive genus. As a consequence, we have

$$\chi(\widehat{S}) \le \chi(S) + \chi(S_0) - 2k,$$

with equality if and only if there were no constant curves in U and U_0, and V consisted of only trivial cylinders (the $-2k$ comes from gluing S and S_0 together at the $k = n_p(\psi) = n_{p_0}(\psi_0)$ points where u and u_0 intersect $\{p\} \times \Delta$ or $\{p_0\} \times \Delta$). Furthermore, if V consists of anything but trivial cylinders, or if there are constant curves appearing in the limit, then

$$\chi(\widehat{S}) \le \chi(S) + \chi(S_0) - 2k - 2.$$

Applying the formula for the Fredholm index from equation (10.57) to u and u_0 implies that

$$\operatorname{ind}(\psi, S) + \operatorname{ind}(\psi_0, S_0) \le 2k + \operatorname{ind}(\psi \# \psi_0, \widehat{S}) - 2 = 2k - 2.$$

By assuming that $\mathcal{M}(\psi)$ and $\mathcal{M}(\psi_0, \mathbf{d})$ are smoothly cut out and of the expected dimension for any $\mathbf{d} \in \mathcal{D}$, if constant components appear in U or U_0, or, if V contains nontrivial curves, then one of the expected dimensions of $\mathcal{M}(S, \psi)$ and $\mathcal{M}(S_0, \psi_0, \mathbf{d})$ is negative. Hence, constant components of U and U_0, as well as nontrivial curves in V, do not arise generically. This concludes step (3).

The above limiting argument shows that the sequence u_{T_i} converges to a pair (u, u_0), where u has Maslov index zero, u_0 has Maslov index $2k$, and $\rho^p(u) = \rho^{p_0}(u_0)$. On the other hand, Proposition 10.50 describes a neighborhood of (u, u_0) in the compactified 1-dimensional moduli space $\overline{\bigcup_T M_{J(T)}(\psi \# \psi_0)}$, from which we conclude that

$$\#\mathcal{M}_{J(T)}(\psi \# \psi_0) = \#\{\, (u, u_0) \in \mathcal{M}(\psi) \times \mathcal{M}(\psi_0) : \rho^p(u) = \rho^{p_0}(u_0) \,\} \tag{10.58}$$

for T sufficiently large. Thus, to prove Proposition 10.32, it is sufficient to count triangles u_0 on $(\Sigma_0, \boldsymbol{\alpha}_0', \boldsymbol{\alpha}_0, \boldsymbol{\beta}_0)$ that satisfy a matching condition with a generic divisor $\mathbf{d} \in \operatorname{Sym}^k(\Delta)$ for some $k \in \mathbb{N}$. For $x \in \mathbb{T}_{\alpha_0'} \cap \mathbb{T}_{\alpha_0}$, we define

$$\mathcal{M}_{(x, \mathbf{a}, \mathbf{b})}(\mathbf{d}) = \coprod_{\substack{\psi_0 \in \pi_2(x, \mathbf{a}, \mathbf{b}) \\ n_{p_0}(\psi_0) = k}} \mathcal{M}(\psi_0, \mathbf{d}).$$

Note that, by using the expected dimension from equation (10.47), for a J achieving transversality, any curve in $\mathcal{M}(\psi_0, \mathbf{d})$ satisfying $(M1)$–$(M6)$ also satisfies $(M7)$ and $(M8)$.

Summarizing, we have shown that, for sufficiently large T, the only Maslov index zero homology classes that have representatives are of the form $\psi\#\psi_0$ with $\mu(\psi) = 0$ and $\mu(\psi_0) = 2n_{p_0}(\psi_0) = 2n_p(\psi)$. Using the fibered product description of the moduli spaces $\mathcal{M}(\psi\#\psi_0)$ in equation (10.58), to prove Proposition 10.32, it is sufficient to show that for each $u \in \mathcal{M}(\psi)$, we have

$$\#\mathcal{M}_{(\Theta,\mathbf{a},\mathbf{b})}(\rho^{\mathbf{p}}(u)) \equiv 1 \pmod 2.$$

This follows from Lemma 10.59 below. □

LEMMA 10.59. *For $\mathbf{d} \in \mathrm{Sym}^k(\Delta)$ not contained in the fat diagonal, the moduli space $\mathcal{M}_{(\Theta,\mathbf{a},\mathbf{b})}(\mathbf{d})$ is a smoothly cut out 0-manifold for a generic choice of almost complex structure J. Furthermore, for J achieving transversality, we have*

$$\#\mathcal{M}_{(\Theta,\mathbf{a},\mathbf{b})}(\mathbf{d}) \equiv 1 \pmod 2.$$

PROOF. Note that, for a fixed $\mathbf{d} \in \mathrm{Sym}^k(\Delta)$ that is not in the fat diagonal in $\mathrm{Sym}^k(\Delta)$ and also includes no points in the neighborhood U of $\partial\Delta\setminus\{v_{\alpha'\alpha}, v_{\alpha\beta}, v_{\alpha'\beta}\}$ appearing in axiom $(J'5')$, transversality of the matched moduli space $\mathcal{M}_{(\Theta,\mathbf{a},\mathbf{b})}(\mathbf{d})$ can be achieved by Proposition 10.48. In fact, since the set of such J is of the second category by the Sard-Smale theorem, one can pick a J achieving transversality simultaneously for any countable collection of submanifolds of $\mathrm{Sym}^k(\Delta)$ that do not intersect the fat diagonal and $U \times \mathrm{Sym}^{k-1}(\Delta)$.

We first show that the quantity $\#\mathcal{M}_{(\Theta,\mathbf{a},\mathbf{b})}(\mathbf{d})$ is independent of $\mathbf{d}$ for generic $\mathbf{d}$. Let $\mathbf{p}\colon [0,1] \to \mathrm{Sym}^k(\Delta)$ be an embedded arc starting at $\mathbf{d}_0$ and ending at $\mathbf{d}_1$ that does not intersect the fat diagonal and $U \times \mathrm{Sym}^{k-1}(\Delta)$. Consider the 1-dimensional moduli space

$$\mathcal{M}_{(\Theta,\mathbf{a},\mathbf{b})}(\mathbf{p}) = \bigcup_{t\in[0,1]} \mathcal{M}_{(\Theta,\mathbf{a},\mathbf{b})}(\mathbf{p}(t)).$$

We will count the ends of $\mathcal{M}_{(\Theta,\mathbf{a},\mathbf{b})}(\mathbf{p})$. First, note that there are ends corresponding to $\mathcal{M}_{(\Theta,\mathbf{a},\mathbf{b})}(\mathbf{d}_i)$ for $i \in \{0, 1\}$.

There are additional ends of $\mathcal{M}_{(\Theta,\mathbf{a},\mathbf{b})}(\mathbf{p})$ that correspond to degenerations of holomorphic triangles into broken holomorphic triangles, in the sense of Definition 10.39. We will show that, by picking J generically, the only degenerations that occur correspond to a sequence of Maslov index $2k$ holomorphic triangles breaking into a Maslov index $2k-1$ triangle that matches a divisor $\mathbf{p}(t)$ and a Maslov index 1 holomorphic disk that has multiplicity zero at p_0.

Let $u_i\colon S_0 \to \Sigma_0 \times \Delta$ for $i \in \mathbb{N}$ be a sequence of triangles in $\mathcal{M}_{(\Theta,\mathbf{a},\mathbf{b})}(\mathbf{p})$ with domain ψ_0. We show first that the sequence u_i cannot degenerate along a closed curve c in the interior of S_0 as $i \to \infty$ for a generic J. Note that a closed surface bubbles off if c is separating, and a source curves with an interior node appears in the limit if c is non-separating. Since the path $\mathbf{p}$ misses the fat diagonal and $U \times \mathrm{Sym}^{k-1}(\Delta)$, the moduli space $\mathcal{M}(\psi_0, S_0, \mathbf{p}) := \mathcal{M}(\psi_0, S_0, \mathrm{Im}(\mathbf{p}))$ (see Definition 10.43) is a manifold of the expected dimension by Proposition 10.48. If S_0 collapses along a simple closed curve in the interior, we note that the limiting curves must map the node corresponding to the collapsed curve to a point in $\Sigma_0 \times \Delta$ or $\Sigma_0 \times [0,1] \times \mathbb{R}$ (as opposed to being asymptotic to a Reeb orbit in the ends) since there are no Reeb orbits in the ends, only Reeb chords. Thus, if S_0 collapses along a closed curve in the interior, the resulting source curve must have a nodal singularity in the interior. Suppose that the resulting node occurs in a curve mapping into $\Sigma_0 \times \Delta$ (the case that the curve maps into $\Sigma_0 \times [0,1] \times \mathbb{R}$ is handled analogously). Let S_0' denote the collection of all source curves in the broken triangle mapping into $\Sigma_0 \times \Delta$, and write ψ_0' for the associated homology class. Since S_0' has at least one nodal singularity,

$$\operatorname{ind}(\psi_0', S_0') \leq \mu(\psi_0') - 2 \leq \mu(\psi_0) - 2 = 2k - 2$$

by equation (10.47). Hence, we have

$$\dim \mathcal{M}(\psi_0', S_0', \mathbf{p}) = \operatorname{ind}(\psi_0', S_0') - (2k-1) \leq (2k-2) - (2k-1) = -1,$$

so $\mathcal{M}(\psi_0', S_0', \mathbf{p})$ is empty for a generic J. We conclude that, for a generically chosen almost complex structure on $\Sigma_0 \times \Delta$, $\mathcal{M}(\psi_0, \mathbf{p})$ contains no curves with nodes along the interior, the limit of any sequence of curves in $\mathcal{M}(\psi_0, \mathbf{p})$ has no nodes along the interior. Note that, in the above argument, we assumed transversality at the limiting curves. By Proposition 10.48, the only possibility for curves to arise that do not achieve transversality are curves with a component u_0 such that $\pi_\Delta \circ u_0$ is constant with value in U. Such curves cannot appear as they would need to match a point in $\mathbf{p}(t)$ for some $i \in [0,1]$, but $\mathbf{p}$ avoids $U \times \operatorname{Sym}^{k-1}(\Delta)$.

The formation of curves with nodes along the boundary is ruled out by a reasoning similar to the proof of [**19**, Proposition 7.1]. By the index calculation of equation (10.47), since boundary nodes contribute $+\frac{1}{2}$ to $\operatorname{sing}(u)$, transversality alone does not prohibit them from forming. Furthermore (and more troublesome), we will not be able to achieve transversality at such curves, as will be clear from the following discussion. To show that such curves cannot form, we instead use the matching condition. Following the proof of [**19**, Proposition 7.1], by applying the maximum modulus principle near the boundary of the curve, one can see that nodes forming along the boundary of the source of a holomorphic curve result in a limiting curve that is mapped entirely into a fiber $\Sigma_0 \times \{x\}$ over a point $x \in e_{\alpha'} \cup e_\alpha \cup e_\beta \subset \partial\Delta$, or over an x in the corresponding edges of $\partial([0,1] \times \mathbb{R})$. As each of $\boldsymbol{\alpha}_0'$, $\boldsymbol{\alpha}_0$, and $\boldsymbol{\beta}_0$ is individually non-separating on Σ_0, any such holomorphic curve must have multiplicity at least 1 on all of Σ_0, but such curves cannot match any non-empty subset of $\mathbf{p}(t)$, as $\mathbf{p}$ does not intersect the boundary of Δ. The remaining curves cannot match any $\mathbf{p}(t)$ since they have combined multiplicity at most $k-1$ over p_0.

Finally, we note that the formation of nontrivial constant holomorphic curves is prohibited by transversality, as in the proof of Proposition 10.32. More precisely, if nontrivial closed components form, then the holomorphic curves obtained by deleting these constant curves would achieve transversality. But, in light of the index formula from equation (10.57), if the constant curves are nontrivial (i.e., there are no spheres with exactly one or two punctures), then the index of the remaining curves drops by at least 2, so the remaining curves do not appear in any moduli space matching $\mathbf{p}$, as $\operatorname{Im}(\mathbf{p}) \subset \operatorname{Sym}^k(\Delta)$ has dimension 1.

The remaining types of degenerations correspond to arcs connecting boundary components of source curves that map to two different types of attaching curves (e.g., collapsing an arc that connects an α boundary component and a β boundary component). These correspond to holomorphic strips breaking off.

Given a sequence of holomorphic triangles u_i in $\mathcal{M}(\psi_0, \mathbf{p}(t_i))$ that converges to a broken triangle with a nontrivial holomorphic strip, we know that the holomorphic triangle u appearing in the limit must also satisfy the matching condition with respect to a divisor $\mathbf{p}(t)$ for some $t \in (0,1)$. The curve u represents a class in $\pi_2(x, \mathbf{a}, \mathbf{b})$ for some $x \in \mathbb{T}_{\alpha'} \cap \mathbb{T}_\alpha$. We now consider separately the contributions to the ends of $\mathcal{M}_{(\Theta,\mathbf{a},\mathbf{b})}(\mathbf{p})$ corresponding to different possibilities for x.

First, suppose that $x = \Theta$. As the class of the whole degenerate curve is ψ_0, we have

$$\mu(u) = 2|\mathbf{d}| = \mu(\psi_0)$$

by Lemma 10.51. The remaining curves have total Maslov index zero and have multiplicity zero at p_0, and hence and hence represent the constant class. By transversality, the curve u is contained in the interior of $\mathcal{M}_{(\Theta,\mathbf{a},\mathbf{b})}(\mathbf{p})$ and so does not contribute to the boundary of the moduli space.

We now consider the case that u represents a class in either $\pi_2(\theta_1^+\theta_2^-, \mathbf{a}, \mathbf{b})$ or $\pi_2(\theta_1^-\theta_2^+, \mathbf{a}, \mathbf{b})$. By Lemma 10.51, the remaining curves have total Maslov index 1, and have multiplicity zero over p_0. The only possibility is that the remaining curve consists of a single Maslov index 1 holomorphic strip. So $\mathcal{M}_{(\Theta,\mathbf{a},\mathbf{b})}(\mathbf{p})$ has additional ends corresponding to

$$\bigcup_{\substack{t\in[0,1] \\ x\in\{\theta_1^+\theta_2^-,\theta_1^-\theta_2^+\}}} \bigcup_{\substack{\phi\in\pi_2(\Theta,x) \\ n_{p_0}(\phi)=0}} \widehat{\mathcal{M}}(\phi) \times \mathcal{M}_{(x,\mathbf{a},\mathbf{b})}(\mathbf{p}(t)).$$

The above space has an even number of points, since the differential on the chain complex $\widehat{CF}(\Sigma_0, \boldsymbol{\alpha}_0', \boldsymbol{\alpha}_0, p_0)$ vanishes by Lemma 10.53.

We now claim that, for a generic choice of almost complex structure, there are no additional ends of $\mathcal{M}_{(\Theta,\mathbf{a},\mathbf{b})}(\mathbf{p})$. To this end, we now claim that, for generic J, the space $\mathcal{M}(\psi, \mathbf{p})$ is empty for any $\psi \in \pi_2(\theta_1^-\theta_2^-, \mathbf{a}, \mathbf{b})$ with $n_{p_0}(\psi) = k$. By Lemma 10.51, $\mu(\psi) = 2k - 2$. Proposition 10.48 implies that $\mathcal{M}(\psi, S, \mathbf{p})$ is a smooth manifold of dimension

$$\operatorname{ind}(\psi, S) - (2k-1) \leq \mu(\psi) - (2k-1) = -1,$$

for a source S. In particular, for generic J, there are no holomorphic triangles u (embedded or nodal) representing a class in $\pi_2(\theta_1^-\theta_2^-, \mathbf{a}, \mathbf{b})$ with $\rho^{p_0}(u) \in \mathbf{p}(t)$, for any t.

Thus, modulo two, the count of the ends of $\mathcal{M}_{(\Theta,\mathbf{a},\mathbf{b})}(\mathbf{p})$ is just

$$0 = \#\mathcal{M}_{(\Theta,\mathbf{a},\mathbf{b})}(\mathbf{d}_0) - \#\mathcal{M}_{(\Theta,\mathbf{a},\mathbf{b})}(\mathbf{d}_1),$$

showing that the quantity $\#\mathcal{M}_{(\Theta,\mathbf{a},\mathbf{b})}(\mathbf{d})$ is independent of $\mathbf{d}$ modulo two for a generic almost complex structure J.

The final step is to show that $\#\mathcal{M}_{(\Theta,\mathbf{a},\mathbf{b})}(\mathbf{d}) \equiv 1 \pmod 2$ for some choice of $\mathbf{d}$. To do this, we adapt an argument of Ozsváth and Szabó [**29**, p. 653] from holomorphic strips to triangles. Let $\mathbf{p}\colon [1,\infty) \to \operatorname{Sym}^k(\Delta)$ be a smooth embedded path not intersecting the fat diagonal (a codimension 2 subset) starting at a divisor $\mathbf{d} \in \operatorname{Sym}^k(\Delta)$, such that the points in $\mathbf{p}(T)$ are spaced at least distance T apart that smoothly approach the vertex $v_{\alpha\beta}$ in Δ as $T \to \infty$. Here, we endow Δ with a metric in which the ends look like strips, as in Figure 61. We further assume that the points in $\mathbf{p}(T)$ avoid the three sides of Δ. We consider the space

$$\mathcal{M}_{(\Theta,\mathbf{a},\mathbf{b})}(\mathbf{p}) = \bigcup_{T\in[1,\infty)} \mathcal{M}_{(\Theta,\mathbf{a},\mathbf{b})}(\mathbf{p}(T)),$$

which has ends corresponding to $\mathcal{M}_{(\Theta,\mathbf{a},\mathbf{b})}(\mathbf{d})$, ends corresponding to degenerations of holomorphic curves, as well as ends corresponding to whatever curves appear in the limit as $T \to \infty$. The same argument as in the previous case shows that the only degenerations that can occur at finite T correspond to a Maslov index 1 strip breaking off. The same argument we used in the previous paragraph shows that the ends corresponding to Maslov index 1 strips breaking off at finite T have total count equal to zero, modulo two.

We now claim that, as $T \to \infty$, the limit is to a Maslov index zero triangle τ on $(\Sigma_0, \boldsymbol{\alpha}_0', \boldsymbol{\alpha}_0, \boldsymbol{\beta}_0)$, and k Maslov index 2 curves on $(\Sigma_0, \boldsymbol{\alpha}_0, \boldsymbol{\beta}_0)$, each satisfying a matching condition to a point $d_i \in \mathrm{Sym}^1([0,1]\times\mathbb{R}) = [0,1]\times\mathbb{R}$. Note that the total Maslov index is $2k$. The limit contains curves mapping into $\Sigma_0\times\Delta$ and $\Sigma_0\times[0,1]\times\mathbb{R}$, all of which could potentially be nodal. Because this configuration is the limit of curves that satisfied the matching condition with the divisors $\mathbf{p}(T)$, and since the points of $\mathbf{p}(T)$ are separating and heading off to the vertex $v_{\alpha\beta}$, in the limit, we must have k curves on $(\Sigma_0, \boldsymbol{\alpha}_0, \boldsymbol{\beta}_0)$ that each satisfy the matching condition with points $d_i \in [0,1] \times \mathbb{R}$. The homology classes of curves on $(\Sigma_0, \boldsymbol{\alpha}_0, \boldsymbol{\beta}_0)$ are all of the form $e_{\mathbf{a}} + s[\Sigma_0]$, where $e_{\mathbf{a}}$ denotes the constant disk at $\mathbf{a} \in \mathbb{T}_{\alpha_0} \cap \mathbb{T}_{\beta_0}$. The Maslov index of such a curve is $2s$. Since there are k curves satisfying matching conditions with d_i (implying $s \geq 1$), each has Maslov index 2. If τ denotes the remaining curves, which map into $\Sigma_0 \times \Delta$, the only way for the total Maslov index to be $2k$ is for $\mu(\tau) = 0$, and for those k curves satisfying matching conditions to be all of the remaining curves in the limit. Furthermore, the holomorphic triangle τ must satisfy $n_{p_0}(\tau) = 0$, since there are $k = |\mathbf{d}|$ holomorphic strips with multiplicity 1 in the limit, and axiom $(J'4')$ implies positivity of intersections with set $\{p_0\} \times \Delta$. As before, we can use equation (10.47), as well as Proposition 10.48 to show that all of the $k+1$ curves must satisfy $(M1)$–$(M8)$ (or the appropriate analogs for curves mapping into $\Sigma_0 \times [0,1] \times \mathbb{R}$).

Hence, by a gluing result of Lipshitz [**19**, Proposition A.1], we conclude that

$$\#\mathcal{M}_{(\Theta,\mathbf{a},\mathbf{b})}(\mathbf{d}) \equiv \left(\#\mathcal{M}_{(\mathbf{a},\mathbf{a})}(d)\right)^k \cdot \sum_{\substack{\psi\in\pi_2(\Theta,\mathbf{a},\mathbf{b})\\ n_{p_0}(\psi)=0}} \#\mathcal{M}(\psi) \pmod 2, \tag{10.60}$$

where $d \in [0,1] \times \mathbb{R}$ is a point and $\mathcal{M}_{(\mathbf{a},\mathbf{a})}(d)$ denotes the moduli space of Maslov index 2 holomorphic curves u on $(\Sigma_0, \boldsymbol{\alpha}_0, \boldsymbol{\beta}_0)$ satisfying $\rho^{p_0}(u) = d$, for a generic almost complex structure satisfying $(J1)$–$(J4)$ and $(J5')$. Note again that our use of almost complex structures satisfying $(J5')$ ensures that any curve appearing in $\mathcal{M}_{(\mathbf{a},\mathbf{a})}(d)$ achieves transversality, as long as each component is somewhere injective. The condition that they match $d \in [0,1] \times \mathbb{R}$ at p_0 immediately implies that they are not multiply covered. Hence, using Proposition 10.48 as well as the expected dimension from equation (10.47), we see the curves in $\mathcal{M}_{(\mathbf{a},\mathbf{a})}(d)$ satisfy $(M1)$–$(M8)$ (and in particular are not nodal).

Hence, it is sufficient to count $\mathcal{M}_{(\mathbf{a},\mathbf{a})}(d)$ for a generic $d \in [0,1] \times \mathbb{R}$. The argument from the proof of stabilization invariance due to Lipshitz [**19**, Appendix A] adapts to our present situation. We consider the Heegaard diagram for $S^1 \times S^2$ shown in Figure 62.

There are exactly two homology classes of disks that contribute to the differential of the hat complex. One is a bigon and has a unique holomorphic representative. The other is a twice stabilized bigon. By invariance of $\widehat{HF}$, we know the twice stabilized bigon must have one holomorphic representative modulo 2 for any almost

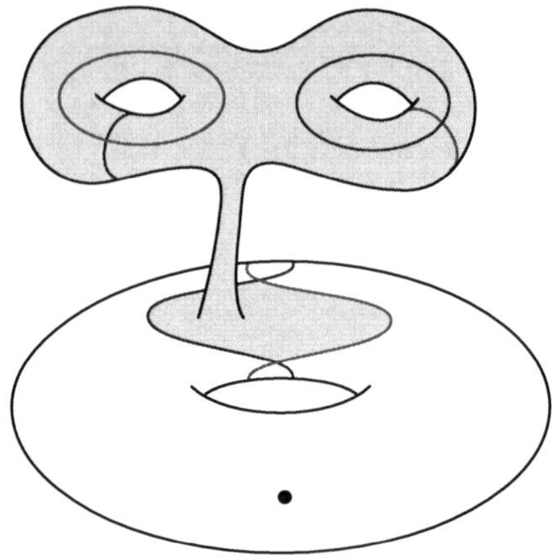

FIGURE 62. A domain on a doubly stabilized diagram for $S^1 \times S^2$.

complex structure achieving transversality. On the other hand, by stretching the neck, we also see that the count for large neck length is equal to the count of strips in $\mathcal{M}(e_{\mathbf{a}} + [\Sigma_0])$ that match the bigon at the connected sum point. We conclude that

$$\#\mathcal{M}_{(\mathbf{a},\mathbf{a})}(d) \equiv 1 \pmod 2.$$

By Lemma 10.54, we know that

$$\sum_{\substack{\psi \in \pi_2(\Theta,\mathbf{a},\mathbf{b}) \\ n_{p_0}(\psi)=0}} \#\mathcal{M}(\psi) \equiv 1 \pmod 2.$$

Hence, from Equation (10.60), we see that

$$\#\mathcal{M}_{(\Theta,\mathbf{a},\mathbf{b})}(\mathbf{d}) \equiv 1^k \cdot 1 = 1 \pmod 2,$$

and the lemma follows. □

10.4. Proof of Theorem 2.33

We now combine the results that we have obtained and together imply Theorem 2.33. As explained at the beginning of Chapter 10, let HF° denote one of the versions of Heegaard Floer homology. In Section 10.1, we showed that HF° is a weak Heegaard invariant in the sense of Definition 2.24. In particular, in Definition 10.20, we associated the $\mathbb{F}_2$ vector space $HF^\circ(H)$ to an isotopy diagram H. We associated the canonical isomorphisms $\Phi_B^{A\to A'}$ and $\Phi^A_{B\to B'}$ to α- and β-equivalences, respectively, in Definition 10.22. For a diffeomorphism d, we constructed an induced isomorphism d_* in Definition 10.24. Finally, to a stabilization $H \to H'$, we assigned the isomorphism $\sigma_{H\to H'}$ in Definition 10.27, and to the destabilization $H' \to H$ its inverse, $(\sigma_{H\to H'})^{-1}$.

The above isomorphisms satisfy the axioms for HF° to be a strong Heegaard invariant in the sense of Definition 2.32. At the beginning of Section 10.2, we showed axiom (1), functoriality, and axiom (2), commutativity. We verified axiom (3), continuity, in Proposition 10.28. Finally, axiom (4), simple handleswap invariance, was proven in Section 10.3. □

APPENDIX A

The 2-complex of Handleslides

In this appendix, we sketch a description of strong Heegaard invariants for classical (i.e., not sutured) single pointed Heegaard diagrams that is equivalent to Definition 2.32, and instead of α-equivalences and β-equivalences, uses more elementary moves: α-isotopies, β-isotopies, α-handleslides, and β-handleslides. The tradeoff is that one has to check the commutativity of the invariant F along a larger number of loops of diagrams. But we do have to impose less on F, and hence strengthen Theorem 2.38. The main tool is a result of Wajnryb [**38**], who constructed a simply-connected 2-complex whose vertices consist of cut-systems, and whose edges correspond to changing just one circle in a cut system. We only sketch the proofs in this appendix.

We start off by looking at those moves that only involve α-circles or β-circles. For these, it is enough to consider only one of the two handlebodies. In particular, we show that any two cut-systems for a handlebody can be connected by a sequence of handleslides. This is in fact a corollary of a result of Wajnryb [**38**]. To state his result, let us first recall some definitions.

Definition A.1. Let B be a handlebody of genus g and boundary $\Sigma = \partial B$. A simple closed curve $\alpha \subset \Sigma$ is a *meridian curve* if it bounds a disk D in B such that $D \cap \Sigma = \partial D = \alpha$. Then D is called a *meridian disk*. We also fix a finite number of disjoint distinguished disks on Σ and we shall assume that all isotopies of Σ are fixed on the distinguished disks.

A *cut-system* on Σ is an isotopy class of an unordered collection of g disjoint meridian curves $\alpha_1, \dots, \alpha_g$ that are linearly independent in $H_1(\Sigma)$ and do not meet the distinguished disks. We denote the cut-system by $\langle \alpha_1, \dots, \alpha_g \rangle$.

We say that two cut-systems are *related by a simple move* if they have $g-1$ curves in common and the other two curves are disjoint.

We construct a 2-dimensional complex $X_2(B)$. The vertices of X are the cut-systems on Σ. Two cut-systems are connected by an edge if they are related by a simple move; this gives the graph $X_1(B)$. If three vertices of X have $g-1$ curves in common and the three remaining curves, one from each cut-system, are pairwise disjoint, then each pair of the vertices is connected by an edge in X and the vertices form a triangle. We glue a face to every triangle in $X_1(B)$ and get a 2-dimensional simplicial complex $X_2(B)$, called the *cut-system complex* of the handlebody B.

The following result is due to Wajnryb [**38**, Theorem 1].

Theorem A.2. *The complex $X_2(B)$ is connected and simply-connected.*

For compatibility with the other moves we consider, we work instead with a 2-complex whose edges are elementary handleslides. To describe the 2-cells, we need another definition.

DEFINITION A.3. A *handleslide loop* is one of the following sequences of cut-systems connected by handleslides.

(1) A *slide triangle*, formed by $\langle\alpha_1, \alpha_2, \vec{\alpha}\rangle$, $\langle\alpha_2, \alpha_3, \vec{\alpha}\rangle$, and $\langle\alpha_3, \alpha_1, \vec{\alpha}\rangle$, where α_1, α_2, and α_3 bound a pair-of-pants.
(2) A *commuting slide square*, involving four distinct α-curves, as in the link of a singularity of type (A1a).
(3) A square formed by sliding α_1 over α_2 and/or α_4, as in case (A1b).
(4) A square formed by sliding α_1 and/or α_3 over α_2, with α_1 and α_3 sliding over α_2 from opposite sides, as in case (A1c).
(5) A square formed by sliding α_1 over α_2 in two different ways, approaching α_2 from opposite sides, as in case (A1d).
(6) A pentagon formed by sliding α_1 over α_2, which is itself sliding over α_3, as in case (A2); see Figure 22.

Now suppose that there is exactly one distinguished disk on $\Sigma = \partial B$. Then let $Y_2(B)$ be the 2-complex whose vertices are cut-systems on B, its edges correspond to handleslides avoiding the distinguished disk, and its 2-cells correspond to the handleslide loops of Definition A.3.

PROPOSITION A.4. *The complex $Y_2(B)$ is connected.*

PROOF. To prove connectivity, it suffices to show that the endpoints of each edge in $X_1(B)$ can be connected by a path lying in the 1-skeleton $Y_1(B)$ of $Y_2(B)$. Suppose we have an edge in $X_1(B)$ connecting $\langle\alpha_0, \vec{\alpha}\rangle$ and $\langle\alpha_1, \vec{\alpha}\rangle$. Then α_0 and α_1 do not intersect. The combined set of circles $\langle\alpha_0, \alpha_1, \vec{\alpha}\rangle$ by hypothesis cuts ∂B into two components, exactly one of which does not contain the distinguished disk; call this component F. Both α_0 and α_1 necessarily appear in ∂F. We can get from $\langle\alpha_0, \vec{\alpha}\rangle$ to $\langle\alpha_1, \vec{\alpha}\rangle$ by sliding α_0 over every component of $\partial F \setminus (\alpha_0 \cup \alpha_1)$. □

PROPOSITION A.5. *The complex $Y_2(B)$ is simply-connected.*

PROOF SKETCH. For simple connectivity, we first show that all the different ways of turning an edge of $X_1(B)$ into a path in $Y_1(B)$ are homotopic inside $Y_2(B)$. This can be done (with some work) using handleslide loops of type (3). For a simple example, see Figure 63.[1]

Next, we show that if we convert the edges e_0, e_1, and e_2 of a triangle Δ in $X_2(B)$ into paths in $Y_1(B)$, we obtain a loop that is null-homotopic in $Y_2(B)$. Let v_i be the vertex of Δ opposite the edge e_i. We distinguish two cases:

- The same circle moves in all three edges of the triangle; i.e., the cut-system $v_i = \langle\alpha_i, \vec{\alpha}\rangle$ for $i \in \mathbb{Z}_3$.
- Two circles are involved; i.e., the cut-system $v_i = \langle\alpha_{i-1}, \alpha_{i+1}, \vec{\alpha}\rangle$ for every $i \in \mathbb{Z}_3$, where $i-1$ and $i+1$ are to be considered modulo 3.

The first case is simple: we end up with a trivial loop even in $Y_1(B)$ for an appropriate choice of resolutions. Indeed, for $i \in \mathbb{Z}_3$, let F_i be the component of the complement of $\langle\alpha_{i-1}, \alpha_{i+1}, \vec{\alpha}\rangle$ that does not contain the distinguished disk. Then $F_i = F_{i-1} \cup F_{i+1}$ for some $i \in \mathbb{Z}_3$. We first convert e_{i-1} and e_{i+1} to paths γ_{i-1} and γ_{i+1} in $Y_1(B)$ using the procedure above, then we choose γ_i to be $\gamma_{i+1}^{-1}\gamma_{i-1}^{-1}$. By the first step, any two choices for γ_i are homotopic, so we can pick this particular one.

[1]To do this properly, note that a minimal path in $Y_1(B)$ corresponding to an edge in $X_1(B)$ gives a pants decomposition of a subsurface of ∂B. To show that two such paths are homotopic in $Y_2(B)$, it suffices to show connectivity of a suitable variant of the pants complex.

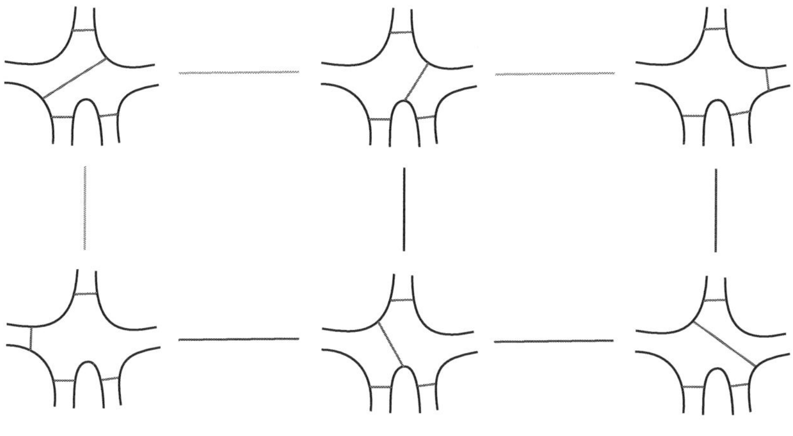

FIGURE 63. A simple example of a homotopy in $Y_2(B)$ connecting two different resolutions of an edge of $X_1(B)$. The lower left and the upper right cut-systems are the vertices of the edge of $X_1(B)$ we are resolving. One resolution is shown in green, the other one in blue. We show half of the component F whose boundary contains α_0, α_1, and no basepoints. In this case, there are $k = 3$ other boundary components of F. The surfaces shown should be doubled along the black boundary to obtain F; in this way the red arcs become red circles.

In the second case, we get a component F with boundary containing α_0, α_1, and α_2. A handleslide loop connects $\langle\alpha_0, \alpha_1, \vec{\alpha}\rangle$, $\langle\alpha_1, \alpha_2, \vec{\alpha}\rangle$, and $\langle\alpha_0, \alpha_2, \vec{\alpha}\rangle$. If there are no other components of ∂F, this is a slide triangle (a handleslide loop of type (1)). Otherwise, if there are k other boundary components of ∂F, let α_0' be the curve obtained from α_0 by sliding over one of the other k components. By induction, the triangle connecting $\langle\alpha_0', \alpha_1, \vec{\alpha}\rangle$, $\langle\alpha_1, \alpha_2, \vec{\alpha}\rangle$, and $\langle\alpha_0', \alpha_2, \vec{\alpha}\rangle$ can be decomposed into allowed two-cells. The remaining region (a quadrilateral with corners at $\langle\alpha_0', \alpha_1, \vec{\alpha}\rangle$, $\langle\alpha_0', \alpha_2, \vec{\alpha}\rangle$, $\langle\alpha_0, \alpha_1, \vec{\alpha}\rangle$, and $\langle\alpha_0, \alpha_2, \vec{\alpha}\rangle$) can be decomposed into $k-2$ commuting slide squares (type (2)) and one slide pentagon (type (6)). The entire large triangle is decomposed into one slide triangle, $k-1$ slide pentagons, and $\binom{k-1}{2}$ commuting slide squares. An example of the end result is shown in Figure 64. □

Let $\mathcal{G}'$ be the graph defined just like in Definition 2.22, but with the word α/β-equivalence replaced by α/β-handleslide. So the vertices of $\mathcal{G}'$ are isotopy diagrams, and its edges correspond to handleslides, stabilizations, destabilizations, and diffeomorphisms. Since every handleslide is an α-equivalence or a β-equivalence, $\mathcal{G}'$ is a subgraph of $\mathcal{G}$.

Similarly, we can modify Definition 2.24. If $\mathcal{S}$ is a set of diffeomorphism types of sutured manifolds and $\mathcal{C}$ is a category, then $\mathcal{G}'(\mathcal{S})$ is the full subgraph of $\mathcal{G}'$ spanned by those isotopy diagrams H for which $S(H) \in \mathcal{S}$. The main result of this appendix is the following.

THEOREM A.6. *Let $\mathcal{S} = \mathcal{S}_{\text{man}}$ be the set of diffeomorphism types of sutured manifolds introduced in Definition 2.25, and let $\mathcal{C}$ be a category. Then every morphism of graphs $F' : \mathcal{G}'(\mathcal{S}) \to \mathcal{C}$ extends to a weak Heegaard invariant $F : \mathcal{G}(\mathcal{S}) \to \mathcal{C}$.*

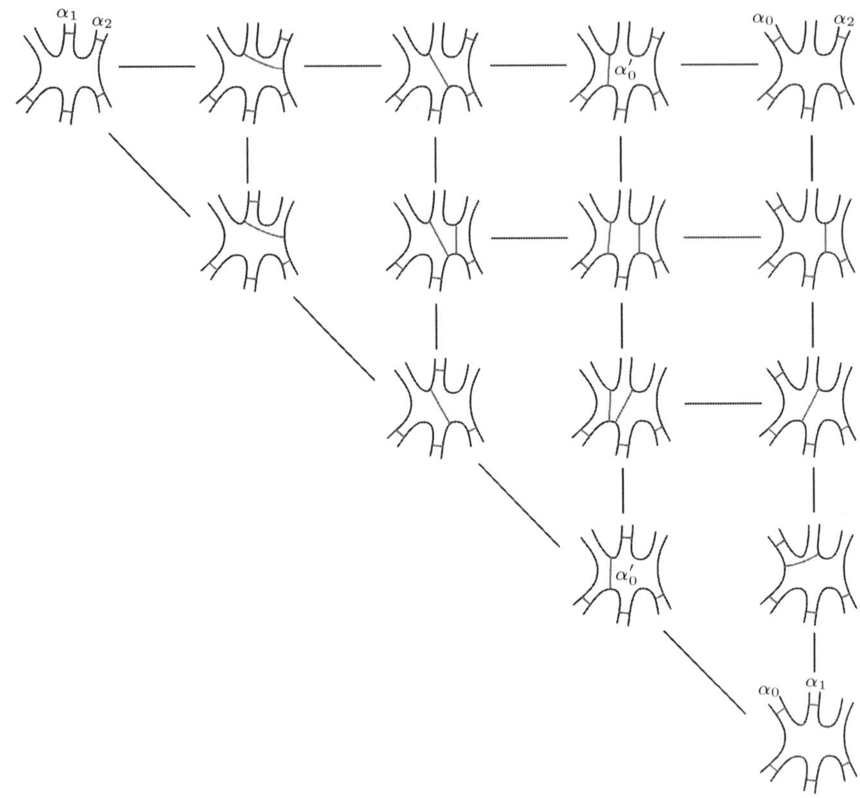

FIGURE 64. Decomposing a large slide triangle. The three vertices are the vertices of a triangle in the complex $X_2(B)$. We show half of the component F whose boundary contains α_0, α_1, and α_2, and no basepoints. In this case, there are $k = 3$ other boundary components of F.

If, furthermore,

- *F' satisfies the second half of axiom* (1) *and axioms* (2)–(4) *of Definition* 2.32*, replacing "α/β-equivalence" with "α/β-handleslide,"*
- *F' commutes along each handleslide loop in Definition* A.3*, and*
- *F' commutes along every stabilization slide (see Definition* 7.7*),*

then F' uniquely extends to a strong Heegaard invariant $F\colon \mathcal{G}(\mathcal{S}) \to \mathcal{C}$.

REMARK A.7. Note that $\mathcal{G}'_\alpha(\mathcal{S})$ and $\mathcal{G}'_\beta(\mathcal{S})$ are not sub-categories of $\mathcal{G}'(\mathcal{S})$, since the composition of two handleslides is in general not a handleslide. The functoriality of F' restricted to the subgraphs $\mathcal{G}'_\alpha(\mathcal{S})$ and $\mathcal{G}'_\beta(\mathcal{S})$ is replaced by the requirement that F' commutes along handleslide loops.

Also note that, in a stabilization slide, we subdivide the α- or β-equivalence into two handleslides, so we view this as a loop of length four.

PROOF SKETCH. To prove the first part, we only have to define $F(e)$ for the edges e of $\mathcal{G}(\mathcal{S})$ that correspond to an α-equivalence or a β-equivalence. Without loss of generality, suppose that e is an α-equivalence between the isotopy diagrams

$H = (\Sigma, \boldsymbol{\alpha}, \boldsymbol{\beta})$ and $H' = (\Sigma, \boldsymbol{\alpha}', \boldsymbol{\beta})$. Let $\overline{\Sigma}$ be the surface obtained by attaching a disk D to Σ along its boundary. This way we obtain two Heegaard diagrams $\overline{H}$ and $\overline{H}'$, containing a distinguished disk D. Let Y be a 3-manifold containing both $\overline{H}$ and $\overline{H}'$ as Heegaard diagrams, and let B be the handlebody lying to the negative side of $\overline{\Sigma}$. By Proposition A.4, the complex $Y_2(B)$ is connected, so $\overline{H}$ and $\overline{H}'$ can be connected by a path of handleslides $\overline{h}_1, \dots, \overline{h}_k$ avoiding D. This gives rise to a sequence of handleslides $h_1, \dots, h_k$ connecting H and H'. Then the isomorphism $F(e)$ is defined to be the composite $F(h_k) \circ \cdots \circ F(h_1)$.

We now prove the second part. According to Proposition A.5, the complex $Y_2(B)$ is simply connected. Together with the fact that F' commutes along every handleslide loop (i.e., along the boundary of every face of $Y_2(B)$), we see that the extension of F' to an α- or β-equivalence edge e is independent of the choice of path $h_1, \dots, h_k$. Functoriality of the restriction of F to $\mathcal{G}_\alpha(\mathcal{S})$ and $\mathcal{G}_\beta(\mathcal{S})$ is clear from the construction.

What remains to show is that F commutes along every distinguished rectangle of type (1), (2), and (3) (see Definition 2.29), with sides e, f, g, and h. First, consider a rectangle of type (1). Write the α-equivalence e as a path of α-handleslides $h_1, \dots, h_k$ and the β-equivalence f as a path of β-handleslides $h_1', \dots, h_l'$. Then we can subdivide the big rectangle into a grid of smaller rectangles with sides h_i and h_j' for $i \in \{1, \dots, k\}$ and $j \in \{1, \dots, l\}$.

Given a rectangle of type (3), let $h_1, \dots, h_k$ be the path of handleslides in the resolution of the α- or β-equivalence e, and let d be the diffeomorphism corresponding to f and g. Then we can subdivide the big rectangle into a row of smaller rectangles with sides h_i and d for $i \in \{1, \dots, k\}$.

Finally, consider a rectangle of type (2). Then let $h_1, \dots, h_k$ be the resolution of the α- or β-equivalence e on the destabilized side. Then, on the stabilized side h, we can choose the stabilizations $h_1', \dots, h_k'$ of the above handleslides. However, the endpoint of h_k' might differ from H_4, the endpoint of h, by a sequence of handleslides over the new α or β-curve appearing in the stabilization. We can correct this by attaching a row of stabilization slides to the row of rectangles with horizontal sides h_i and h_i'. □

Bibliography

[1] C. Abbas, *An introduction to compactness results in symplectic field theory*, Springer, Heidelberg, 2014, DOI 10.1007/978-3-642-31543-5. MR3157146

[2] D. V. Anosov, S. Kh. Aranson, V. I. Arnold, I. U. Bronshtein, V. Z. Grines, and Yu. S. Il′yashenko, *Ordinary differential equations and smooth dynamical systems*, Springer-Verlag, Berlin, 1997. Translated from the 1985 Russian original by E. R. Dawson and D. O'Shea; Third printing of the 1988 translation [*Dynamical systems. I*, Encyclopaedia Math. Sci., 1, Springer, Berlin, 1988; MR0970793 (89g:58060)]. MR1633529

[3] V. I. Arnold, V. V. Goryunov, O. V. Lyashko, and V. A. Vasil′ev, *Singularity theory. I*, Springer-Verlag, Berlin, 1998. Translated from the 1988 Russian original by A. Iacob; Reprint of the original English edition from the series Encyclopaedia of Mathematical Sciences [*Dynamical systems. VI*, Encyclopaedia Math. Sci., 6, Springer, Berlin, 1993; MR1230637 (94b:58018)], DOI 10.1007/978-3-642-58009-3. MR1660090

[4] G. Arone and M. Kankaanrinta, *On the functoriality of the blow-up construction*, Bull. Belg. Math. Soc. Simon Stevin **17** (2010), no. 5, 821–832. MR2777772

[5] F. Bonahon, *Cobordism of automorphisms of surfaces*, Ann. Sci. École Norm. Sup. (4) **16** (1983), no. 2, 237–270. MR732345

[6] F. Bourgeois, Y. Eliashberg, H. Hofer, K. Wysocki, and E. Zehnder, *Compactness results in symplectic field theory*, Geom. Topol. **7** (2003), 799–888, DOI 10.2140/gt.2003.7.799. MR2026549

[7] M. J. D. Carneiro and J. Palis, *Bifurcations and global stability of families of gradients*, Inst. Hautes Études Sci. Publ. Math. **70** (1989), 103–168 (1990). MR1067381

[8] J. Cerf, *The pseudo-isotopy theorem for simply connected differentiable manifolds*, Manifolds–Amsterdam 1970 (Proc. Nuffic Summer School), Lecture Notes in Mathematics, Vol. 197. Springer, Berlin, 1970, pp. 76–82. MR0290404

[9] S. Eilenberg and N. Steenrod, *Foundations of algebraic topology*, Princeton University Press, Princeton, New Jersey, 1952. MR0050886

[10] Y. Eliashberg and N. Mishachev, *Introduction to the h-principle*, Graduate Studies in Mathematics, vol. 48, American Mathematical Society, Providence, RI, 2002, DOI 10.1090/gsm/048. MR1909245

[11] D. Gabai, *Foliations and the topology of 3-manifolds*, J. Differential Geom. **18** (1983), no. 3, 445–503. MR723813

[12] J. E. Grigsby and S. M. Wehrli, *On the colored Jones polynomial, sutured Floer homology, and knot Floer homology*, Adv. Math. **223** (2010), no. 6, 2114–2165, DOI 10.1016/j.aim.2009.11.002. MR2601010

[13] K. Hendricks and C. Manolescu, *Involutive Heegaard Floer homology*, Duke Math. J. **166** (2017), no. 7, 1211–1299, DOI 10.1215/00127094-3793141. MR3649355

[14] J. Johnson and D. McCullough, *The space of Heegaard splittings*, J. Reine Angew. Math. **679** (2013), 155–179, DOI 10.1515/crelle.2012.016. MR3065157

[15] A. Juhász and M. Marengon, *Concordance maps in knot Floer homology*, Geom. Topol. **20** (2016), no. 6, 3623–3673, DOI 10.2140/gt.2016.20.3623. MR3590358

[16] A. Juhász, *Holomorphic discs and sutured manifolds*, Algebr. Geom. Topol. **6** (2006), 1429–1457, DOI 10.2140/agt.2006.6.1429. MR2253454

[17] A. Juhász, *Cobordisms of sutured manifolds and the functoriality of link Floer homology*, Adv. Math. **299** (2016), 940–1038, DOI 10.1016/j.aim.2016.06.005. MR3519484

[18] P. Kronheimer and T. Mrowka, *Monopoles and three-manifolds*, New Mathematical Monographs, vol. 10, Cambridge University Press, Cambridge, 2007, DOI 10.1017/CBO9780511543111. MR2388043

[19] R. Lipshitz, *A cylindrical reformulation of Heegaard Floer homology*, Geom. Topol. **10** (2006), 955–1096, DOI 10.2140/gt.2006.10.955. [Paging previously given as 955–1097]. MR2240908

[20] R. Lipshitz, *Correction to the article: A cylindrical reformulation of Heegaard Floer homology [MR2240908]*, Geom. Topol. **18** (2014), no. 1, 17–30, DOI 10.2140/gt.2014.18.17. MR3158771

[21] R. Lipshitz, P. S. Ozsváth, and D. P. Thurston, *Bordered Floer homology and the spectral sequence of a branched double cover II: the spectral sequences agree*, J. Topol. **9** (2016), no. 2, 607–686, DOI 10.1112/jtopol/jtw003. MR3509974

[22] R. Lipshitz, P. S. Ozsvath, and D. P. Thurston, *Bordered Heegaard Floer homology*, Mem. Amer. Math. Soc. **254** (2018), no. 1216, viii+279, DOI 10.1090/memo/1216. MR3827056

[23] D. McDuff and D. Salamon, *J-holomorphic curves and symplectic topology*, 2nd ed., American Mathematical Society Colloquium Publications, vol. 52, American Mathematical Society, Providence, RI, 2012. MR2954391

[24] J. Milnor, *Lectures on the h-cobordism theorem*, Notes by L. Siebenmann and J. Sondow, Princeton University Press, Princeton, N.J., 1965. MR0190942

[25] P. Ozsváth and Z. Szabó, *Holomorphic disks and knot invariants*, Adv. Math. **186** (2004), no. 1, 58–116, DOI 10.1016/j.aim.2003.05.001. MR2065507

[26] P. Ozsváth and Z. Szabó, *Holomorphic disks and three-manifold invariants: properties and applications*, Ann. of Math. (2) **159** (2004), no. 3, 1159–1245, DOI 10.4007/annals.2004.159.1159. MR2113020

[27] P. Ozsváth and Z. Szabó, *Holomorphic disks and topological invariants for closed three-manifolds*, Ann. of Math. (2) **159** (2004), no. 3, 1027–1158, DOI 10.4007/annals.2004.159.1027. MR2113019

[28] P. Ozsváth and Z. Szabó, *Holomorphic triangles and invariants for smooth four-manifolds*, Adv. Math. **202** (2006), no. 2, 326–400, DOI 10.1016/j.aim.2005.03.014. MR2222356

[29] P. Ozsváth and Z. Szabó, *Holomorphic disks, link invariants and the multi-variable Alexander polynomial*, Algebr. Geom. Topol. **8** (2008), no. 2, 615–692, DOI 10.2140/agt.2008.8.615. MR2443092

[30] P. S. Ozsváth and Z. Szabó, *Knot Floer homology and rational surgeries*, Algebr. Geom. Topol. **11** (2011), no. 1, 1–68, DOI 10.2140/agt.2011.11.1. MR2764036

[31] J. Palis and F. Takens, *Stability of parametrized families of gradient vector fields*, Ann. of Math. (2) **118** (1983), no. 3, 383–421, DOI 10.2307/2006976. MR727698

[32] J. A. Rasmussen, *Floer homology and knot complements*, ProQuest LLC, Ann Arbor, MI, 2003. Thesis (Ph.D.)–Harvard University. MR2704683

[33] S. Sarkar, *Maslov index formulas for Whitney n-gons*, J. Symplectic Geom. **9** (2011), no. 2, 251–270. MR2811652

[34] S. Sarkar, *Moving basepoints and the induced automorphisms of link Floer homology*, Algebr. Geom. Topol. **15** (2015), no. 5, 2479–2515, DOI 10.2140/agt.2015.15.2479. MR3426686

[35] M. Schwarz, *Morse homology*, Progress in Mathematics, vol. 111, Birkhäuser Verlag, Basel, 1993, DOI 10.1007/978-3-0348-8577-5. MR1239174

[36] P. Seidel, *π_1 of symplectic automorphism groups and invertibles in quantum homology rings*, Geom. Funct. Anal. **7** (1997), no. 6, 1046–1095, DOI 10.1007/s000390050037. MR1487754

[37] G. Vegter, *Global stability of generic two-parameter families of gradients on three-manifolds*, Dynamical systems and bifurcations (Groningen, 1984), Lecture Notes in Math., vol. 1125, Springer, Berlin, 1985, pp. 107–129, DOI 10.1007/BFb0075638. MR798085

[38] B. Wajnryb, *Mapping class group of a handlebody*, Fund. Math. **158** (1998), no. 3, 195–228. MR1663329

[39] I. M. Zemke, *TQFT Constructions in Heegaard Floer Homology*, ProQuest LLC, Ann Arbor, MI, 2017. Thesis (Ph.D.)–University of California, Los Angeles. MR3697764

[40] I. M. Zemke, *TQFT Constructions in Heegaard Floer Homology*, ProQuest LLC, Ann Arbor, MI, 2017. Thesis (Ph.D.)–University of California, Los Angeles. MR3697764

[41] I. Zemke, *Quasistabilization and basepoint moving maps in link Floer homology*, Algebr. Geom. Topol. **17** (2017), no. 6, 3461–3518, DOI 10.2140/agt.2017.17.3461. MR3709653

[42] I. Zemke, *Link cobordisms and functoriality in link Floer homology*, J. Topol. **12** (2019), no. 1, 94–220, DOI 10.1112/topo.12085. MR3905679

Index

Editorial Information

To be published in the *Memoirs*, a paper must be correct, new, nontrivial, and significant. Further, it must be well written and of interest to a substantial number of mathematicians. Piecemeal results, such as an inconclusive step toward an unproved major theorem or a minor variation on a known result, are in general not acceptable for publication.

Papers appearing in *Memoirs* are generally at least 80 and not more than 200 published pages in length. Papers less than 80 or more than 200 published pages require the approval of the Managing Editor of the Transactions/Memoirs Editorial Board. Published pages are the same size as those generated in the style files provided for $\mathcal{A}_{\mathcal{M}}\mathcal{S}$-LaTeX or $\mathcal{A}_{\mathcal{M}}\mathcal{S}$-TeX.

Information on the backlog for this journal can be found on the AMS website starting from `http://www.ams.org/memo`.

A Consent to Publish is required before we can begin processing your paper. After a paper is accepted for publication, the Providence office will send a Consent to Publish and Copyright Agreement to all authors of the paper. By submitting a paper to the *Memoirs*, authors certify that the results have not been submitted to nor are they under consideration for publication by another journal, conference proceedings, or similar publication.

Information for Authors

Memoirs is an author-prepared publication. Once formatted for print and on-line publication, articles will be published as is with the addition of AMS-prepared frontmatter and backmatter. Articles are not copyedited; however, confirmation copy will be sent to the authors.

Initial submission. The AMS uses Centralized Manuscript Processing for initial submissions. Authors should submit a PDF file using the Initial Manuscript Submission form found at `www.ams.org/submission/memo`, or send one copy of the manuscript to the following address: Centralized Manuscript Processing, MEMOIRS OF THE AMS, 201 Charles Street, Providence, RI 02904-2294 USA. If a paper copy is being forwarded to the AMS, indicate that it is for *Memoirs* and include the name of the corresponding author, contact information such as email address or mailing address, and the name of an appropriate Editor to review the paper (see the list of Editors below).

The paper must contain a *descriptive title* and an *abstract* that summarizes the article in language suitable for workers in the general field (algebra, analysis, etc.). The *descriptive title* should be short, but informative; useless or vague phrases such as "some remarks about" or "concerning" should be avoided. The *abstract* should be at least one complete sentence, and at most 300 words. Included with the footnotes to the paper should be the 2010 *Mathematics Subject Classification* representing the primary and secondary subjects of the article. The classifications are accessible from `www.ams.org/msc/`. The Mathematics Subject Classification footnote may be followed by a list of *key words and phrases* describing the subject matter of the article and taken from it. Journal abbreviations used in bibliographies are listed in the latest *Mathematical Reviews* annual index. The series abbreviations are also accessible from `www.ams.org/msnhtml/serials.pdf`. To help in preparing and verifying references, the AMS offers MR Lookup, a Reference Tool for Linking, at `www.ams.org/mrlookup/`.

Electronically prepared manuscripts. The AMS encourages electronically prepared manuscripts, with a strong preference for $\mathcal{A}_{\mathcal{M}}\mathcal{S}$-LaTeX. To this end, the Society has prepared $\mathcal{A}_{\mathcal{M}}\mathcal{S}$-LaTeX author packages for each AMS publication. Author packages include instructions for preparing electronic manuscripts, samples, and a style file that generates the particular design specifications of that publication series. Though $\mathcal{A}_{\mathcal{M}}\mathcal{S}$-LaTeX is the highly preferred format of TeX, author packages are also available in $\mathcal{A}_{\mathcal{M}}\mathcal{S}$-TeX.

Authors may retrieve an author package for *Memoirs of the AMS* from `www.ams.org/journals/memo/memoauthorpac.html`. The *AMS Author Handbook* is available in PDF format from the author package link. The author package can also be obtained free of charge by sending email to `tech-support@ams.org` or from the Publication Division, American Mathematical Society, 201 Charles St., Providence, RI 02904-2294, USA. When requesting an author package, please specify $\mathcal{A}\mathcal{M}\mathcal{S}$-LaTeX or $\mathcal{A}\mathcal{M}\mathcal{S}$-TeX and the publication in which your paper will appear. Please be sure to include your complete mailing address.

After acceptance. The source files for the final version of the electronic manuscript should be sent to the Providence office immediately after the paper has been accepted for publication. The author should also submit a PDF of the final version of the paper to the editor, who will forward a copy to the Providence office.

Accepted electronically prepared files can be submitted via the web at `www.ams.org/submit-book-journal/`, sent via FTP, or sent on CD to the Electronic Prepress Department, American Mathematical Society, 201 Charles Street, Providence, RI 02904-2294 USA. TeX source files and graphic files can be transferred over the Internet by FTP to the Internet node `ftp.ams.org` (130.44.1.100). When sending a manuscript electronically via CD, please be sure to include a message indicating that the paper is for the *Memoirs*.

Electronic graphics. Comprehensive instructions on preparing graphics are available at `www.ams.org/authors/journals.html`. A few of the major requirements are given here.

Submit files for graphics as EPS (Encapsulated PostScript) files. This includes graphics originated via a graphics application as well as scanned photographs or other computer-generated images. If this is not possible, TIFF files are acceptable as long as they can be opened in Adobe Photoshop or Illustrator.

Authors using graphics packages for the creation of electronic art should also avoid the use of any lines thinner than 0.5 points in width. Many graphics packages allow the user to specify a “hairline” for a very thin line. Hairlines often look acceptable when proofed on a typical laser printer. However, when produced on a high-resolution laser imagesetter, hairlines become nearly invisible and will be lost entirely in the final printing process.

Screens should be set to values between 15% and 85%. Screens which fall outside of this range are too light or too dark to print correctly. Variations of screens within a graphic should be no less than 10%.

Any graphics created in color will be rendered in grayscale for the printed version unless color printing is authorized by the Managing Editor and the Publisher. In general, color graphics will appear in color in the online version.

Inquiries. Any inquiries concerning a paper that has been accepted for publication should be sent to `memo-query@ams.org` or directly to the Electronic Prepress Department, American Mathematical Society, 201 Charles St., Providence, RI 02904-2294 USA.

SELECTED PUBLISHED TITLES IN THIS SERIES

1323 **Shu Kawaguchi and Kazuhiko Yamaki,** Effective Faithful Tropicalizations Associated to Linear Systems on Curves, 2021

1322 **D. Bulacu and B. Torrecillas,** Galois and Cleft Monoidal Cowreaths. Applications, 2021

1321 **Christian Haase, Andreas Paffenholz, Lindsey C. Piechnik, and Francisco Santos,** Existence of Unimodular Triangulations–Positive Results, 2021

1320 **Th. Heidersdorf and R. Weissauer,** Cohomological Tensor Functors on Representations of the General Linear Supergroup, 2021

1319 **Abed Bounemoura and Jacques Féjoz,** Hamiltonian Perturbation Theory for Ultra-Differentiable Functions, 2021

1318 **Chao Wang, Zhifei Zhang, Weiren Zhao, and Yunrui Zheng,** Local Well-Posedness and Break-Down Criterion of the Incompressible Euler Equations with Free Boundary, 2021

1317 **Eric M. Rains and S. Ole Warnaar,** Bounded Littlewood Identities, 2021

1316 **Ulrich Bunke and David Gepner,** Differential Function Spectra, the Differential Becker-Gottlieb Transfer, and Applications to Differential Algebraic K-Theory, 2021

1315 **S. Grivaux, É. Matheron, and Q. Menet,** Linear Dynamical Systems on Hilbert Spaces: Typical Properties and Explicit Examples, 2021

1314 **Pierre Albin, Frédéric Rochon, and David Sher,** Resolvent, Heat Kernel, and Torsion under Degeneration to Fibered Cusps, 2021

1313 **Paul Godin,** The 2D Compressible Euler Equations in Bounded Impermeable Domains with Corners, 2021

1312 **Patrick Delorme, Pascale Harinck, and Yiannis Sakellaridis,** Paley-Wiener Theorems for a p-Adic Spherical Variety, 2021

1311 **Lyudmila Korobenko, Cristian Rios, Eric Sawyer, and Ruipeng Shen,** Local Boundedness, Maximum Principles, and Continuity of Solutions to Infinitely Degenerate Elliptic Equations with Rough Coefficients, 2021

1310 **Hiroshi Iritani, Todor Milanov, Yongbin Ruan, and Yefeng Shen,** Gromov-Witten Theory of Quotients of Fermat Calabi-Yau Varieties, 2021

1309 **Jérémie Chalopin, Victor Chepoi, Hiroshi Hirai, and Damian Osajda,** Weakly Modular Graphs and Nonpositive Curvature, 2020

1308 **Christopher L. Douglas, Christopher Schommer-Pries, and Noah Snyder,** Dualizable Tensor Categories, 2020

1307 **Adam R. Thomas,** The Irreducible Subgroups of Exceptional Algebraic Groups, 2020

1306 **Kazuyuki Hatada,** Hecke Operators and Systems of Eigenvalues on Siegel Cusp Forms, 2020

1305 **Bogdan Ion and Siddhartha Sahi,** Double Affine Hecke Algebras and Congruence Groups, 2020

1304 **Matthias Fischmann, Andreas Juhl, and Petr Somberg,** Conformal Symmetry Breaking Differential Operators on Differential Forms, 2020

1303 **Zhi Qi,** Theory of Fundamental Bessel Functions of High Rank, 2020

1302 **Paul M. N. Feehan and Manousos Maridakis,** Łojasiewicz-Simon Gradient Inequalities for Coupled Yang-Mills Energy Functionals, 2020

1301 **Joachim Krieger,** On Stability of Type II Blow Up for the Critical Nonlinear Wave Equation in $\mathbb{R}^{3+1}$, 2020

1300 **Camille Male,** Traffic Distributions and Independence: Permutation Invariant Random Matrices and the Three Notions of Independence, 2020